T0187705

THIS WILD ABYSS
The Story of the Men Who Made Modern Astronomy

IN THE PRESENCE OF THE CREATOR
Isaac Newton and His Times

FOX AT THE WOOD'S EDGE
A Biography of Loren Eiseley

WRITING LIVES IS THE DEVIL!
Essays of a Biographer at Work

EDWIN HUBBLE

MARINER OF THE NEBULAE

E D W I N

CRC Press
Taylor & Francis Group
Boca Raton London New York

CRC Press is an imprint of the
Taylor & Francis Group, an **informa** business

GALE E. CHRISTIANSON

HUBBLE

MARINER OF THE NEBULAE

CRC Press
Taylor & Francis Group
6000 Broken Sound Parkway NW, Suite 300
Boca Raton, FL 33487-2742

First issued in paperback 2019

ISBN-13: 978-0-7503-0423-8 (hbk)
ISBN-13: 978-0-367-40103-0 (pbk)

British Library Cataloguing-in-Publication Data
A catalogue record for this book is available from the British Library.

Library of Congress Cataloging-in-Publication Data are available

First edition published 1995, by Farrar, Straus & Giroux Inc.
This edition published 1997, by Institute of Physics Publishing

Designed by Fritz Metsch.

FRONTISPIECE
Plate H335H. Hubble's discovery of variable stars in Andromeda, the first proof of the existence of galaxies beyond the Milky Way.
(*The Observatories of the Carnegie Institution of Washington*)

Visit the Taylor & Francis Web site at
http://www.taylorandfrancis.com

and the CRC Press Web site at
http://www.crcpress.com

TO R.P.

CONTENTS

Photographs follow page 244

A NOTE TO THE READER

Throughout his lifetime as an astronomer, Edwin Hubble refused to embrace the term "galaxy" when referring to the great stellar systems beyond our own Milky Way. Keeping faith with my subject, I have employed his term "nebula" in all but a few instances.

Hubble was an atrocious speller, a trait often commented on by those with whom he corresponded. Operating on the premise that idiosyncrasies are an important part of the historical record, I have rarely corrected his prose nor have I employed the inelegant and intrusive [sic].

EDWIN HUBBLE

MARINER OF THE NEBULAE

Amongst those who go to sea there are
the navigators who discover new
worlds, adding continents to the
earth and stars to the heavens:
they are the masters, the great,
the eternally splendid.

—GUSTAVE FLAUBERT

CHAPTER ONE

MARSHFIELD

I

Fifty years earlier, a young and powerfully built Martin Jones Hubble had first ridden into what was then the village of Springfield, Missouri, with the aura, almost, of a magus. Now, puffing on an ancient pipe made from a real cob, with a stem more than a foot long, the seventy-year-old pioneer was bent low over his oak desk, deeply absorbed in a much contemplated labor of love. In the interests of authenticity, Martin had chosen to write out each of the ten dinner invitations with a goose-quill pen on a sheet of foolscap, which he then carefully folded into the shape of an envelope, as was the custom before envelopes became commonplace. These he sealed with red wax and posted in late March 1906. They read:

> On the last day of March, 1856, I rode into Springfield on a red sorrel horse having four white feet and a white nose, a flax mane and tail. The tail touched the ground and his mane reached his knees. I

sold him to Hugh T. Hunt, who knew his stock, for $250. I was twenty years old, and now at the end of fifty years, I want all of the men who lived in the city or county then, and live in the city now, to take dinner with me on that anniversary.

There will not be many of you, so I urgently ask you to dine with me at my house at 12 o'clock noon next Saturday, the 31st day of March, 1906.

MENU

Turnip Greens	*Hog's Jowl*
Corn Bread	*Buttermilk*
Boiled Custard	*Pound Cake*[1]

Martin later described the thirty-first of March as a "beautiful, sunshiny day," with all of the invited guests in attendance. The next morning the host was met on the street by the Reverend J. J. Lilly, who suggested that those at the dinner preserve the early history of the city and county. Reverend Lilly was added to the guest list as toastmaster the following year, and a stenographer was engaged to take down verbatim the personal reminiscences of those present. After the eighth such gathering of old men, who by 1914 ranged in age from seventy-eight to ninety-four, Martin Hubble saw the oral history of Springfield, Missouri, through the press, then sadly declared an end to the round-table discussions which the participants had affectionately dubbed "pulling taffy."[2]

Broader at his bearded jaw than at his forehead, with a nose rather too large and eyes wide-set, Martin was an imposing, if not a particularly handsome, presence. He stood six feet and, along with his impressive height, had inherited his father's massive shoulders, bull neck, and expansive girth. In addition to his abiding sense of history, he was remembered by his grandson Edwin as a man of unshakable opinions: the finest novel ever written was *Great Expectations*; the loveliest song was "Annie Laurie"; the only political party worth its salt was the Democratic Party; the consumption of alcohol was the direct road to perdition; the most beautiful things in the universe were the planets and the stars.[3]

Martin's upbringing and aspirations had been quintessentially American. His forebears came from England, Ireland, and Wales, with "no strains," as one family member later wrote, of foreign blood. The first among them to reach American shores five generations earlier was Richard Hubball, an officer of the Royalist army who fled to Connecticut in the early 1640s, after the beheading of Charles I. The colonial squire fathered fifteen children by three successive wives and died rich in land and other property. Hubbells (the "e" having replaced the "a") served on both sides in the Revolutionary and Civil Wars; in times of peace they moved west, laying claim to an abundance of land never dreamed of by their yeoman ancestors. They lived and worked in the old America of Jeffersonian democracy—a rural community of cooperating individualists, with a passion for self-government, combined with pure and fervent patriotism. Later generations, equally patriotic but more urbanized, entered the professions, including law and medicine, while others became public servants. Martin liked to stress the fact that the family had never known the stigma of a divorce.[4]

It was to Justice Hubbell, a signer of the Articles of Association protesting the closing of the port of Boston by the British in 1775, that Martin traced his ancestry when applying for membership in the Sons of the American Revolution, as would his daughters on joining the D.A.R. He also knew that it was Justice Hubbell's son Joel, a Virginia slaveholder of Seven Mile Ford, who changed the spelling of the family name to Hubble for reasons unknown. Martin's father, John B. Hubble, was Joel's son, and had migrated from Virginia to Boone County, Missouri, where he practiced medicine near Columbia until his early demise in 1847, at age thirty-six. Twelve-year-old Martin, together with his four younger brothers, was taken by his mother, Sarah Lavinia Jones Hubble, to Giles County, Tennessee. There the boys grew up on the plantation of Martin's grandfather and namesake Martin Jones, a prominent cotton farmer and slave owner.

Having been reared in a culture where human bondage was taken for granted, Martin underwent a crisis of conscience on the eve of the Civil War. The Missouri to which he had returned following his mother's death had been admitted to the Union as a slave state in 1821. In 1854 the issue of slavery was exacerbated by the passage of the Kansas-Nebraska Act, leaving the question of slavery in these territories to the settlers themselves. Proslavery forces in Missouri became active in trying to win "bleeding Kansas" for the slave cause and contributed to the violence and disorder that tore the territory apart in the years just prior to the Civil War. Nevertheless, Missouri also had leaders opposed to slavery, including its prominent senator, Thomas Hart Benton, and Free-Soil congressman, Francis P. Blair, who later distinguished himself as a major general of the Union Army in the Vicksburg, Chattanooga, and Atlanta campaigns. A state convention meeting in March 1861 voted against secession, and in August the Union general John C. Frémont issued a proclamation instituting martial law and freeing all slaves. An appalled Abraham Lincoln immediately countermanded this order with another bringing Missouri into conformity with existing federal law.

While their Grandfather Jones yielded to the inevitable and freed his slaves, the five sons of John and Sarah Hubble were deeply split on the issue of war. One died while fighting in the Confederate ranks; the youngest, redheaded George Washington Hubble, became a standard-bearer of the Stars and Bars. Shot through the leg, he was wrapped in the same flag he had carried into battle by his friend the regimental drummer boy. The youths were subsequently captured by Union troops and spent the remainder of the war in Libby prison. George carried his battle scars for life, walking with a limp, a disability which sparked the imagination of his great-nephew Edwin.[5]

Though a Democrat and no supporter of a besieged Lincoln, Martin eventually opted for the Union. In February 1861, the Greene County Court appointed him a member of the three-man "patrol" for Campbell township, responsible for keeping order among the slaves for the next twelve months. He was elected clerk of the circuit court in 1863 and began taking an active part in the Union League, whose stated objective was to "aid and abet by all honorable means the Federal government in its efforts to put down the rebellion." Emboldened by his modest political success, Martin ran for Congress on the Democratic ticket in November 1864, only to be crushed by his Republican foe, 1,129 to 288. Months later, the Confederates briefly reappeared in force under the leadership of Missouri native Sterling Price, causing enlistments to soar. In Springfield, over the vehement protests of Confederate sympathizers, the Enrolled Missouri Militia was organized in August 1864. Martin was commissioned captain and quartermaster on September 28, and served until March 12 of the following year, when the Confederacy was in its death throes. Though he saw little or no action, he was destined to remain Captain Hubble in the minds of his fellow citizens until his death fifty-five years later.[6]

II

The only woman known to have contributed to the reminiscences of the Springfield pioneers was the diminutive but gritty Mary Jane Hubble. Little is known of the background of this Springfield matriarch, who, during a span of twenty-one years, had given birth to four sons and five daughters, eight of whom survived infancy. Mary Jane was born on a farm four miles east of Springfield in 1840, the daughter of General George Joseph Powell, a veteran of the early Indian Wars, and Jane Massey, whose father, Captain James Massey, had emigrated from Ireland and fought in the War of 1812. Mary Jane was orphaned at an early age and moved into the household of her uncle, W. H. McAdams, one of Springfield's leading pioneer merchants. After completing grammar school she attended the local Christian College and was baptized by immersion in the cold waters of Fulbright Spring, just west of town. Her attraction to Martin must have been instantaneous and compelling, for their marriage had taken place less than six months after his grand entrance on horseback, and only three weeks after Mary Jane's sixteenth birthday.

The Hubbles' first child, Mary Lavinia, was born in May 1858, but died a short time later. Mary Jane's second pregnancy culminated in the birth of a healthy son on April 3, 1860, little more than a year before Confederate commander P. G. T. Beauregard ordered the firing on Fort Sumter. Keeping with tradition, John Powell Hubble, father of the future astronomer, was given a previously used Christian name, followed by the maiden name of his mother.

Tradition also dictated that the first son of a Hubble receive special consideration in terms of education and moral grounding, an attenuated form of entitlement reaching back to the Middle Ages, when the family was known by its menacing crest bearing three leopard heads and a wolf. And while John would grow up in the new Missouri that rose out of the war, its semi-Southern atmosphere and river life of steamboating in decline, the flavor of antebellum nostalgia lingered in the works of his favorite American writer and Missouri's most celebrated son, Mark Twain.

The earliest surviving picture of John seems to have been taken during his college or university days, sometime in the late 1870s or early 1880s. Handsome and dashing, he has perfectly trimmed hair parted in the middle and complemented by an equally well-groomed mustache, adding a touch of hauteur to a refined, aesthetic face. Though more slender than Martin, at six feet three inches John towered over his father and three younger brothers, Marshall, Levi, and Joel. As in later photographs, he is meticulously arrayed in a crisply starched collar, fashionable suit, and deftly knotted tie.

John attended Springfield public schools and then enrolled in Drury College, within walking distance of the family home. The newly founded institution boasted an enrollment of 225 undergraduates and a faculty of eleven full-time instructors. Though Drury was nondenom-

inational, its mission was unequivocally set forth in the college catalogue: "The founders of Drury College do not wish to disguise the fact
that instructing youth in the Sacred Scriptures and the principles of
the Christian religion has been a ruling motive in undertaking the
work." Education divorced from the inculcation of Christian principles
"is, at best, only half done, and likely to prove dangerous to society."
To this end, students were required to attend morning chapel, in addition to Sunday services. "The use of profane, vulgar, or unbecoming
language" was punishable by expulsion, as were the consumption of
intoxicating liquors and the patronage of billiard halls and saloons. The
principles of nonsectarian puritanism also barred "scuffling, noisy sport,
and disorderly company"; above all, gentlemen were not to visit the
rooms of the lady students, nor the ladies the rooms of the gentlemen.[7]
If John was ever in danger of a moral lapse, his father was present to
remind the youth of his duty. Martin had not only pledged $150 toward
Drury's fledgling endowment fund in 1875, he was also celebrated as
the man who had single-handedly saved Drury for Springfield. When
a financial misunderstanding resulted in a threat that the institution
would be located in Neosho, Martin saddled his horse and rode back
and forth among the quarreling parties, negotiating a modus vivendi.[8]

Fifteen-year-old John opted for the three-year Classical Preparatory
Course, which, if successfully completed, would gain him admission to
the department of his choice and, eventually, a degree. Arithmetic,
English grammar, and Latin filled the first term; United States history,
natural philosophy, and more Latin the second. He next marched and
died with Caesar before graduating to Greek grammar, history, and
Latin prose composition; algebra, quadratics, Virgil, Cicero, Xenophon,
and Homer's *Iliad* rounded out the course of instruction.[9]

Martin's ambitions for his son exceeded the then limited financial
benefits that could be expected from a Bachelor of Arts degree. The
burgeoning city of Springfield, which boasted a population of some
8,000 in 1875, promised great opportunity to a young man in the right
profession. After attending Drury for three years, a degreeless John
headed northeast to St. Louis in September 1878. At eighteen, he was
a year too young to qualify for entry into Washington University's prestigious law school. Thus, John spent the year enrolled in the advanced
class at Smith Academy, one of two preparatory schools run by the
university.[10] He was able to meet the age requirement one year later,
and succeeded in passing the law school entrance exam. The aspiring
attorney was now faced with a curriculum requiring only two years of
study.

Exactly what happened next is impossible to determine. The name John Powell Hubble does not appear on the list of first-year (junior) law students in the 1879–80 catalogue, as would be expected. However, John is listed among the members of the senior class the following year. One thing is certain: he left Washington University, as he had Drury College, without a degree, perhaps having failed his final exams in June 1881. Had he finished, he would have been awarded the L.L.B. and granted automatic admission to the Missouri bar. If John still wanted to practice law he would have to satisfy a judge of his qualifications during a public examination in circuit court.[11]

III

Despite his respected position in the community, Martin Hubble had not had an easy time providing for the ten members of his household. Political connections had gained him appointment as county land commissioner, at a salary of $1,000 per year in 1867, but a change in administration led to his ouster two terms later. Nor did he benefit significantly from masterminding the 1872 election of Democrat John S. Phelps to the governorship of the state. His stock and membership on the board of directors of the Robberson Avenue Street Railway, while prestigious, did not prove very lucrative.[12] It was not until the early 1880s, after the growth of Springfield had led to the rise of a stable middle class, that the "street car magnate" and most influential Mason in Greene County discovered a way to turn his political and social contacts to financial advantage.

Martin, together with Mary Jane and sons John and Joel, entered the insurance business, establishing an office in their residence at 460 South Campbell Street, near the center of the city. They specialized in plate-glass and fire insurance, a necessity for anxious merchants investing heavily in new construction.[13]

Family ambitions were not confined to Springfield. Twenty miles east of the city, on tracks completed by the San Francisco–St. Louis Railroad Company in 1870, lay the Ozark town of Marshfield, with its twenty stores, two hotels, livery stable, weekly newspaper, and two wagon shops clustered around the traditional town square. The seat of Webster County, Marshfield boasted a population of nearly a thousand, an estimated 150 of whom were described as "mostly good coloreds," confined to the north side of town.

Located on one of the highest points in Missouri, Marshfield was surrounded by rolling countryside, part prairie and part timber, inter-laced by a flourishing network of rivers, creeks, and springs. Agricultural production centered on wheat, rye, corn, oats, and hay. Tobacco thrived on the uplands, and was cultivated more intensively than anywhere else in the region. Several kinds of fruit prospered in the rich, well-drained soil, including grapes, strawberries, peaches, and plums. However, the most favored of the fruit crops was apples, whose millions of spring blossoms covered the orchard grounds like giant, perfumed quilts. It was partly for the sake of apples and the promise of high profits at harvesttime that the Hubbles came to Marshfield.

The Hubble Land and Fruit Company, known locally as the "Hubble Ranch," covered all of section 22, a mile and one-half south of town on Fordland Road. Stone entrances graced the middle of each mile-long boundary, providing four points of access to the winding dirt road that connected the fields, orchards, and spring-fed ponds. In the center stood the two-story house, with fireplaces both above and below. The nearby bungalow was let to the ranch manager as part of his wages, which were further supplemented by pork raised in the large hog house. The Hubbles' was the only 640 acres in Ozark Township wholly owned by a single family. Legal documents list the independent-minded Mary Jane as company president, Joel as vice president, and John, who han-dled most of the financial affairs, as secretary. All three, together with Martin, were members of the board of directors. Theirs was also some of the most encumbered land in the county; a note in the amount of $3,500 was owed to one Lena Guggenheim.[14]

For a time, Martin moved his family from Springfield to the ranch, making it easier for the two younger boys remaining at home to perform a share of the labor. Their father rode the train out during the week-ends, his comings and goings regularly noted in the Friday editions of *The Marshfield Chronicle*.[15] He also took an active part in community affairs by joining the Immigration Society and serving on the building committee of the Marshfield Christian Church, located on the south-west corner of the square and dedicated in September 1882. A district grand deputy of the Masonic Order, Martin became the founding fa-ther of Marshfield's Doric Lodge No. 300.[16]

What proved a fateful accident had occurred on the ranch while John was home from college one summer. Without warning, the team of horses he was working bolted, entangling him in the harness lines and dragging the helpless youth across field and furrow. He was driven into Marshfield by wagon to the office of Dr. William H. James, who immediately sent for his daughter, Virginia Lee, to assist with the agonized patient. Virginia retched when she walked into the room. John was covered in blood, his shredded and muddy bib overalls completely soaked through. She later recalled that the gory sight convinced her that she "never wanted to see John Hubble again."[17] John felt quite differently about the reluctant Florence Nightingale, whose most attractive features were her mesmerizing blue eyes and blond, curly hair. The courtship blossomed with the passing months, leading to the prospect of marriage, but not before the tall, stately "Jennie," who was four years younger than her ardent suitor, completed two years of study at a nearby women's college.

The Jameses, like the Hubbles, were deeply rooted in the Old South. Jennie's father was born in Blount County, Tennessee, the grandson of Scotch-Irish Virginia slaveholders. Her mother, Lucy Ann Wade, came from Virginia and traced her ancestry back eight generations to colonial icon Miles Standish. According to Jennie's older brother Jefferson Beauregard, whose Rebel name bespoke volumes, their mother was an autocratic lady, who "didn't know that the Confederate war was over."[18]

William Henderson James had acquired his medical skills by apprenticing himself to Johnson County physician William Huff in 1844. He established his own practice in Johnson County three years later, but it was short-lived. The discovery of gold at Sutter's Mill early in 1848 brought more than 40,000 fortune seekers into western California, including William James. The doctor found no gold, but he established other mineral rights and soon opened a quartz mine, which together with the practice of his profession and some general merchandising afforded him a more than comfortable life in the halcyon Napa Valley. Like many speculators, James eventually came to grief by overextending himself. He lent large sums to the holders of Spanish land grants, only to see the promising wheat crop fail. Then the quartz vein petered out. Risking the vagaries of violence and vigilante justice in hopes of recouping his losses, he moved his wife and four young sons to Virginia City, Nevada, the brawling, lawless mining camp and "capital" of the Comstock Lode, the country's richest silver deposit. It was here, on the side of Mount Davidson, that Lucy James gave birth to Virginia, on May 16, 1864.[19]

Having failed in his dream of joining the ranks of the fabulously wealthy silver barons, James returned to Missouri in 1869, along with his wife and family of seven. He settled permanently in Marshfield in 1877, gaining a reputation as an expert on the dreaded typhoid fever. The physician was soon back in business as well. W. H. James & Son Drug Store, located on the northwest side of the square, did a brisk trade in perfumes, ventilated trusses, paints, oils, varnishes, watches, fancy articles, and window glass, in addition to patent medicines and homemade pills of every description. If the business ads appearing in the town's newest weekly, *The Marshfield Mail*, cannot be described as factual, they were invariably creative:

SHE WAS TOO LAZY
To steep up herbs and so she tried
pills and cathartics till she was sallow
and bilious as an opium eater. Then
she took a friend's advice and began
using Park's Tea. Now she is fresh
and blooming "as a daisy." Sold by
W. H. James & Son.[20]

I V

After leaving Washington University without a law degree in 1881, John was employed by the Old American Insurance Company of Chicago as their state adjuster for farm losses. Meanwhile, he continued reading the law. In March 1882, a month before his twenty-second birthday, John became a member of the bar by passing scrutiny before Judge Robert W. Fyan of the Webster County Circuit Court, thus living up to at least a part of his father's high expectations.[21] A history of the period describes his budding Springfield practice as a "good" one, adding that it "is of such stuff as men are made."[22] Like all families, the Hubbles developed their own cherished legends. John was later described as insurance commissioner for the state of Missouri and prosecutor for Webster County, neither of which is borne out by the records.

John Powell Hubble and Virginia Lee James were wed in Marshfield on August 10, 1884. The first of their eight children was a withdrawn, sensitive boy christened Henry James, born in October 1886; their second, the demure and rather plain Lucy Lee, arrived two years later. Like her mother and mother-in-law, Jennie became pregnant as soon as she weaned her children from the breast and started ovulating again. It was this practice that accounted for the lockstep two- to three-year spacing between babies.

As the time for Jennie's third delivery approached in November 1889, John brought his wife and children back from Springfield to his in-laws' home in Marshfield, so that the baby could be born under the watchful eye of Dr. James. The single-story house, comprised of seven rooms and two large porches, no longer stands, but locals who remember it believe it likely that Edwin Powell, who arrived by kerosene lamp near midnight on the twentieth, was delivered in the living room, which contained the only fireplace. The "fine, healthy infant," as he was described by one of his sisters, received his paternal grandmother's maiden name, James having already been appropriated by Henry, the firstborn.[23] Edwin, which was Jennie's choice, was taken from her nineteen-year-old brother. According to the *Chronicle*, Wednesday's child was first "interviewed" by his father two days later. After recuperating for another two weeks, Jennie returned to Springfield with her husband and children in time to celebrate Christmas and the new year.[24]

In June 1890, six months after Edwin's birth, John moved his wife and children to Marshfield.[25] The exact location of their home on Maple Street, which the newspapers refer to only as the "Hubble property," is not known, owing to the fact that many of the town's tax and real estate records were destroyed when fire leveled the first courthouse.

There is some doubt whether John Hubble's legal practice ever came close to fulfilling the expectations with which it was launched, for his career as an attorney ended within three or four years of its inception. The decision to close the law office was subsequently attributed to failing eyesight, an explanation given credibility by a later photograph that shows him wearing thick, wire-rimmed spectacles.[26] Yet the fact that John had never given up the insurance business, which also requires considerable reading and the continuous filing of written reports, suggests that the law may not have been his strong suit.

During the coming decade, John would hold various positions with a succession of insurance companies, including Home, Hamburg-Bremen, Atlas, and, finally, Greenwich of New York. The peripatetic father of three had a penchant for traveling far and wide. It was not unusual for him to be away for a month to six weeks at a time, riding trains and frequenting the lobbies of small hotels scattered the length and breadth of Missouri, Kansas, Illinois, and Wisconsin. Thus Jennie and the children would be safer and better cared for by living near her father and mother. Moreover, Martin's family was still ensconced at the Hubble Ranch just outside town, though they would soon return to Springfield for good, the children having grown up and the tedium of country life wearing thin with a socially active Mary Jane.

Jennie's feelings about these long separations are not known. But twice, before Edwin was old enough for school, she and the children joined John elsewhere after renting out the family home. William Martin Hubble was born on Lydia Avenue in Kansas City, Missouri, in January 1892, where the family resided for two years. Then, in March 1894, they moved from Kansas City to St. Louis when John changed employers.[27] Virginia, the fifth child and her mother's namesake, was delivered in their new home in November.

One of Edwin's earliest memories dates from this period. He recalled walking with his parents in a big city feeling very tired and muttering, "I'm sick, I'm sick." He was taken somewhere and left to play alone in a garden. A large dog approached him, and he was bitten when he accidentally stumbled against the beast. Feeling more sorry for the startled dog than for himself, he approached it carefully and patted it until the animal became friendly.[28]

By October 1895 renter Ward Harmon had vacated the Hubble house on Maple Street, and Jennie and the children were back in Marshfield. Edwin finally started school. He had learned to read and cipher before entering first grade, when he had protested not being allowed to attend classes with his older brother and sister. Precocious and bored, he now had a behavior problem, giving his elementary teachers "quite a deal of misery." The situation was made worse by a crisis of deeper proportions. Fourteen-month-old Virginia, the baby of the family, was persona non grata to her older brothers, who resented the toddler's habit of calmly knocking over the castles and bridges Bill and Edwin painstakingly constructed out of blocks. The boys agreed that the only way to cure her was to step on her fingers, for which they were swiftly punished. Days later the baby came down with an unidentified childhood disease and steadily declined despite the best efforts of Grandfather James. Virginia succumbed on January 14, 1896, plunging Edwin, and to a lesser degree Bill, into depression. In their minds baby Virginia had died because of the injury they had inflicted on her. "Edwin," a sister wrote, "became psychologically ill and had it not been for his very understanding and intelligent parents, this paranoia might have caused another tragedy in the family."[29]

John Hubble was the product of a puritanical upbringing and strict education. A stern, hard-bitten blend of moral high-mindedness and relentless ambition, he became a demanding taskmaster who "ruled the roost" in no uncertain terms. He was also an inveterate student of the Bible who lived his religion. John had little time for the Christian Church attended by his parents and in-laws, whose tenets he found wanting. As a young man he had chosen to become a Baptist, having been attracted to the denomination's modified form of Calvinism. Following Martin's example, he renounced alcohol and rarely, if ever, cursed. The only vice inherited from his father was a love of tobacco; John smoked both a pipe and large cigars, blowing perfectly formed rings at the insistence of his children, who attempted to make bracelets of the ethereal bands.

Jennie was described by her daughters as a lovely lady in every way. "How she put up with Papa we don't know."[30] Her sense of humor doubtless helped. When John, who was taller than almost every man he encountered, took one of the girls to be vaccinated, they both fainted. Jennie was not hesitant about reminding her abashed husband of his failed example when he came down too hard on the children. She, too, was religious and a "fine Bible student" who put her knowledge to use by teaching the women's Sunday school class. But unlike John, the more serene Jennie did not wear her faith "on her sleeve." The belief in salvation held a more prominent place in her thinking than the threat of damnation. She was always someone the children could talk to; it was to their father that they listened.

John's stature, combined with his overbearing demeanor, was normally sufficient to keep the children in line, thus sparing the rod. Edwin remembered being spanked only rarely, and said that he had probably deserved it. Yet it was a sister who observed that "Edwin felt Papa blighted his life."[31] A family photograph shows him as an ungainly, sad-eyed boy, possessed of his father's body and his mother's face. As he was too tall for his clothes, his wrists hang bare from beneath the sleeves of his ill-fitting shirt and suit coat. As the tallest among his peers he was always assumed to be older than his age, causing him to seek the company of boys ahead of him in school. So he also seemed to his father, a man whose brief and sporadic appearances were fraught with a mixture of anticipation and dread. John, the absentee censor, was suddenly transformed into the strict disciplinarian and rival for Jennie's affections. Edwin's graciousness toward his mother was remembered as a gift,[32] a recognition that both had been victimized by circumstances beyond their control. Aside from Jennie, it was to his grandfathers that Edwin turned for adult companionship.

v

High in the black firmament above the Missouri village, where no streetlights glowed, the stars were numerous, white, and cold. During winter especially, it was difficult to think of them as great suns diminished by vast distances. To a boy contemplating their wonders they may have appeared as holes in the sky, stab wounds of the gods after whom the constellations are named, diamonds that could sever glass, or even alien intelligences. So they may have also appeared to the romantic William James, who had followed the invisible glimmer of gold all the way to California as a young man. Thanks to the kindly doctor, Edwin is said to have gotten his first look through a telescope on his eighth birthday. Working with materials largely available at W. H. James & Son, his grandfather built the instrument with his own hands. Instead of the usual party, Edwin asked that he be allowed to stay up late to look at the heavens. The wish was granted, and he and his grandfather had what was described as a great time.[33]

In all likelihood, Martin Hubble was invited by his grandson to gaze upon the planets and stars when he next visited Marshfield. The two talked a good deal about astronomy and harbored a mutual fascination for the planet Mars. It was about this time that the dapper, self-confident astronomer Percival Lowell was swept up in the belief made popular by Italian Giovanni Schiaparelli that Mars was covered by an intricate network of crisscrossing lines, the so-called *canali*. On the basis of his own observations conducted on Mars Hill in Flagstaff, Arizona, Lowell concluded that the canals did indeed exist, and were used for the purpose of conveying great amounts of water from the planet's melting polar caps to the inhabitants of the parched equatorial cities, members of an older and wiser race than ours. Family tradition holds that Edwin wrote a detailed letter about the planet to Martin, who was so impressed with the boy's grasp of the subject that he got it printed in one of Springfield's three newspapers.[34]

Martin was a fine storyteller, and Edwin remembered absorbing much of American history and that of the Hubbles at his grandfather's knee. He dreamed of visiting the Confederate museum containing the bloodstained flag that had stanched the wound of lamed George Washington Hubble, who had been little older than Edwin when he was shot through the leg and fell into enemy hands. The youth was taken to visit Hazelwood Cemetery on Springfield's east side. There, in terrible but poignant symmetry, were the graves of hundreds who had fallen in the war, the white stone markers of Yankees and Rebels separated in death, as in life.[35]

The tie between John "Bud" James, one of his mother's distant relatives, and the infamous Jesse James, an early recruit of Confederate captain William Clarke Quantrill, leader of the infamous Missouri Bushwhackers, was no less intriguing. Accompanied by his brother Frank, Jesse arrived at John's cabin near Niangua, Missouri, unannounced, hoping to find out if they were related. The brothers invited themselves to stay the night and requested a supper of ham with biscuits and red-eye gravy. John's wife became terrified; while there was plenty of meat in the smokehouse, she had no idea how to make the gravy. Frank calmly took over, mixing cream with the ham grease to achieve the desired result. After talking late and concluding that their grandfathers were probably cousins, the brothers bedded down on the floor, their unholstered guns within arm's reach. Following a quiet night, Jesse and Frank rode out of the gate and into family mythology.[36]

Edwin had "settled down" considerably by the time he entered fourth grade, and began earning the high marks that would distinguish the remainder of his school years. The restlessness and boredom he had experienced were replaced by worlds brought to Marshfield in books that were part of the new golden age of children's literature. Among his favorites were the escapist *Alice in Wonderland* and *Through the Looking Glass*, Charles Kingsley's *Water Babies*, and Kipling's *Jungle Books*. He traveled to the center of the earth and the bottom of the sea via the imagination of Jules Verne, trekked a perilous Africa in quest of King Solomon's mines with H. Rider Haggard, rafted down the Mississippi with fellow spirit Huckleberry Finn. Like all boys for whom the dread chore bell tolls, Edwin was intrigued by the psychology that enabled Tom Sawyer to persuade Hannibal's innocents that it was a privilege to whitewash Aunt Polly's fence, and one worth paying for. Arthur Conan Doyle was at the height of his popularity then, and someone gave the youth *The Hound of the Baskervilles*. Thinking it the story of a dog, he began reading the book alone in the house at night. Before the end he rushed upstairs to his bedroom and hid beneath the covers. He was also very well read in the King James version of the Bible, a development looked upon with an approving eye by both his parents.[37]

Edwin was no less at home outdoors. He wandered much of the open countryside, along streams, through woods, observing birds and animals. His father and Grandfather Hubble taught him to shoot, but always at targets rather than at living things. Indeed, Martin ran regular "notices to the public," warning that all persons caught hunting on the Hubble Ranch would be prosecuted to the full extent of the law.[38] Edwin learned to swim the old-fashioned way: bigger boys threw him in the local pond and as often as he scrambled out they threw him in again, until he mustered the courage to turn and swim to the opposite bank. He also developed a large collection of flint arrowheads, still numerous in the verdant fields, which he eventually gave to a museum.[39]

Boyhood companion Sam Shelton remembered that the Hubble

home was located on a large lot, with many trees, about three blocks from the red-brick schoolhouse, which contained all grades through high school. Sam often stopped by to pick up Edwin on his way to class of a morning. Marshfield's Negro children attended a separate facility on North Marshall Street, not far from the Colored Methodist Church. About the only contacts with Negro boys their own age were the Sunday-afternoon corncob fights that pitted blacks against whites in a barn on the town's north side.

Edwin was a frequent visitor in the Shelton home, which held a special appeal. It was located on a five-acre plot, almost qualifying it as a small farm in the 1890s. The family of eight cultivated extensive vegetable gardens and kept milk cows and horses, in addition to hogs and laying hens. Edwin, whom Sam described as a "nature lover," liked to observe the stock and hunt for the dens of wild animals. A small stream crossed the property where they sometimes found muskrats, which the boys attempted to trap, without success. "Neither of us really wished to catch them anyway." Nearby was the large pond where the children of Marshfield learned to skate. Sam classified Edwin as one of the better athletes on blades, "almost an exhibitionist." Unfortunately, their fun was interrupted just when the ice was at its peak. Townsmen armed with huge saws cut the pond's inert surface into large blocks, which were stacked away between thick layers of sawdust in the old icehouse.[40]

Sam remembered that he and Edwin were almost ten when his excited friend informed him of an impending event of great importance —a total lunar eclipse. It was scheduled to begin after midnight on June 23, 1899, and Edwin wanted to stay outdoors all night to be certain not to miss a moment of it. He had received his parents' permission, but they wanted Sam to join him. The Sheltons were dubious. They thought their son too young for this unorthodox enterprise, but were finally worn down by the boys' incessant pleading. Sam remembered it as a brilliantly clear night: "We spent most of [it] out in the open spaces where the view of the sky was unobstructed. To us it was a magnificent show." The eclipsed moon finally set in the west, its blackened surface fading in the light of dawn. Many years later, as the executive assistant to the publisher of the *St. Louis Post-Dispatch*, Sam Shelton would think back on this experience and wonder whether it had contributed to Edwin's decision to become an astronomer.[41]

VI

The sting of baby Virginia's death diminished over time and was further relieved by the birth of John and Jennie's sixth child, Helen, in December 1898. By now John was traveling farther and was away longer than ever before. His latest employer, Greenwich of New York, gave the insurance adjuster and field manager four states to oversee: Missouri, Kansas, Michigan, and Wisconsin. He also spent considerable time at the company's regional office in Chicago, and occasionally visited the home office in distant New York City.

Still, John had not given up his dream of joining the ranks of the landed gentry. He was elected secretary of the newly formed Farmer's Club and was reported to be superintending the erection of a new barn on his farm south of town in August 1898.[42] Oddly, this positive development was the result of family misfortune—his parents had gone bankrupt a few months earlier. On the basis of the large number of property transfers bearing their names during the 1880s and 1890s, it is clear that Martin and Mary Jane were speculating heavily in real estate, both in and around Marshfield. John often served as the middleman in these deals, though there is no way of knowing how much, if any, of his own money may have been involved. At times he would purchase land in his own name, then sell it to his mother, or both parents, a few months later. These transactions are explained by the fact that the elder Hubbles had returned to Springfield, leaving their son, with his knowledge of the local market and his legal training, in a better position to broker on their behalf.[43]

Whatever might happen in the long run, Jennie had reached the point where she could no longer cope single-handedly with five children, the oldest of whom, Henry and Lucy, had entered adolescence. The Jameses, her mainstay during John's peregrinations, were aging, a fact of life made worse by the diagnosis that her father was going blind. Nor can the possibility be dismissed that Jennie was more than a little embarrassed over her awkward position at a time when, short of war or working on the railroad, a husband's presence at home was taken for granted.

As reported in the paper, John's final return from Chicago occurred on October 8, 1899. He spent only a day "looking after business affairs" before heading north again. Three weeks later, Jennie shipped the last load of her family's belongings to Evanston, Illinois, after which she walked the children over to her parents' home so that the latest renter could take possession of the house. On Tuesday morning, November

14, relatives and friends gathered at the railroad station for a tearful farewell.[44] Jennie, a still-suckling Helen in her arms, boarded the hissing train, followed by William, Edwin, Lucy, and Henry. The conductor signaled the engineer and swung aboard as the great drive wheels first spun, then suddenly took hold. Just as suddenly, life in Marshfield was transformed into a memory.

CHAPTER TWO

"AN AWFUL MOMENT"

I

Very little is known about the year the Hubbles spent in Evanston, a quiet residential community on Lake Michigan's southwest shore. The event that marked their lives more than any other was the sudden death of Jennie's father in September 1900. William and Lucy James had only recently returned to Marshfield from an extended visit to their daughter's new home. According to the paper, the seventy-three-year-old doctor had greatly benefited from the trip. Though his eyesight was failing, he saddled his horse and rode out into the Missouri countryside on his accustomed Monday-morning rounds. After returning in the afternoon, he began vomiting and complained of severe chest pains. A fellow physician was summoned and on his arrival began preparing a mustard plaster. As Dr. James was pulling up his shirt and telling his colleague how he wanted the plaster put on, he gasped and died. The funeral was held in the family home three days later. Jennie received the news by wire and arrived in time for the service. At her side was

Martin Hubble, who came over from Springfield to bid farewell to his old friend.[1]

In 1892, Graham Burnham, a Chicago real estate developer and brother of Daniel H. Burnham, then directing the massive construction for the World's Columbian Exposition of 1893, published a pamphlet titled *Wheaton and Its Homes*. "It is the purpose on the part of the citizens," Burnham wrote, "to make Wheaton a safe place in which to bring up and educate children." The flag that waved perpetually over the public school stood for more than a mere ornament. "It means that Wheaton people are patriotic, and that those who do not reverence the stars and stripes can have no place among us." Newcomers who were American citizens, believed in American principles, and had a high regard for the canons of morality would be cordially received. "Without taking on airs or appearing pharisaical the people of Wheaton may justly claim for their town that its moral atmosphere is equaled by few cities in the United States."[2] Whether John Hubble read Burnham's widely circulated pamphlet is uncertain, but it was almost certainly a touch of the "pharisaical" that piqued his interest in this seemingly charmed community of 2,400, located twenty-four miles west of downtown Chicago.

The seat of Du Page County, one of the wealthiest in Illinois, Wheaton had only recently been incorporated as a city. It boasted a streetlight on every corner, and by 1900 most of its several hundred homes had been wired for "incandescents," the vernacular for the new electric lamps. Thanks to the recently completed waterworks, indoor plumbing rendered cisterns and outhouses obsolete. The town's original earthen footpaths were being replaced with miles of sidewalk, leading to the removal of unsightly fences and the extensive planting of tall hedges and great hardwoods, which in the suffused light of October displayed a natural incandescence of their own. Property was a bargain when compared with other suburban enclaves such as Oak Park, River Forest, and Elmhurst; a cottage could be purchased for $500, a two-story home with a large yard for a few thousand. Rent payments of $30 to $40 a month provided comfortable housing for a sizable family with a live-in maid. Burnham, though prone to exaggeration, was not far off the mark when he noted that a family man with an income of $900 to $1,000 a year could enjoy a standard of living that would make him the equal of nearly all. Homogeneity was indeed the dominant theme. Social and racial differentiations, "the bane of the majority of urban centers," were declared alien to Wheaton, whose residents were dedicated to keeping it that way. Like those who made the decisions in ancient Athens,

Wheaton's male citizenry boasted that a more democratic society would be difficult to find.[3]

The moral virtues of this new middle-class existence were underscored by the city's long-standing ban on taverns and the sale of alcohol, reaffirmed annually at the ballot box. The fear of backsliding was met with frequent admonishments from the pulpits of the town's many churches, including the Methodist, First Baptist, Episcopalian, Congregational, Unitarian, and Lutheran. President Blanchard of Wheaton College was famed for his animated lectures on temperance, the texts of which sometimes appeared on the front page of the *Wheaton Illinoian*.[4] It was proclaimed that church membership included a larger percentage of the population than in any other community of Wheaton's size west of New England.[5] When the Hubbles moved into their rented home at 225 Franklin Street and joined the First Baptist Church in the summer of 1901, that proud number increased by seven.

By most standards of measure, including his own, the son of Martin Hubble had arrived, or was about to. In his latest photograph, the general agent of the Western Department of Greenwich Insurance Company looks every inch the successful businessman. Now forty-one and balding at the hairline, John has compensated by growing a thicker mustache, his owl-like glasses and knitted brow contributing to the aspect of a severe professor gazing into some ominous future. Waxing sartorial, he wears a suit of the latest cut, the raised collar of his gleaming white shirt heavily starched and perfectly pressed.

Starting at six in the morning, thirty-two trains a day plied the double set of Chicago & Northwestern tracks linking Wheaton and the Loop, completing the twenty-four-mile run through the flat countryside and a half dozen small towns in an average of forty-three minutes. When John was not in the field checking on one of the 600 agents in his division, he could usually be found at his desk in downtown Chicago. Widely known for his expertise as an underwriter, which had gained him membership in the Blue Goose, the fraternity's exclusive inner circle, he churned out long articles and thick pamphlets on all aspects of fire insurance, infusing the mundane subject with a moral urgency bordering on evangelical fervor: "The best definition we have found for civilization is that a civilized man does what is best for all, while the savage does what is best for himself. Civilization is but a huge mutual insurance company against human selfishness." While he conceded that fire insurance companies are not philanthropic institutions, "we do claim that the fire insurance industry is, perhaps, more than any other modern activity, the creature of civilization, and that it is impossible to conduct this business in such a manner as to reap a profit without bringing beneficent results to the nation as a whole."[6]

In the evenings John returned to Wheaton and the grand Victorian house on Franklin Street, a comfortable six-block walk from the downtown depot. First occupied in 1874, the house has changed little over the years and still retains much of its original flavor. Rich in oak, leather, and heavy wallpaper, the front parlor, dining room, and rear parlor were separated by great sliding doors, which controlled the flow of heat and kept the children from eavesdropping when their parents entertained. Eight bedrooms with large windows graced the second floor. The ornately lathed porches and the cavernous attic were for playing on rainy days; the full basement, with its echoing cistern and raging coal furnace, for snowbound afternoons. The favorite attraction was the turreted second-story bedroom containing a trio of circular windows topped by an exotic onion dome. The imaginative child who occupied the center pane was the master of all he or she surveyed—a Guinevere, a Launcelot, the Lady of the Lake, D'Artagnan, Galileo.

II

In September 1901, Henry, Lucy, and Edwin all entered Central School, dubbed the "Old Red Castle" by locals for its Victorian Gothic tower. Along with their classmates, they were watched intently from the ramparts every morning by Gus, Central's amiable custodian, who was responsible for ringing the tardy bell. If Gus saw that a student was a bit late but hurrying, he would continue ringing until the winded child had reached the safety of his classroom. But if he was seen dawdling, the ringing abruptly ceased and a detention slip followed, which explains why Gus occasionally arrived in the morning to discover that the tower rope had been cut.

Elementary and junior high school students occupied the building's first floor, Henry and his high school classmates the second. Though the tall, angular Edwin would not turn twelve until November 20, he was placed in the eighth-grade class of Minnie M. Riddell, two to three years ahead of the other students his age. He posted his best average of the year, a 90, at the end of the first term ending in October, good enough to rank him sixth in a class of about thirty. But this was before he began taking spelling, a skill he would never master despite a deep love of history and literature. Reading, with grades in the low 90s, was his best subject, followed closely by arithmetic, United States history, grammar, and writing. By the end of the year he had slipped to tenth in his class with a final term average of 85, barely good enough to rank him in the upper third.[7]

A more telltale sign of Edwin's feelings was his declining marks in deportment, a disturbing reminder to his parents of the emotional turbulence of his early school days in Marshfield. Miss Riddell, who was quite generous with E's (Excellents), gave him a G (Good) in the first term, which by midyear slipped to an M (Mediocre), then plummeted to a virtually unacceptable P (Poor) by the time summer vacation finally arrived in May. Albert Colvin, a classmate who lived on Main Street a block away from the Hubbles, remembered that "Ed" did not have any close friends in grade school. This apparently had little to do with his status as a newcomer. "[It] was largely his fault because he did not show any desire to be particularly friendly with any of us." He seemed naturally arrogant and had "a way of acting as though he had all the answers, so there was no intimate contact with him." Nor did Albert recall the youth being very much closer to his brothers and sisters.[8]

There was doubtless some pretty straight talk after John received his son's latest report card. It was becoming clear that Henry, although four years Edwin's senior, could not be counted on to advance the family name. Once described as a "totally carefree boy," Henry had taken what his sisters guardedly characterized as an "impractical" turn, his tastes "running to exotic things."[9] Rumors of mental instability later circulated, although these were denied by members of the family.[10] Henry's few surviving pictures are ones of a melancholy, abstracted presence bordering on the effeminate, favoring neither parent in looks. Despite his relative youth the years appear to have been unkind, as if he is already resigned to a rendezvous with disappointment.

Edwin responded as John knew he could. Much to the freshman's relief, spelling was absent from the ninth-grade curriculum, which was weighted toward the classics that his father had studied so assiduously at Drury College. With few exceptions, his term grades ranged from the low to high 90s in a series of demanding courses that included English, civics, physiology, biology, algebra, rhetoric, beginning Latin, and rhetoricals. He did equally well the following year, receiving helpful pointers in mythology and Caesar from John. The change in his deportment scores was even more marked. The eight E's and two G's stand in bold relief against the many M's and the notorious P of his first year at Central.[11] What is more, Edwin had gained an admirer on the faculty. Miss Harriet Grote, a teacher who was about to marry

Schuyler Colfax Reber, the son of Wheaton's mayor, saw something in her pupil that remained hidden from others. She later gained a measure of local fame for her prediction that "Edwin Hubble will be one of the most brilliant men of his generation."[12] Harriet's son, Grote Reber, would become famous in his own right by constructing the world's first radio telescope in his hometown.

Edwin was fourteen when he suffered the only serious illness of his youth, an appendicitis.[13] He was bedridden for some weeks following surgery, during which time he read books on astronomy to his heart's content and grew a reputed three and one-half inches.[14] Otherwise, little is known of the astronomy of his dreams. The letters written to Grandfather Hubble concerning the planets and the stars are lost, and the subject was not taught in any depth at Central School. The one person who might have fueled his interest in the universe was the superintendent and principal, John B. Russell, a former professor of science at Wheaton College. Described as "a prince of a man" by faculty and students alike, Russell often went out of his way to help those whose interests exceeded the bounds of the curriculum. Moreover, Russell had maintained close ties with Wheaton College, site of the only observatory within miles. Its sturdy 12½-inch reflector was housed in a conical building sixteen feet in diameter aptly dubbed the "Lemon." Wheatonian Herman A. Fischer, Jr., who worked summers as a computing assistant at the University of Chicago's Yerkes Observatory, remembered Ed Hubble as a loose-jointed kid famous for his high jumping, but he had no recollections of him hanging around the local observatory.[15] Still, Fischer was away much of the time, and it is possible that the youth visited the Lemon without Fischer's having known of it.

Edwin did much of his skyward looking on Sunday mornings while singing in the choir and attending services at the ivy-covered First Baptist Church, which boasted the finest location of any house of worship in the city, as well as the most scholarly pastor, the Reverend Robert Bryant. An activist in civic and religious affairs, John was chosen to serve as a church trustee; Jennie, who taught Sunday school, represented the church at the Baptist State Convention. Both took part in the decision to remove the simple frame building and erect a new structure in its place. On September 18, 1903, after the old church had been hauled away in sections, the foundations were laid, followed by the cornerstone the first week of November. By April the massive stone tower had risen above the surrounding buildings and trees. During the interim, the congregation met in the structure recently vacated by the Methodists, who boasted an elegant new church of their own. A diversion from the staid and ritualistic was also available on occasion, courtesy of a onetime professional baseball player and spellbinding evangelist named William "Billy" Ashley Sunday. The Chicago YMCA employee and recently ordained Presbyterian minister preached many of his early sermons in Wheaton, the home of his sister, and in surrounding communities, becoming the master of classic one-liners such as "Isadora Duncan is a Bolshevik hussy who does not wear enough clothes to pad a crutch."[16]

Removed though Wheaton was from the evils of city life at the opposite end of the tracks, all was not well in Arcadia. Graham Burnham's effusive pamphlet contained at least one sentence that John Hubble would have gladly deleted: "The county fair grounds located close to the city contain a fine track for speeding horses, which is in common use during a good portion of the summer and fall."[17] As a Missourian with an eye for horseflesh, John had no quarrel with racing, or, for that matter, with the betting that accompanied it. But the way to and from the fairgrounds ran close to the Hubble residence, and therein lay the rub. As a draw for recreational commuters from Chicago, the track encouraged patrons to evade Wheaton's ban on the sale of alcohol by carrying their own bottles with them on the train. By the end of the day, many a boisterous reveler returned to the depot half intoxicated. To ensure his children's safety on race days, John confined them to the back yard, much to the displeasure of Edwin and Bill, who yearned to join their buddies in the "den of iniquity." According to Helen, "Papa was sure they were paving their way to the lower regions by their low tastes."[18]

John went on the offensive by heading a small but vociferous group of reformers bent on closing down the operation. Their families were serenaded by drunken jockeys and threats were exchanged. Nicknamed "Little John" by some clever local, Hubble, like the gruff but compassionate giant of Sherwood Forest, held his ground; indeed, waited in vain for the threatened attack that never came.[19] Bristling with self-righteousness, he undermined his position with the already nervous business community by taking the local druggists to task for selling liquor to their friends under the counter. When a minister voiced a similar opinion, he was brought up short by someone who quipped, "I wish you would bump heads with 'Windy' Hubble."[20] The disparaging appellation stuck.

Toddlers abounded in the Hubble household. Three-year-old Helen had been replaced as the baby of the family by Emily Jane, who was born in the home on Franklin Street in June 1902. "Janie" was followed in turn by Elizabeth, the last of the seven Hubble children. By the time of "Betsy's" birth in February 1905, the family had moved to 301 Union Street, where they stayed little more than a year. The Hubbles then moved into the last residence they would occupy in Wheaton, remembered by Helen as "the house we loved."[21]

The rambling, two-story clapboard home at 606 North Main Street had been built in 1887 by John Colvin, co-owner of a downtown dry-goods store located across from the depot. When Colvin died in 1902 the property was left to his widow, Florence, and would eventually pass into the hands of their only child, Albert, Edwin's classmate and some-time chum. The Hubbles rented the sprawling first floor and three large bedrooms on the second, the adjoining four rooms having been made into a separate apartment. The floors, balustrades, and trim were all glistening hardwood, the ceilings of the living room, parlor, dining room, and downstairs bedroom rimmed with molded plaster. Huge bay windows, reaching from floor to ceiling in every room but the kitchen, provided excellent views of the lawn, one of the largest in the town, and the colorful beds of nasturtiums, asters, and roses planted by Jennie. Hinged wooden shutters and heavy drapes provided a barrier against the sun in summer and the cold in winter. The crackling living-room fireplace was banked only in the small hours between late autumn and early spring.[22]

John insisted that dinner be served at six-thirty, with every one of his children present and accounted for. Yet Helen remembered that Papa always waited a minute or two after the dinner bell rang to say grace, "thereby allowing a little discretion for the tardy arrival, but that was all. We all knew the rules and we understood very well that we were to follow them or pay the price." It was permissible for the children to ask their friends to dine, so long as Mama was warned in advance. Papa served the plates from where he sat at the head of the table, and then everyone burst into conversation about the events of the day. The meal had something of a "town hall" atmosphere, with Papa and Mama acting as referees, never allowing anger to manifest itself. After the main course was finished and the dishes were cleared, Jennie served dessert from her place at the foot of the table, whose leaves were never removed. There was homework to be done, and the children returned with their books and tablets to work quietly under John's watchful but approving eye.[23]

Jennie was never without domestic help, provided for the most part by the daughters of immigrant Scandinavian families. Helen remembered them as being fine cooks and fastidious housekeepers, but their presence did not mean that the children were permitted to shirk their own responsibilities. Everyone had specific duties to perform, including making the bed, hanging up clothes, and keeping their bedrooms tidy. "Those rooms were our own territories and Heaven help the one who invaded [them]. This was not our parents' decision but ours." Throughout it all an unruffled Jennie was remembered to have lost her temper only once. As was the custom when guests were invited to dinner, the children had been fed ahead of time in the kitchen. The dining-room table was beautifully arrayed and the little nut cups filled. Just before company arrived Jennie discovered to her horror that the cups were empty, and it was too late to send for more nuts. Deeply angered, she set off in pursuit of the culprits. "We were all so amazed to see our placid, calm mother in such a state." It turned out that Henry and Lucy, the oldest, were the guilty parties, and the duo reaped a desert they hadn't bargained for.[24]

Aside from the inevitable Sunday-morning trek to church, weekends took on the aura of holidays. The many games played by the children included flinch, checkers, charades, and a spelling contest that involved the adding of one letter person by person while trying not to finish the designated word. In the heat of battle the dictionary became more sacred than the Bible. Depending on the season, the hall closet was raided for bats, balls, tennis rackets, skates, and croquet mallets. The children followed the example set by their father, whose fine set of golf clubs sat amid their large store of sports equipment. John had gained membership in the exclusive Chicago Golf Club on Wheaton's southwest side. The first 18-hole course in the country, it became a major stop on the Aurora & Elgin Railroad, which carried wealthy Chicagoans to their summer homes lining nearby Merrill Drive and Hawthorne Lane. John regularly brought home engraved cups and ribbons, as did Bill, who demonstrated a natural flair for the game and might have gotten even better had he not been forbidden to play on Sundays for religious reasons. Edwin also scored well, but had little enthusiasm for the game. He was more interested in the good money to be picked up caddying. It was a standing joke among the children that Father had moved the family from Evanston to Wheaton because it had the finer golf course.[25]

Social life expanded as the children grew older. There were hay rides in the summer, sleigh rides in winter, parties of all kinds, dancing, taffy pulling, and apple bobbing. Together with friends, they often walked downtown to browse the row of modest shops lining Front Street, a dozen stores selling a wide variety of food, clothing, toys, and household goods. The nickelodeon claimed much of their spending money, as did the makeshift movie theater set up in the rear of the drugstore on weekends. "Popcorn Brown," son of the postmaster, had a shop on Front Street from which he dispensed heaping five-cent bags of his well-buttered product. At sixteen, a blond and musically inclined Lucy was old enough to accompany her mother to meetings of the W.C.T.U.; Henry, Edwin, and Bill attended the stage shows put on by the Fire Department, whose equipment continued to be pulled by teams of horses until 1917. The boys also went to evening lectures and occasional scientific demonstrations at Wheaton College. Stereopticons, the famed magic lanterns, were in vogue and often used by itinerant speakers for dramatic effect. Sensational advertising played upon the active imagination: an escaped convict from Andersonville prison would tell how he had tunneled his way to liberty during a lecture in the high school hall—adults twenty-five cents, children ten cents.[26]

Idyllic as they may seem from the distance of many decades, those days were no less enchanted to the children who lived them. Lora Fox, a classmate of Bill's and a friend of his sister's, remembered turn-of-the-century Wheaton as a good clean town with "no really bad children." The most serious trouble anyone got into was being suspended for cutting the school-bell rope.[27] Helen Hubble simply recalled it as a time when life was less hurried—"a time for the little things that make life so pleasant."[28]

IV

Edwin reached his full stature during his junior year in high school. His siblings liked to claim that he was six feet three inches, the same height as their father.[29] In fact, he was an inch shorter, but fully a head taller than every other boy in his class with the exception of friendly giant Jack Alexander, who measured six feet four. Broad at the shoulders but narrow through the hips and almost perfectly proportioned, Edwin displayed the graceful carriage of an athlete, though it would be some time before the rather lanky youth reached his mature weight of

185 pounds. His brown, slightly wavy hair retained a glint of auburn from Jennie's lineage and was parted on the left, the same side he reserved for the camera, a self-conscious pose seemingly practiced in the mirror. His well-shaped ears were set close to a broad, full head, the nose was high and straight. Clear hazel eyes gazed out from beneath level brows and heavily shadowed lids; the sculpted chin was dimpled, the teeth strong and straight. Youthfully handsome, Edwin would grow even more so as character and time deepened the contours of his patrician face.

While Helen described her brother as "very much a typical boy," those closer to Edwin's own age harbored rather different memories. Lora Fox, who lived across the street from the Hubbles for several years, remembered Bill as being much more friendly and outgoing than his older brother.[30] Albert Colvin, who knew Ed better, felt that his friend had changed very little since grade school, for he was still acting as if he had all the answers. "He always seemed to be looking for an audience to which he could expound some theory or other." At a class party given by twins Catherine and Gretchen French, Ed decided to demonstrate his acrobatic skill. He picked up an expensive dining-room chair and placed it on his chin, leaning far back. The chair suddenly slipped and crashed to the floor, breaking off a leg, much to the annoyance of those present. Yet such an awkward bid for attention did not translate into a desire for intimacy; Ed remained as standoffish as before. A puzzled Albert had the feeling that "he sought more of a challenge than we could provide. His goals were high and . . . he thought we would drag him down."[31] More than this, Albert believed Ed to be a dreamer; like Jules Verne, he seemed to have a scheme for everything, an assessment with which the Hubbles all agreed.[32]

Helen noted that even before Edwin's college days he had become interested in the recently established Rhodes Scholarships. He may have gotten the idea from a series of large ads placed in the local paper by President Blanchard of Wheaton College, which was caught up in a financial crisis. The central message was that a college is better equipped than a large university to prepare students for the life ahead of them. Wheaton had produced no Rhodes Scholars, but three other Illinois colleges had, the implication being that lightning might yet strike locally.

Edwin continued to pursue the classical curriculum he had opted for while a freshman, with its emphasis on Latin and German. He turned fifteen in the middle of his junior year, and his report card contained nothing but 90s for final marks. As a measure of his application he earned his highest grades—100, 99, and 97—in spelling, his Achilles' heel in the past. He did almost as well in eleven other subjects, which, in addition to languages, included history, chemistry, English, and geography. Conspicuously absent from his records are courses in advanced mathematics; the future scientist would graduate from high school without taking anything beyond the required algebra and geometry.

Once again he was found wanting in deportment. In contrast to his sterling academic marks he received low 80s, a full ten points below most of his peers.[33] Class valedictorian Olive Stark, who stands between Edwin and Jack Alexander in the back row of their senior picture, was able to shed some light on this mystery more than sixty-five years later. Edwin never accepted what his teachers said, Olive recalled, and it was clear that they did not like being questioned by a smart aleck.[34]

Edwin's emotional immaturity in the classroom was offset by his prowess in the gym and on the athletic field, where his exploits put him in an elite class second only to Wheaton's immortal Galloping Ghost, Red Grange. Central School suited up its first men's basketball team in the 1904–5 season, Edwin's junior year. Scheduling proved difficult because there were so few area schools to compete with, but *The Wheatonia*, Central's annual, reported that only a scattering of defeats were entered against a long string of victories. Expectations were high the following November when Coach Howard Pinckney's squad of seven took the floor, six of whom were starters in the days before rule changes reduced the number to five. Will Leonard, Larmon Smith, and Harry Brooks played the forward positions, Ed Hubble was at center, and Ray Matter and Lawrence Fischer were guards, with Ernest Guild the lone reserve. Community interest in the team expanded exponentially as the victories accumulated. The Wheaton College gymnasium, where the home games were played, was usually sold out, and many of the players' families accompanied them when they went on the road, often spending a good portion of the night traveling by train. After going through the season undefeated, the maize-and-blue Wheatonians met the Ottawa High six for the state championship. In what proved a disappointing mismatch, Wheaton drubbed their outmanned opponents by a score of 46–10. Edwin was singled out in the paper as the hometown hero for having scored a game-high five baskets. It was also reported that the team would "mingle" at the home of senior Rose Mills. "The girls are so proud of the boys that a reception can only give vent to their feelings."[35]

Wheaton's football program had an even shorter history than basketball. According to Albert Colvin, the boys sometimes got together after school or on Saturdays to scrimmage in a vacant field. One day they became conscious of a rather unusual spectator, a Negro named William Henry Newburn, who, at eighteen, had recently dropped out of the University of Chicago's Morgan Park Academy, where he had been a scholarship student. Newburn was impressed by some of the raw talent on display and offered to mold the group into a team by drawing up plays and scheduling games. The boys accepted and volunteered among themselves to chip in the not inconsequential sum of one dollar per player per game as a form of compensation. They also agreed to pay for their own uniforms, consisting of padded sweaters and knee pants, dark socks, and leather shoes and helmets. The squad of fifteen, accompanied by grade school mascot Dink Weldon, played what may have been its first game against an experienced Elgin Academy in October 1905. They lost by a score of 6–5 on their opponent's home turf, but the local paper extolled it as a moral victory. Edwin, who loved physical contact, was in his element at the position of tackle, where he more than held his own. But like many of his teammates' parents, Jennie and John were fearful that their son might be seriously injured and nearly forbade him to play. The reward came at the end of the season when, with a proud Coach Newburn looking on, the members of the squad accepted the first school letters ever awarded by Wheaton High.[36]

One of the few surviving articles from this period is a small paperbound notebook containing entries on the track team during Edwin's senior year. Elected captain, the dauntless all-arounder turned in the kind of performances that gave rise to the legends so revered by the Greeks. While he did well during the tryouts for upcoming meets, he was one of those special athletes who exceed all expectations once the preliminaries are over. In a dual meet against archrival Elgin High, he virtually carried the day single-handedly by winning the pole vault, shot put, standing high jump, running high jump, discus, and hammer throw, then followed up by running the second leg for the victorious mile relay team. Only in the broad jump did he falter, placing third with an effort of 18 feet 4 inches. Though the youth would never experience quite such a day again, he came close in a number of meets held later in the season, writing himself into the school record books in the high jump, pole vault, hammer throw, and discus. He capped a brilliant spring at the Northwestern University Interscholastic Meet on May 6, 1906, clearing the high bar at 5 feet 8½ inches. Not only was the effort good enough to claim victory; Hubble was hailed in the papers for establishing a new state record in the event.[37]

V

Summer jobs were scarce, but members of the athletic teams were given preferential treatment. Edwin would struggle out of a warm bed before daybreak and ride his bike to the train station to pick up the morning papers from Chicago, then deliver them to the residents of the still-slumbering town.[38] The most sought-after jobs were on the horse-drawn ice wagons. Edwin and Bill both joined crews that ambled through the streets watching for "ice cards" in the windows. They chipped off the designated amount with a long pick and then, with a set of heavy metal tongs, hoisted their freezing burden onto a heavily padded shoulder and carried it inside, where it was deposited in the top of the icebox. Aside from building muscle, the main fringe benefits were provided by housemaids and doting matrons who plied the familiar youths with freshly baked rolls, pie, and cookies rarely shared with the have-nots on the crew, occasionally giving rise to ill feelings. Less desirable were their jobs as "water boys" on the local road crews. Captives of the blazing sun, the brothers had no free goodies or refreshing chunks of ice to offset the workmen's incessant demands for more buckets of fresh water.[39]

During the summer between his junior and senior years, Edwin was allowed to join a surveying crew in northern Wisconsin, a job arranged by his father, who knew the head surveyor. This experience, according to Helen, was the high point of her brother's early life. When he returned, the boy in him seemed to have vanished; "he was a man in his own right."[40]

Part of the summer earnings were saved; the remainder went for "frills," such as clothing that did not fit John's definition of "commonsense needs." Girls were sometimes treated to ice cream at the downtown soda fountain, but aside from an occasional flirtation Edwin showed little romantic interest in the opposite sex. When he began walking Lora Fox's older sister Mayme home from basketball games, no one took it seriously, and neither did they.[41] A restless Edwin must have gained a degree of perspective while staring for hours into the campfires of the Wisconsin woods. Now that he had pondered the lure of the wider world, the insularity of Wheaton held no lasting appeal, especially to one who dreamed of navigating the stars.

The academic record compiled during his senior year was as respectable as that of the year before. He earned A's in every one of his eight subjects, finishing with an overall average of 94.5. After starting the fall term with an 85 in application and an even lower 80 in deportment, he raised these marks to 90s, which held steady the rest of the year.[42]

Graduation exercises were scheduled for June 14, 1906, a Thursday evening, in Gary Memorial Church. Even those who had a child in the class of twenty-seven, the largest in Wheaton's history, were in for a long ordeal. In addition to multiple selections by the Wheaton High School Girl's Trio and a commencement address, every student in the class delivered a customary oration, ranging from the proposed "Annexation of Canada" to "The Value of Discomfort," which must have seemed comically ironic to many in the cramped audience."[43] While waiting his turn, a terrified Albert Colvin spent the evening repeating the merits of "The Irish Land Question" to himself. He remembered being so distracted that he missed most of Ed's recitation on the history of the senior class, but Albert felt certain that his self-possessed friend did well. "He probably enjoyed the experience."[44] Near the end of the evening, the tables were abruptly turned when Superintendent Russell rose from his seat to announce the class honors. "Edwin Hubble, I have watched you for four years and I have never seen you study for ten minutes." The stunned audience, which included his crestfallen parents, was deadly silent, and a mortified Edwin still remembered it years later as "an awful moment." But Russell had only paused for effect, a kind of final repayment to the bright but headstrong youth for the trials of the past four years. "Here," the superintendent continued, a smile inching its way across his roguish face, "is a scholarship to the University of Chicago."[45]

CHAPTER THREE

"A THING
SO OUTLANDISH"

I

At eight minutes past noon on May 1, 1893, a massively rotund Grover Cleveland completed his brief address to an estimated 200,000 chilled spectators silhouetted against the fogbound shores of Lake Michigan. His chubby index finger simultaneously depressed a large ivory button, setting into motion the mechanical ballet that most had anticipated since midmorning. Old Glory broke forth 300 feet above the President's head, trailed a second later by the properly respectful banners of Christopher Columbus and his adopted Spain. Some 700 additional flags unfurled brilliantly across the 600 acres that comprised the neoclassical White City; fountains sprang to life, shooting so high that umbrellas began popping up everywhere; the great Allis electrical generating engine, god of the Palace of Mechanic Arts, sputtered and took hold, while huge folds of drapery fell from the giant Statue of the Republic, her gilt arms gesturing toward the now mythical western frontier.[1] The World's Columbian Exposition, popularly known as the Chicago

World's Fair, was officially open; before its scheduled closing in No-vember, some 27 million visitors would come to gaze upon the future, one that so bewildered the normally poised historian Henry Adams that the universe itself seemed to be tottering.[2]

In this still-infant age of the skyscraper, one of the best views of Chicago was to be had from atop the exposition's giant Ferris wheel. Looking down and to the north, riders glimpsed four great limestone structures ornamented with what, in the wavering distance, appeared to be live gargoyles. This was no mirage. The seemingly old medieval bastions housed the new University of Chicago, whose visionary pres-ident, William Rainey Harper, a man of Theodore Roosevelt's propor-tions in figure, face, and temperament, was planning "to build the city gray."[3]

Thirteen years later, on October 1, 1906, a matriculating Edwin Hub-ble stood shoulder to shoulder with his fellow University of Chicago freshmen in the massive convocation room of Hutchinson Commons, one of four neo-Gothic structures making up the so-called Tower Group. Led by the president, faculty, and trustees, the service consisted of hymns, prayers, and readings. As in almost every university ceremony invoking pomp and circumstance, the program concluded with the singing of the "Doxology," with no thought for the scattering of Jews and other non-Christians in the audience.

The setting could not have been more appropriate for a youth whose sights were already fixed on England and a Rhodes Scholarship. Chi-cago's building committee had chosen Gothic to represent what they hoped to achieve with their university, founded in 1890 with a com-bination of Rockefeller millions, hundreds of thousands in Baptist sub-scriptions, and Harper grit. Gothic arching spires and vertical lines were seen as glorious and majestic, as striving upward toward the divine, reflecting Harper's belief that "the university is the keeper, for the church of democracy, of holy mysteries, of sacred and significant traditions."[4] Soaring Mitchell Tower, the new axis of the campus, was modeled directly after the famous Magdalen Tower of Oxford. And it was also to Oxford that architect Charles A. Coolidge had traveled for the plan of the great dining hall, which he had found in the original at Christ Church. The buildings of Oxford and Cambridge had lasted five hundred years; it was believed that those of Chicago would endure equally well.

In the assessment of many, Chicago was the finest institution of higher education west of the New England seaboard from the day it admitted its first class in October 1892. The initial roll of faculty included eight college presidents and a future Nobel laureate, Albert Michelson, head of the Department of Physics. Also on the staff were John Nef, a former professor of chemistry at Clark University; Eliakim Hastings Moore, who came down from Northwestern as professor and head of mathematics; and Rabbi Emil Hirsch, the first Jewish scholar to teach in a Christian university. Philosopher-educator John Dewey would join the faculty within two years, lured, as were many of his colleagues, by Harper's policy whereby promotion and salary were directly related to research and publication. Looking ahead, a euphoric Albion Small, head of sociology, believed that Chicago was uniquely qualified to realize what he saw as the main goal of his own field: "to organize a *Novum Organum* of all the sciences which contribute to an understanding of life." In the words of a reflective President Robert Maynard Hutchins in his 1929 inaugural address: "If the first faculty had met in a tent this still would have been a great university."[5]

II

As an educated businessman and active member of his country club, John Hubble knew the intellectual value of a university education as well as the practical benefits of developing social contacts. It was decided that Edwin should live on campus rather than commute from Wheaton every day. While the institution of fraternities and sororities had no place in President Harper's plans, as he thought them anti-intellectual, his proposal that they be barred from campus had been roundly defeated by the faculty at its very first meeting. Among the fraternities to form a Chicago chapter was Kappa Sigma, which had received its national charter in 1904. The social organization quickly gained favor among star athletes and other young men with high profiles on campus. Edwin, whose accomplishments in football, track, and basketball were well known, joined Kappa Sigma's autumn pledge class, along with fellow Wheatonian and athletic giant Jack Alexander. The two roomed together their freshman year after settling with their brothers in Hitchcock Hall pending the acquisition of a chapter house.

Constructed on the Oxford plan, Hitchcock was unabashedly proclaimed "the most distinctive and unique . . . of dormitories in the whole category of American institutions, contributing much to the beauty of the far-famed 'battlement towers' of the 'City Gray.' "[6] The usual monotonous row of cells was dispensed with, and five separate sections were built in their place, each with its own stairs and entryway but joined with its neighbors by a long, intricately tiled corridor. The five units had individual charters and their own elected officers, yet all were dedicated to furthering what was termed a "home-pervading atmosphere" and a "strong, virile fellowship among the members of the hall."[7] In the mornings, before nine, the residents met in private dining rooms for breakfast, which typically consisted of grapefruit, cream of wheat, and hot chocolate, while lunch and dinner were served in the great hall of Hutchinson Commons, overlooked by the censorious and universally ignored portraits of John D. Rockefeller, President Harper, and some of Chicago's most influential trustees. For snacks and conversation there was the Shanty at the corner of Fifty-seventh and Ellis, a board-and-batten vestige of the Columbian Exposition run by the motherly Mrs. Ingham. Hitchcock's well-appointed library doubled as a party and reception room for distinguished campus visitors, who spoke on subjects ranging from evolution and belles lettres to aeronautics and intercollegiate athletics. The club room down the corridor often rocked with late-night piano and vocal renditions of popular "rags." It was probably in the dormitory's small, dome-shaped card room that Edwin mastered the sleight of hand with which he sometimes dazzled his sisters and their friends.

Social life created divisions as well as community. The young men and women with whom Edwin associated tended to be of the upper middle class, possessing the time and resources to devote to frivolity. Working-class students and the commuters—numbering more than half the student body of 5,000—concentrated harder on getting an education and, on average, earned higher grades, although the sixty "plebeians" who occupied rival Snell Hall were lampooned in the university yearbook, *Cap and Gown*, for exhibiting mental traits resembling a cross between a barrel of monkeys and a balloon.[8] This division was heightened by a running conflict with the administration, which had little patience for "sophomoric traditions." Hazing, rushing, and similar practices were discouraged, but for students struggling to create a heritage the only models were the high jinks of the all-male schools of the East Coast. Having just turned seventeen, Edwin loathed his fraternity's more civilized requirement that he attend university dances. He soon learned that popular girls could look out for themselves, while a mere freshman was left with the wallflowers and a bad conscience. He usually took the gentlemanly course, though regretting it later: "They all danced badly and they had pins in their belts that stuck," he lamented.[9]

Equally humiliating was the fledgling custom of the "green stamp," which had been hatched by fellow student and gadabout "Win" Henry. Despite much protesting, every freshman was required to don a bright green beanie as a means of promoting class spirit. This debate was rekindled in the councils of student government each autumn, and always with the same result; dissenters were gleefully overridden by upperclassmen who had experienced the same embarrassment and were set on revenge.[10]

Edwin would remember his undergraduate days as a time of great financial stress. The Entrance Scholarship he had received, which was one of dozens awarded to high school seniors throughout the country, supposedly had been given to a second student by mistake, leaving him with only half the money he had counted on.[11] Yet the letter confirming his windfall included a voucher good for full tuition during all three quarters of his freshman year. The remaining costs, including room and board, books, laundry, and matriculation fees, totaled about $280.[12]

Strong disciplinarian though he was, John Hubble hardly qualified as a skinflint; his children were well dressed; the large home he rented was tastefully furnished; there was money for dinner parties, music lessons, and books. Having placed his son on campus, it is unlikely that he would have left him to his own financial devices, thus risking failure, which was unthinkable. While Edwin stated that he worked his way through the university by tutoring fellow students and holding down summer jobs, his record of participation in a variety of sports and other extracurricular activities casts doubt on this version of his personal history. That he worked and was less well off than some of his friends may have been true, but he also developed the habit early on of taking full credit for everything he accomplished.

Betsy remembered that her brother had "only one thought in his mind, and he wasn't going to let anyone else bother that." Edwin was determined to become an astronomer. But "Papa," Helen insisted, "wouldn't have let him go through school if he was going to be a thing so outlandish."[13] Well aware of his father's resolve in this matter, Edwin decided to play a waiting game. He would take the scientific and technical courses required for advanced study in astronomy while satisfying the curricular prerequisites for admission to law school. He later admitted that he was biding his time as much as anything, hoping against hope that some turn of events would prompt a change in his father's attitude.[14]

His first-year courses included algebra, analytic geometry, trigonometry, two quarters of inorganic chemistry, surveying, English, physical culture, and descriptive astronomy. A friend and fellow athlete observed that "Ed was brilliant in math and lightning fast in his work."[15] But this was long after Hubble had gained renown as an astronomer and greatness could be read back into his early life. Edwin himself was always sensitive to his limitations as a mathematician and later regretted not having taken more than one course per quarter in the subject.[16]

The undergraduate teacher who left the deepest impression on him, indeed the only one he ever spoke about with fondness, was Forest Ray Moulton, associate professor of astronomy. The eldest son of eight children, Moulton was named Forest Ray by an intemperately romantic mother who thought her son a "perfect ray of light and happiness in that dense forest."[17] Moulton received his Ph.D. in astronomy summa cum laude from Chicago in 1899, three years after joining the faculty as a combination assistant professor and graduate student. While working toward his doctorate he was invited by Thomas Chrowder Chamberlin, the grizzled chairman of geology, to participate in an investigation of Earth's origins.

In the early nineteenth century, the French astronomer Pierre Simon Laplace had hypothesized that the planet had been born out of a vast cloud of fiery gas thrown off by the spinning sun. Some of this gas had liquified into a molten sphere, which has been slowly cooling ever since. After the primordial sun, which was 5.5 billion miles in diameter, had ejected what would become Earth and its sister planets, it contracted, reaching its present diameter of 865,000 miles and a rotational velocity of 270 miles per second. In testing this theory, Moulton imagined that all the planets were reabsorbed by the sun. His calculations indicated that, immense though the star is, it could not have had sufficient momentum to hurl off Laplace's gigantic rings of matter. Following an exhaustive examination of photographs of the solar eclipse of May 1900, Moulton and Chamberlin developed their planetesimal hypothesis, published in 1904.[18] According to this theory, the gases given off by the sun and the other stars form small chunks of matter, which are then collected and fused together in dynamic regions or "knots" of the spiral nebulae. Although the Moulton-Chamberlin hypothesis no longer stands by itself, it provided a basis for current theories and was just coming into its own when an eager Edwin took his first of several courses with Moulton. His professor's undergraduate textbook, *An Introduction to Celestial Mechanics* (1902), became badly outdated over the years, but it always retained a cherished place on Hubble's library shelf.[19] Less convincing, so far as Edwin was concerned, were Moulton's public lectures devoted to reconciling astronomy and religion.

III

On an autumn afternoon in 1906, Edwin answered a knock on his door and found himself staring into the face of a living legend. Although he was only forty-four years old, weighed a mere 150 pounds, and was without a hint of gray in his dark, wavy hair, Amos Alonzo Stagg had long been dubbed the "Grand Old Man" of the U. of C. campus. The director of the Division of Physical Culture had come to Chicago in 1892 at the insistence of William Rainey Harper, his onetime professor of Semitic languages at the Yale Divinity School. Harper's dislike of fraternities was balanced by a love of intercollegiate athletics; he had watched in awe from the sidelines while Stagg, as both an undergraduate and a divinity student, pitched Yale to five consecutive baseball championships, and so excelled on the gridiron that the diminutive end was named a member of Walter Camp's first All-American football eleven in 1889. Lacking oratorical skill, Stagg voluntarily left divinity school, took a degree in physical education, and was ready when Harper summoned him. As Chicago's football coach, the nonswearing teetotaler, who celebrated each victory by downing a pint of ice cream, displayed a genius for both the practical and the bizarre. The inventor of the tackling dummy was also the architect of heart-stopping plays such as the end around, the hidden-ball trick, the double reverse, and the flea-flicker. Starting in 1896, Stagg's teams won six Western Conference (later Big Ten) titles and tied for another; five of his teams went undefeated, nor did he ever fall to nearby upstart Notre Dame.

Stagg demanded to know why Hubble wasn't out on Marshall Field. Surely he was aware that this was the first afternoon of practice. Edwin explained that he had wanted to play college football more than anything, but he had promised his mother, who considered the sport brutally dangerous, that he would give it up after high school. Stagg understood. He invited a distraught Edwin to come down to the field anyway, where he could watch from the sidelines until the coach had a chance to speak with his father. Sports reporters and photographers were present to look over the fall prospects; the next day Edwin's picture, which had been taken without his knowledge, appeared in the paper. "Coach Stagg," the *Record-Herald* trumpeted, "has landed another star 'prep' for the maroon forces. Hubble, the Wheaton High School 'phenom,' is the latest acquisition."[20] An incensed John Hubble happened to see the photo and immediately came down from his office in the Loop, breathing fire and insisting on an explanation. Stagg attempted to run interference, but his pleadings were useless. A promise had been made; Edwin would not play football, and that was that![21]

Stagg was not a man easily dissuaded. Edwin started bringing home weekend guests, who happened to be members of the team. John was polite to the earnest youths, but unyielding. After hearing them out and reaffirming his position, he made it clear that there would be no sleeping late the next morning, for church awaited. The chastened players reported back to Stagg, who dispatched a different pair the next weekend, with the same result. A desperate Edwin finally decided on a more cerebral approach. His father was a close follower of professional baseball, so he began keeping track of all the injured players. After satisfying himself that the sport was no less dangerous than football, he approached John with his findings. If what he had found was true, John countered, then his son had better not play baseball either.[22] While this episode had its comic side, the pain and humiliation lingered. No one but himself, Edwin recounted, could truly understand what the renunciation of his favorite sport had cost.[23]

With his manhood on the line, he had to do something to redeem himself in the eyes of his fraternity brothers and friends. Ironically, his parents' prohibition against playing football did not extend to boxing, which had recently been dominated by the flamboyant and iconoclastic Jack Johnson, the son of an ex-slave and the first Negro to claim the world heavyweight championship. Boxing was not a university sport, but the Fifty-third Street Y.M.C.A., which had several rings and a training room, was only blocks from campus. Edwin, who weighed 175 pounds when he enrolled in the university, fought in the amateur heavyweight division, and, by his own account, soon sparked the interest of Chicago promoters, who urged him to turn professional and fight the champion.[24]

While the youth may have been the victim of some wishful thinking concerning his potential in the ring, he soon participated in his first athletic competition for the university. At six feet two inches, Hubble was touted in *The Daily Maroon* as the natural understudy for "Long John" Schommer, star center of the varsity basketball team.[25] Practice began on December 4, and hopes were high for a conference championship. However, when the athletes returned from Christmas vacation, Edwin and two other freshmen who had made the team received some depressing news. According to head coach Dr. Joseph Raycroft, a new conference rule barring first-year students from participating on varsity teams was to go into effect immediately. Only when playing noneducational institutions, such as Amateur Athletic Union and Y.M.C.A. squads, could freshmen join upperclassmen on the court. "The rulings of the Conference," mourned the sports editor of *The Maroon*, "have practically shot the Varsity basketball squad to pieces."[26]

This same rule extended to track, relegating Edwin to the freshman team during the 1907 indoor season. The Wheaton "phenom" raised some eyebrows during the first practice of the year by clearing the high bar at 5 feet 10 inches, setting an unofficial Bartlett gymnasium record. This also happened to be John Schommer's event, as was the shot put, in which Edwin worked out. The track reporter for *The Maroon* anticipated a fierce struggle of wills when Edwin became eligible for the varsity, and predicted that he would press the talented Schommer "hard for his laurels" in the coming year.[27]

In a major reversal of form, Edwin continued to score better in practice than in the heat of competition. Not once in the five meets of his freshman year did he come close to equaling the mark posted during his first practice, nor would he match his state high school record of 5 feet 8½ inches. His two victories, one a jump of 5 feet 4½ inches, the other 5 feet 6, came against nonconference teams, the weakest on the schedule. His two best efforts, both 5 feet 8 inches, were posted in back-to-back meets with the freshmen squad from Illinois, garnering him a second and a third. In the shot put, his other major event, he finished no higher than second with a season best of 37 feet.[28] The Old Man, who doubled as varsity track coach and kept meticulous records on the progress of the upcoming freshmen, must have wondered whether something was ailing young Hubble.

I V

Sam Shelton, who had walked downtown to take part in the local Fourth of July celebration, was stunned when he glanced across the street. There was Edwin Hubble striding across Marshfield's courthouse square in the direction of some female celebrants. Sam looked on while Edwin placed his hands under the arms of a pleasantly startled Edith Case, the young woman Sam would marry after the two finished college, and lifted her high in the air. He was amazed by the change in his childhood companion, who had been transformed from a tall, gawky ten-year-old into "a towering and handsome young man."[29]

Edwin had come back to visit relatives and spend part of the summer working on his father's ranch. A Springfield newspaper clipping from a family scrapbook describes him as "eighteen years of age, six feet three inches in height, and one of the best all round athletes of the University of Chicago." He got his name in the paper for undertaking what was considered a "record walk." Starting from the Hosmer dairy farm seven miles northwest of Marshfield on a warm Saturday morning, Edwin covered the approximately twenty-two miles to the Springfield home of Grandfather Hubble in four hours.[30]

Much had changed in the eight years since the Hubbles had left Marshfield. Both of Edwin's maternal grandparents were gone, and the homes in which he had been born and reared were occupied by strangers. Still, the James family remained prominent, thanks largely to Edwin's favorite uncle, Jefferson Beauregard, who owned a prospering jewelry store and was now serving his second term as mayor, an office he would hold longer than anyone else in the town's history. Edwin was developing a flair for stylish clothes, stimulated in part by his Uncle Jeff's well-earned reputation as a dandy. The gregarious mayor wore custom-tailored three-piece suits laden with gold chains and elegant fobs in winter, exchanging them for bold sport coats, white ducks, and a boater with the onset of summer. While His Honor's taste in clothing may have raised some eyebrows among Marshfield's no-nonsense citizenry, Jeff James was a man revered for his compassion and generosity. Edwin's stepcousin Virgil, Jeff's oldest son, had been plucked, crying and terrified, from one of the "orphan trains" funded by the federal government after the Civil War. The jeweler had walked the floor that stormy night with the one-year-old clinging to him and began adoption proceedings the next morning.[31]

The retired Martin and Mary Jane Hubble, with whom Edwin stayed during weekend visits in Springfield, were as spry as ever, and deeply involved in compiling the oral history of the city's early days. The family insurance business was now in the hands of Edwin's Aunt Emily Jane, "Janie," a merry, plump, auburn-haired spinster whose voice, together with the family Christmas tree, formed some of Edwin's earliest memories of Springfield.[32] Aunt Janie was involved in all things pertaining to the D.A.R.; what she didn't know about the city and the background of its residents was deemed inconsequential. Janie's younger sister, Louisa May, was almost equally well informed, and had enhanced the family's social position by finally snaring a shy local bachelor, one Mr. Dickerson, who took his December bride on frequent trips to Europe, where they traveled in high style on the interest from his lucrative investments.

Doting aunts that they were, the two could not restrain themselves when Edwin formed his first and, according to his sisters, only serious romantic attachment before marrying years later. Little is known of the young woman named Elizabeth whom Edwin met that summer in Springfield, except that Helen remembered her as being "a very lovely, charming girl," with whom her brother had fallen in love. Unfortunately, Edwin's aunts "had a severe case of D.A.R. ideas about any of us seeing boys and girls that did not meet their standards," Helen wrote. For whatever reason, Elizabeth did not measure up and the relationship was doomed. She apparently saw things more clearly than did Edwin and decided to go her own way, using his plans for the future as her excuse; never, she told him when they parted, could she hope to rival his love of Mars and the distant nebulae.[33]

V

Having shed the green stamp, stigma of his freshman year, Edwin began the first of another three quarters largely devoted to the study of mathematics and science. The atmosphere in the Department of Physics was electric; in November 1907, Albert Abraham Michelson, professor and head of the department, had been informed by the Swedish Academy of Sciences that he was the winner of the Nobel Prize for his highly accurate means of measuring the speed of light. Michelson's was the first Nobel ever awarded to an American, and it so happened that Edwin was enrolled in the department's Electricity and Light course, taught by the hard-driving associate professor and Michelson protégé Robert Andrews Millikan, a Nobel laureate in the making. Edwin had already taken Millikan's class in Mechanics, Molecular Physics, and Heat, thus positioning himself for special consideration when Millikan was ready to select his student laboratory assistant for the coming academic year.

Still in search of a chapter house, the young men of Kappa Sigma again occupied rooms in Hitchcock Hall. Studious but not to a fault, a fun-seeking Edwin was photographed with several roistering fraternity brothers and fellow athletes. Grinning like Cheshire cats, all, including basketball stars John Schommer and team captain William Georgen, are arrayed in boaters and sport coats, the teddy bears on their laps doubtless having been won from some unwitting concessionaire.

It was during Edwin's sophomore year that he took part in a classic act of mischief whose potential consequences would haunt him for months to come. Alienated by the cloud of austerity that perpetually enveloped the students of the neighboring Divinity School, the boys of Kappa Sigma took it upon themselves to change the atmosphere. After laying in an ample supply of fresh eggs, the brothers waited at their windows for the cleaners to deliver the dignified black suits that were the trademark of the aspiring theologians. The yellow-and-white bombs had barely found their targets before the culprits were called on the carpet by the administration. Their fathers were also contacted, leaving a shaken Edwin to face a "terribly embarrassed" John Hubble during his next visit home. Worse still was the fear that the prank might have done irreparable damage to his chances for a Rhodes Scholarship. The only person who seemed capable of putting the incident into perspective was Martin Hubble, who thought that his grandson showed promise.[34]

Contrary to the predictions of disaster after freshmen had been declared ineligible for league competition, the varsity basketball team had gone on to tie Minnesota and Wisconsin for a share of the Western Conference championship. All but one of the starters were back for the 1907–8 season, and this, coupled with the addition of four talented sophomores, including Hubble, gave rise to hopes for one of the most successful campaigns ever. Competition began with a tightly scheduled swing through the Midwest in late December. Five games were played in less than a week, including a lopsided blistering of the Des Moines Y.M.C.A. squad on Christmas Day. After routing visiting Columbia University in early January, the undefeated Maroons launched their conference season by more than doubling the scores on Indiana, Iowa, and Purdue. Then they went on the road to face archrival Wisconsin on a bitter evening in January. Nothing Chicago did could crack the Badgers' tenacious defense, and the weary team returned home at 3 a.m. on the short end of a 27–17 score.

From this point on the season was all Chicago's. After downing Illinois, the squad was slated to entertain Northwestern, the perennial conference pushover. With the regulars watching the action from the bench, the reserves pummeled the Methodists 41–6. For the first and only time during the season, Hubble stepped out from behind Long John Schommer's shadow. The substitute center responded by "throwing" seven baskets, tops among all scorers, while at the same time shutting out Culbertson, his woefully overmatched counterpart.[35]

In a nip-and-tuck rematch against Wisconsin, the screaming Bartlett gymnasium crowd of 2,000 was rewarded with a 24–19 victory, tying the two squads for the conference lead. With a possible national championship at stake, touted as the first in the history of intercollegiate athletics, the Maroons returned to Madison for the rubber game. The five starters played the entire contest without a substitution. With the final whistle about to blow and the score tied 16–16, Schommer, in a display of his All-American form, hit the deciding basket from long range, and Chicago was glory bound.

The University of Pennsylvania, the reigning Ivy League powerhouse with over twenty victories to its credit, was heavily favored to claim the national championship by besting the Maroons in their upcoming three-game series. Besides being underrated, Chicago had the added advantage of playing the first game on its home floor. The March 21 contest was as hard-fought as those against Wisconsin. The score remained deadlocked until near the very end, when the worn-down Quakers suffered two critical defensive lapses that resulted in a 21–18 Chicago victory.

The second and, as it turned out, final game of the series took place in Philadelphia four days later. Chicago was represented by a small but vocal contingent of rooters that included university president Harry Pratt Judson, recent successor to William Rainey Harper. "From the first blow until the last," *The Maroon* reported, "the contest took on every aspect of a championship struggle." As usual, Schommer was everywhere; his work on both offense and defense was described as flawless. Even the Quaker fans went wild when the broad-shouldered Achilles made two sensational baskets, one with his back toward the goal, the other from midcourt. In a virtual replay of the first contest, the score was knotted numerous times, with Chicago's greater strength and endurance asserting themselves late in the second half. Following their 16–15 victory, the new national champions climbed aboard a train and headed south "to recuperate." Edwin paid his first visit to the Capitol, where both houses of Congress were in session. A boat excursion 200 miles down the Potomac River and Chesapeake Bay to Old Point Comfort followed, as did a specially guided tour of the navy yards at Newport News. The much feted team finally returned to campus on March 31, a Monday morning, where they were met by a crowd of cheering students, faculty, and alumni. A week later, championship medals were awarded to the top six players on the squad, leaving Edwin, who had played in only two of twenty-three games, empty-handed.[36] Ironically, he had not even accumulated enough time on the court to earn a varsity "C," prize emblem of athletic accomplishment. Still, he was a member of the nation's greatest basketball team and must have consoled himself with the thought that Schommer's eligibility had come to an end; no longer would he be forced to play Patroclus to Achilles.

The recently completed winter track season, which paralleled basketball, had done did little to bolster Edwin's self-confidence. Competing under the legendary Stagg for the first time, he had again gone head to head with the more physically mature Schommer. He did not capture his first point until the team's sixth meet, a 55–31 trouncing of Illinois. As reported in *The Maroon:* "Maddigan and Schommer did the expected in taking one-two in the shot put, the interest in this event being furnished by Hubble, who beat out both McCord and Litt of Illinois [for third]." "Schommer," the article continued, "displayed rare form in the high jump despite his strenuous activities in basketball." Long John took first by clearing the bar at 5 feet 9⅜ inches, an event in which Edwin did not place.[37]

Then disaster struck, or so it seemed. While preparing for a dual meet with Purdue in early May, Schommer "split" a leg muscle and apparently was lost for the season. Thrust into the limelight, Edwin responded by garnering a total of 8 points in his three events. After tying teammate Roy Maddigan for first in the high jump with an unimpressive mark of 5 feet 2 inches, he went on to place second to Maddigan in the shot and third in the hammer throw, helping to seal a narrow 62–55 Chicago victory.[38]

Chafing from inactivity, and perhaps a little concerned over these developments, Schommer defied the predictions of his athletic demise and came charging back. He captured first place in both the high jump and shot put during the next conference meet against Illinois, giving him a total of 10 points to Edwin's 1½. Only in the high jump against Wisconsin during the last regular meet of the season did the sophomore gain a measure of satisfaction by equaling Schommer for the first and only time. When the individual scores were tallied at the end of the year, Long John led his team with 68½ points, while Edwin, who had hardly pressed Schommer "for his laurels," came in ninth with a total of 15.[39] Although he had not scored a point during the all-conference meet held at Marshall Field in June, Chicago had prevailed, owing largely to Schommer and future Olympic sprinters Noah Merriam and Clare Jacobs, making Edwin a member of another championship team. Yet, as in basketball, he was not recommended by the Old Man for a varsity "C."

VI

June 9, 1908, marked the official completion of Edwin's sophomore year. He was awarded a two-year Associate in Science degree, which was then standard at Chicago. While grades for individual courses are kept private by the university, he had studied hard enough to earn honorable mention for his achievements and was granted the Junior College Scholarship in Physics, plus the coveted slot he had been angling for as Robert Millikan's laboratory assistant.[40]

The astronomy course titled Introduction to Surveying, which Edwin had taken under Assistant Professor Kurt Laves, a native German and University of Berlin Ph.D., would soon pay practical dividends. During a reunion with Wheaton classmate Albert Colvin years later, Edwin told of working the summers of 1908 and 1909 for the Chicago, Burlington & Quincy Railroad. He was assigned to an Iowa-based engineering crew responsible for conducting extensive surveys in preparation for the C.B.&Q.'s westward expansion.[41]

Although the chronology is muddled, he seems to have worked one other summer in the "Big Woods" around the Great Lakes, the genesis of much Hubble apocrypha. According to Helen and Betsy, one of their brother's duties required that he ride several miles to the nearest town on horseback to obtain camp supplies. A large bear crossed his path during one such trip, attracted by the unfamiliar scent of the packhorse. The curious bruin proceeded to follow Edwin and, as he told the story, was getting dangerously close. Unarmed except for a small pocketknife, he punctured a sack of sugar, leaving an irresistible trail of sweet powder on the ground. The bear was soon distracted and Edwin made good his escape. "He tried to do things," Betsy noted, "to prove he was capable of doing them."[42]

In a later, more heroic, version of the story, Edwin claimed that the packhorse bolted on seeing the bear, scattering its load. Long after sunset, guided only by the faint glow from the stars, he led the horse into camp, the supplies, though badly battered and mixed, still intact. The tent he was sharing with two others was supposedly crushed by a falling tree during a thunderstorm not long thereafter, killing the man sleeping next to him.[43]

He described the towns along the border as "pretty rough." At dusk, while walking through the rail yards of one such community, he was accosted by two Italians who demanded his money. When he laughed and walked on, a knife was thrust into his shoulder. He spun round, knocking out one of the would-be robbers while the other took to his heels. Only after examining his fallen attacker and determining that he was not seriously hurt did he seek attention for his wound. This Bunyanesque summer of close calls ended with yet another. While the rest of the crew headed south, Edwin stayed in camp with a second man to finish up some work, after which they were to be met at the railhead by the final train of the season. But no train came, and their supplies were depleted during the futile wait. Having no choice, the two walked for three days before reaching an outpost of civilization. "We could have killed a porcupine or small game," a nonchalant Edwin remarked, "but there was no need, and there was plenty of water."[44] These were deeds his ancestors would have been proud of, deeds in which John Hubble, for all his commitment to the manly virtues, had never taken part.

VII

With the egging incident behind him and a Rhodes Scholarship still paramount in his thinking, Edwin began balancing his work in the sciences and mathematics with courses in French, Latin, Greek, political economics, and public opinion. Because Virgil, Cicero, and many of the other classics were already familiar territory, this breathing room enabled him to become more involved in a range of campus organizations and activities that could prove critical to the selection process.

Among these was the Blackfriars, founded in the winter of 1904 by Frank Adams, a frustrated student actor. Named after the ebony-cloaked Dominican order of Elizabethan London, the Blackfriars was inspired by other early collegiate musical comedy groups such as Harvard's Hasty Pudding and Princeton's Triangle. Membership was restricted to male students, and included one man from each fraternity as well as any campus actors "fitted for amusing themselves and others." With their first musical comedy, *The Passing of the Kahn*, an annual tradition was born.

Ranked below the superiors of the order—the abbot, the prior, the scribe, the hospitaler—were the forty-eight brothers, whose company "Friar Hubble" entered his junior year. Possessed of a reasonably good singing voice, he donned the garb of a Saxonian chorister for the burlesque "The Sign of the Double Eagle," which was billed in *The Maroon* as "the scream of the year." As in Shakespeare's day, all female roles were played by men, whose performance was summed up for the 1909 *Cap and Gown* by the show's press agent: "The costumes were effective and sometimes almost daring; and the stage evolutions of the large and pulchritudinous chorus were startlingly clever."[45]

Even as the production was going on, Edwin, whose accomplishments had come to the attention of the administration, was chosen to take part in a vaunted university tradition. "It gives me great pleasure to inform you," the letter from Harry Pratt Judson's secretary began, "that the President has appointed you as a Marshal of the University. . . . I hope that you will find it possible to accept this great honor; and trust that you will hold it highly."[46] Attired in full academic regalia, the nine marshals, together with their female counterparts, the aides, represented the student body at all convocations until their own graduation a year later. Edwin, who had inherited the family love for pageantry, immediately accepted the president's invitation. His proudest moment as a marshal came when, to the strains of Elgar's "Pomp and Circumstance," he joined the academic procession that led the class ahead of his on its final, most nostalgic walk across the commons.

Edwin recorded nothing of his experience as Millikan's laboratory assistant. Four decades in the future, long after both men had become world-famous and were reunited in California, Millikan would publish his autobiography, which likewise contains no mention of his erstwhile student. Thus, it will never be known whether Edwin was aware that the year of his apprenticeship marked a turning point in the physicist's life. The consummate teacher and author of textbooks had "decided to gamble" in 1908 by attempting to measure the electronic charge. Working first with a charged cloud of water vapor and subsequently with drops of oil—a medium that led to much "ragging" because of the Rockefeller connection with Chicago—Millikan developed the most persuasive evidence yet that electrons are fundamental particles of identical charge and mass. By late 1910 he was able to publish the value for an electronic charge; three years later, owing to refinements in his experimental work, Millikan set forth a revised value which would serve the world of science for a generation.[47] The Nobel Prize in Physics followed in 1923.

Amid everything else, Edwin had not given up his elusive dream of athletic stardom. It lasted through the fall quarter and into the winter term, until the appearance of this front-page headline in *The Maroon:* " 'LONG JOHN' SCHOMMER ELIGIBLE FOR BASKETBALL." During a special meeting of Western Conference officials, a vote had been taken in favor of waiving the three-year rule in certain instances, thus granting "the greatest center in basketball history" a reprieve. If Edwin had remained sufficiently composed to read the rest of the article, he would have come across his own name far down the page: "Of last year's substitutes, Hubble, Cleary and Kelley will be back."[48]

The dream was rekindled, if only for a short while, with the beginning of practice. Coach Raycroft, who some must have thought certifiable, moved Schommer to a guard position and installed "Big Hubble" at center. In responding to a swarm of queries about his actions, the embattled coach reminded the press that practice was not over and that the lineup for the opening contest against Indiana was still to be named.[49] When the whistle sounded against the Hoosiers two days later, Hubble and Schommer had exchanged positions. "Hubble played a strong game at guard," noted the reporter who covered the contest, "but the big fellow's size seemed to be in his way at times."[50] Coach Raycroft must have agreed, for he was demoted to the role of sixth man for the remainder of Chicago's first and only undefeated conference season. As against Northwestern the previous year, he turned in one stellar performance while substituting for injured left guard Pat Page. Eight of his season-total 28 points came during a 27–2 rout of the Minnesota Gophers. Besides scoring four baskets, he was described as a "tower of strength on defense."[51] When the 1909 conference champions were photographed for the Cap and Gown, Hubble was sporting his first varsity "C."[52]

His second came in track, where Schommer again dominated the high bar and other field events until the engineering major quit the team at the end of the winter quarter to play baseball. All might have gone well for Edwin had it not been for the sudden appearance of Schommer's doppelgänger in the form of underclassman William Lucas Crawley, an almost mythic all-arounder. Besides running both the 120 and 220 hurdles, Crawley entered the discus, shot put, pole vault, and high jump. The fleet youth chalked up an individual 21 points in a meet with Wisconsin, scored 16 against Illinois his next time out, and added another 21 against Purdue a week later, each time eclipsing Hubble's combined total of 11½ points for the entire season. After twice tying Crawley for first in the high jump, the onetime phenom claimed his letter and decided to forgo track his senior year, never having recaptured his record-setting high school form in three years of competition.[53]

VIII

"Study has been 'my middle name,' as the slang has it," Edwin wrote Grandfather Hubble in August 1909. "This summer I have taken nothing but Latin in preparing, as I am, for the Rhodes Scholarship Exams.

I have less than two months left now, as they come Oct. 19th and 20th."[54] He was writing from the nearly deserted Kappa Sigma house on Hyde Park's Kimbark Avenue, a three-story brick Victorian structure within easy walking distance of campus and the pleasures of the Lake Michigan shoreline. The charms of ancient Rome were wearing thin as the summer wound down, and he looked forward to four weeks of rest before the beginning of the fall quarter in October, hoping, somehow, to squeeze in a trip to Springfield, his nostalgia for Missouri piqued by a recent visit from Aunt Janie. Meanwhile, he would do his best to answer Martin Hubble's query about a seemingly unfamiliar celestial object.

The brilliant "star" visible in the southeast late in the evening was actually the planet Mars. Only once every fifteen years does it come this close to Earth, "so I reckon an immense amount of work will be done this fall on the Canals of Mars. The real problem will, I think, be in photographing these baffling specters." Edwin accounted for this cyclical phenomenon by noting that as the two planets revolve around the sun their gravitational attraction for each other causes perturbations, or disturbances in their orbits, sometimes drawing them nearer, sometimes driving them farther apart. As this cosmic rumba repeats itself, observers on Earth are treated to a spectacular show some six or seven times a century.

In truth, such gravitationally induced changes are so slight that they cannot explain this unusually close opposition of the spheres. Rather, it is a matter of what astronomers call sidereal periods, the revolutions of the planets around the sun. Eight sidereal periods of Mars (8×1.88089 years $= 15.047$ years) are nearly the same as 15 sidereal periods of Earth. Hence fifteen years after a close opposition—or a distant one, when Mars is at its faintest—the situation is repeated, or nearly so.

How Edwin could have made such a fundamental mistake is not entirely clear. Most probably it had to do with the astronomy curriculum itself. Aside from an early course in descriptive astronomy and another titled Observatory Work, everything taken through his junior year was almost pure mathematics, with no attention given to the study of the natural forces that would have been relevant to this problem. Only after entering Forest Moulton's introductory course on celestial mechanics during the winter quarter of his senior year would he realize that he had erred, something Martin Hubble would never know.

Edwin also alerted his grandfather to the anticipated arrival of Halley's comet, which was fast closing on the sun to keep its only rendezvous in the normal life span of a man. He offered the first evidence of his extensive reading in the history of science by noting that a second comet tracked by Edmond Halley in 1680 was thought by some to be the same harbinger of fate that had flared over England during the Norman Conquest of 1066. "Some identify it with the great comet which blazed forth at the Death of Caesar—B.C. 43—while a few declare this was none other than the Star of Bethlehem." Edwin went on to explain that it was William Whiston, a disciple of the great Isaac Newton and his successor as Lucasian Professor of Mathematics at Cambridge, who believed that the comet of 1680 had literally grazed the earth after the fall of Eden, triggering the Noachian Deluge in 2346 B.C.—"the year of sin." As far as Edwin was concerned, Whiston's theory "makes as good reading for me as the 'Blue Fairy Book' does for Emma Jane."[55]

IX

Cecil John Rhodes, the adventurous son of a Hertfordshire clergyman, had matriculated at Oxford in 1873, the same day as George Parkin, a young man of melancholy countenance but great personal charm. When the archimperialist passed away suddenly in May 1902, leaving a fortune of 6 million pounds, the fifty-six-year-old Parkin, who had become a recognized leader in the Church of England and an impassioned colonialist as well, was given the responsibility for putting the huge benefaction creating the Rhodes Scholarships into effect. A year later Parkin addressed a convention of college and university presidents in Chicago, telling them that they would form the committees responsible for screening the candidates from each of the then forty-four states and four territories which had been assured of admission to the Union. "What, exactly, were the committee members to look for?" someone asked. "Candidates likely to become President of the United States, Chief Justice of the Supreme Court, or American Ambassador to Great Britain would almost certainly satisfy the Rhodes Trustees," Parkin replied with a straight face.[56]

The requirements set forth by Rhodes in his will were anything but conventional. The diamond king was emphatic in his desire that those selected for scholarships must not be mere bookworms. Besides literary and scholastic attainments, the successful candidate should exhibit specific qualities of manhood—"truth, courage, devotion to duty, sympathy for and protection of the weak, kindliness, unselfishness and fellowship." "Moral force of character" and "instincts to lead and to take an interest in his schoolmates" were considered almost as important. Finally, a Rhodes Scholar must have exhibited a "fondness for and success in manly outdoor sports such as cricket, football and the like."[57]

As Edwin had written his grandfather, the qualifying examination, only the first of several hurdles, took place on October 19. "For a full year," he told a reporter from *The Maroon*, "I have looked forward to the competition for the Rhodes Scholarship, and have arranged some of my courses in order to get as much knowledge as possible of the subjects which are given in the qualifying examination."[58] The equivalent of Oxford's Responsions, the test was used to determine whether the candidates were capable of doing university work in mathematics, Latin, and Greek. Though it was not particularly difficult, less than half of those who sat for the examination passed, their downfall resulting from little or no training in the language of Homer and Socrates.

Since the outcome would not be announced until mid-December, Edwin was in for an excruciating two months. So was his classmate Winston "Win" Henry, the controversial architect of the green beanie, who had also taken the examination and was hoping to move on to the state competition in January 1910. Henry was none other than the younger brother of Huntingdon Henry, holder of the first Rhodes Scholarship from Illinois. A fellow university marshal and a member of the baseball team, Henry was also president of the Reynolds Club, a prestigious gathering place for undergraduate males, and a brother in Chi Psi fraternity.

In a brilliant flanking maneuver that Rhodes himself would have admired, Edwin ran unopposed for vice president of his senior class in December 1909, only one month before the interviews that would determine the scholarship winner.[59] That this step was well calculated is further underscored by the fact that he had accumulated enough credits to graduate in March rather than June, limiting his term of office to a brief four months but effectively checkmating Henry's accomplishments in the event both made it to the finals. With this honor came another—membership on the class executive committee—an appointment announced just in time for him to include it in his updated application to the state selection committee.[60]

If Win Henry and the other four Rhodes candidates from around Illinois were much on Edwin's mind, so too was John Schommer, who had undergone still another incarnation. The 1909 graduate and future member of the Basketball Hall of Fame had been hired to coach the varsity five.

Handicapped not only by his youth and by his membership on the team during the previous year, Schommer was faced with instituting rule changes that would affect the entire complexion of the game. No longer would a player be allowed to renew his dribble once the ball had come to a stop in his hands, nor could he raise his pivot foot off the floor once it was planted. As an attempt to replace physical contact with finesse, both charging and the use of hands and arms to block an opponent would be whistled as fouls.[61]

The team's early practices were described in *The Maroon* as a "real gloomy bear story." When the "scrubs" upset the varsity during a scrimmage on November 27, Long John called a "council of war" with his predecessor, Joe Raycroft. After much debate, the decision was made to suspend workouts until the beginning of the winter quarter. Although Edwin had taken part in the varsity's humiliation, privately his spirits must have soared, thanks to words he had been longing to read in *The Maroon* ever since his sophomore year: "Hubble seems to have the center's position clinched."[62]

Word suddenly came down concerning the results of the scholarship examinations. Edwin had passed, but Win Henry and four of the five other candidates had not, victims all of a poor acquaintance with Greek. The other successful candidate, one L. E. Elan of Greenfield College, would join him during final interviews before the committee on selections, whose membership must have given him further cause for hope. Only President Ramelkampf of Jacksonville College represented a small institution; the balance of power rested in the hands of Presidents James of Illinois, Harris of Northwestern, and Judson of his own university.[63]

He began soliciting letters of recommendation from his professors soon after learning that he was a finalist. Mathematician Herbert E. Slaught, from whom he was taking his fourth course, characterized the senior "as a gentleman in every respect; a young man of fine physical bearing, a good athlete, and all round fine fellow." Slaught specifically mentioned his student's recent election to the vice presidency of his class, "which shows that he has proved popular with his mates."[64] Robert A. Millikan was even more laudatory, his remarks bordering on an encomium: "I find [Hubble] a man of magnificent physique, admirable scholarship, and worthy and loveable character." As a laboratory assistant in physics he had proven himself to be "thorough, dependable and popular as a teacher, as well as able." "Seldom," Millikan concluded, "have I known a man who seemed to me to be better qualified to meet the conditions imposed by the founder of the Rhodes scholarships than is Mr. Hubble."[65]

The interviews were conducted on the campus of the University of

Illinois at Urbana-Champaign on January 10. Edwin left by train for Chicago immediately afterward, arriving at Bartlett gymnasium in time to take his place as starting center against nonconference foe Lewis Institute. He played the first half without scoring a point, giving way after intermission to Clark G. Sauer, a burly sophomore who split the nets for 12 points, which equaled Chicago's margin of victory. The conference opener, a 31–4 riddling of hapless Northwestern, took place five days later, with Edwin scoring his first three shots of the year. Playing his best basketball ever, the senior center racked up another 12 goals during a three-game run in which the Maroons outscored their opponents by a combined total of 79 points, laying low the "gloomy bear story."[66]

While the new team hero was savoring his hard-won spot in the limelight and the prospect of another conference championship and varsity "C," Edmund J. James, chairman of the committee on selections, announced that Edwin P. Hubble had been voted the 1910 Rhodes Scholarship from Illinois. The holder would spend the next three years at Oxford studying the subject of his choice, with an annual allowance of $1,500.[67] When Edwin telephoned Wheaton with the news, Betsy, who was only five, thought her big brother had won a boat on which the entire family would soon cross the Atlantic Ocean.[68] Asked about his plans by a campus reporter the day following the announcement, he stated that he had known of the award for too short a time to be certain. "However, I will select a tentative list of the colleges which I would most desire to enter, and matriculate in whichever one I can." As to the question of what he would study: "Although I have diverted most of my attention while at Chicago to the sciences, especially physics, I expect to take up law and international law at Oxford. Most excellent courses in both these subjects are offered in the English institution."[69] Circumspect and diplomatic, Edwin was already sounding like an Englishman.

CHAPTER FOUR

A SON OF QUEEN'S

I

Jennie Hubble was both shocked and amused by the contents of the unmarked package she had discovered on the doorstep early one morning. Some anonymous prankster, bent on having the last laugh on "Little John," had delivered a bottle of fine whiskey to the home of Wheaton's most notorious teetotaler. Wishing to depart the community on peaceful terms, she disposed of the liquor and never said a word to her temperamental husband, who was already in Louisville, Kentucky, preparing for the family's impending move.[1]

Besides, Jennie was preoccupied by more serious matters, the most vexing of which was her husband's declining health. A victim of chronic malaria, which he likely contracted in Missouri before the move north, the fifty-year-old insurance agent was occasionally bedridden by bouts of chills and fever lasting several hours. More recently, these symptoms were complicated by unexplained headaches, a puffiness around the eyes, high blood pressure, and a discoloration of the urine. His normally taut frame gradually took on a flaccid and bloodless aspect; he was often drowsy and easily fatigued, and was forced to cut back on his frequent rounds of golf. The doctors were at a loss and could only prescribe more rest and a reduced work schedule.

Having served as Chicago city manager for National of Hartford since 1908, John applied for and was granted a transfer to Kentucky, where he became the company's state agent in April 1909. Striving, as before, to swaddle his family in the protective membrane of small-town America, he rented a large white frame home at 928 Bland Avenue in Shelbyville, a thirty-mile commute by interurban to his Louisville office on West Main. Henry, who had attended Wheaton College for one year before dropping out and trying his wings without much success, soon joined his father's agency.

The Hubbles sensed that the summer of 1910 would probably be their last together as a family, and they drew close in the few months remaining before Edwin's departure for England. Helen remembered that her brother enjoyed the leisurely pace of life in Kentucky and spent much of the time brushing up on his Greek. A giggling Betsy rode around the neighborhood on her brother's broad shoulders and assumed the same perch when Edwin took his three youngest sisters to the circus. He also bought tickets to Maurice Maeterlinck's *The Blue Bird*, the dreamy, melancholy fantasy based on the quest for the Blue Bird of Happiness by Tyltyl and Mytyl, the children of a poor woodcutter. An entranced Betsy, Janie, and Helen memorized several lines and could be heard repeating them ad infinitum in the weeks to come. Yet it was an occasionally sentimental Edwin who seemed to have enjoyed the performance most.[2] Henry, who was described by his sisters as "sweet and nonaggressive," temporarily escaped the mundane world of insurance premiums, indemnities, and claims adjusting by giving the girls "fancy books," which he read to them late into the night.[3]

Bill, considered Edwin's intellectual and athletic equal by the family, had also come home for the summer. Dubbed Aeneas by his high school classmates for his adventurousness and love of Latin, the compact, ruggedly handsome "sea captain" of class prophecy had recently completed his freshman year at the University of Missouri. He was especially concerned about his ailing father and planned to transfer to the University of Wisconsin at Madison in the fall. Perhaps sensing that the family would soon be in need of another breadwinner, he had decided to major in agriculture in hopes of acquiring a farm of his own near the family property outside Marshfield.

Beginning at dusk, all eyes scanned the horizon's purple rim for the first glimmer of Halley's comet, destined to be at its brightest when Earth passed through the giant snowball's streaking tail late in May. Two years earlier, astronomers had detected cyanogen gas, a known poison, in the tail of comet Morehouse, stirring public fear despite assurances to the contrary. A French writer added to the mounting hysteria by speculating that hydrogen in Halley's tail might react with Earth's atmosphere, overwhelming the planet in a gigantic explosion. Next followed claims that the orbital calculations of the scientists were wrong; Halley's head would strike Earth somewhere between Boston and Boise, knocking the planet into the depths of outer darkness and dooming every creature on its surface. Some sealed their windows against the predicted fallout while others saved the comet the trouble by committing suicide. The more daring, like the Hubbles, stood outside and watched in wonder as Earth floated into Halley's luminous tail. Humorist James Thurber, who was sixteen at the time, remembered that "nothing happened, except that I was left with a curious twitching of my left ear after sundown and a tendency to break into a dog-trot at the striking of a match or the flashing of a lantern."[4] Had he not been otherwise occupied, Mark Twain would have relished the sentiment. As it was, John Hubble's favorite American author had his own prophecy to fulfill: "Here are those unaccountable freaks," Twain wrote of the comet and himself. "They came in together, they must go out together,"[5] and so they did.

On Sunday afternoons, the Hubble home was open to anyone who cared to stop by. Blue-eyed Lucy, now twenty-two and still living with her parents, took her accustomed place at the piano, while John played the violin and Bill the mandolin. Guests sometimes brought their own instruments; those, like Edwin, who didn't play gathered around the piano and harmonized. Virginia, ever the gracious hostess, made certain there was plenty of lemonade, tea, and freshly baked sweets. Looking on from afar, Helen and Betsy studied their towering older brother, who was about to disappear from their lives for three years. Both realized that Edwin had been singled out for special attention almost from the beginning; he was bright, good-looking, confident, and nice to be with. Yet it was also apparent that he did not relate to the family in the same way the others did. "With Edwin," Betsy thought to herself, "it was out of sight, out of mind. When he was with you, you were the only person in the world, but if you were away, he would forget you. His head was in the stars."[6]

II

On September 7, 1910, Edwin said his goodbyes and left Shelbyville by train, bound for Montreal. There he would join his fellow Rhodes Scholars and board the steamer *Canada* for the eleven-day voyage up the St. Lawrence and across the Atlantic. Though his son was about to reach his majority and had never been in serious trouble, John had felt it his duty to lecture him on the temptations and pitfalls awaiting him in the Old World. The tried-and-true Baptist also exacted a promise that would be virtually impossible to keep: Edwin was not to so much as taste liquor in any form, which presumably included French cooking.[7]

Towering in the confidence of twenty-one, Edwin was photographed on deck, his hands thrust nonchalantly into the pants pockets of his three-piece suit, the collar of the heavy cardigan under his jacket turned up, accenting the bow tie and stylish tweed cap crowning his broad, unlined forehead. There were other passengers besides the Rhodes Scholars, but the young men, who represented most of the great universities of the country, took little notice. In years to come, Edwin would reflect on the dubious sense of security that had marked the passage of the blessed to England. International suspicions had been dulled by nearly a century of peace and established order. Economic stability was taken for granted. There was faith in government, pride in independence. Almost every man believed he could better his condition by dint of hard work and honest thrift. Life, by its very nature, was competitive, and it was the role of government to ensure fair play when corruption threatened. Society, like science, seemed progressive, and youth looked eagerly forward to a future filled with promise. No one had dared suggest that the comet's return boded ill for England, as it had in 1066. Edward VII inherited Victoria's empire, and while the King had never been on good terms with his mother, he meant to keep every square inch of British territory. Edward, too, lent his name to an age—flamboyant, ostentatious, at times vulgar and strident, with picturesque contrasts of fortune and circumstance. Yet the last decade of the nineteenth century and the first decade of the twentieth had much in common. Most important of all, God was still an Englishman.

Once Edwin had received word of his selection, he had been responsible for getting himself admitted to one of the twenty colleges whose constitutions provided for the acceptance of undergraduates of the type specified in Rhodes's will. Like other scholars-elect, he had submitted a list of six colleges in order of preference, along with testimonials to his character, a college transcript, a statement of religious preference, and his proposed course of study. His grades had tailed off during his last two years at Chicago, and he finished with a B minus average.[8] Whether this had any bearing on his eventual placement is unknown. More significant was his father's inflexibility regarding his career; he must study law as opposed to astronomy, which meant that every college on his list had to conform to this fiat.

Queen's College, Oxford's sixth oldest, had been founded in 1340 by chaplain Robert de Eglesfield under the patronage of his mistress Queen Philippa, wife of Edward III. Eglesfield's plans called for a provost and twelve scholars, representing Our Lord and his apostles, thirteen chaplains, and seventy-two "poor boys" or "disciples," some of whom were to constitute the choir. Every day pea soup was to be served to the indigent at the gate while the Fellows, attired in fine blood-red gowns symbolizing the Crucifixion, were to be summoned to dinner by trumpet, a custom—minus the red robes—still honored in term time. Income had never met expectations during the late Middle Ages and the Renaissance, and there was always plenty of room to spare. To ensure its survival, Queen's admitted paying guests and thus acquired some of the most illustrious names on its rolls: religious reformer John Wyclif, cardinal and papal legate Henry Beaufort, and Beaufort's inscrutable nephew Henry V, the mythic hero of Agincourt. In 1726, well before a new building scheme was completed, Daniel Defoe described Queen's as "without comparison the most Beautiful College in the University."[9]

The *Canada* reached Liverpool in late September, and the scholars boarded a special railroad car provided by the government. When Edwin arrived at Queen's, he was assigned second-floor rooms on the farthest side of the inner quadrangle next to the library, his back to New College, which also dates from the fourteenth century. Chicago's architectural attempt to create instant history had suddenly given way to history itself. The corridors he walked echoed the footsteps of Joseph Addison, Jeremy Bentham, Walter Pater, and, in an ironic twist, the astronomer Edmond Halley. Within a stone's throw of his desk a host of other worthies—Thomas Sydenham, Christopher Wren, William Blackstone, Percy Shelley, and John Galsworthy—had begun the intellectual toil that would lead to fame, and in some cases fortune.

In what may have been his first letter home, written to his mother on September 30, Edwin thanked his "lucky stars" for having drawn Queen's. "The more I learn about Queen's the better I like it. My rooms are very comfortable and when I get them fixed up a bit will be ideal." His money was holding out and he was prepared to "live high" until the release of his first seventy-five pounds. He was even planning a trip to London for the purpose of buying "reproductions I wish on my walls."[10]

Warren Ault, a historian who had made the same journey in 1907, wrote vividly of the young man who had oriented the Americans during the train ride to Oxford:

> He was dressed in plus-fours, a Norfolk jacket with leather buttons, and a huge cap. He also sported a cane and spoke in a British accent I could scarcely understand. He was, in fact, an American Rhodes Scholar who had been at Oxford for two years. He had come to clue us in. Those two years had transformed him, seemingly, into a phony Englishman, as phony as his accent. I was put off by the sight of him and the sound of him and I resolved that Oxford would not do that to me.[11]

In addition to some atrocious spelling and erratic punctuation, Edwin's earliest letter reveals that he had undergone an equally remarkable transformation during his first few days in Oxford. As if by design, he began speaking and writing in that curiously affected "-egger, -ugger" jargon, which survived the prewar period, to die out gradually in the twenties and thirties. One did not talk the hours away; one jawed. One had a ripping good time; one's clothes were ripping, as were one's digs and the flowers one bought to make them more human. He had met Stolz of California, Bland of Ohio, and Crossland of Alabama—"all splendid fellows." Though not a Rhodes man, Edwards of New Zealand "is a mighty good sort." He had already lent Edwin his "wheel" (bicycle). "As soon as I get on a sound financial footing I shall straitway purchase one for my own." Having taken his preliminary exams, he was heading for a country inn along with three others to await the beginning of term: "[T]he ten days there will be the finest kind of time."[12] So alien was the prose that a bewildered Jennie might have double-checked the handwriting to confirm that it was her son's.

Longing for acceptance, Edwin quickly became acquainted with most of the Americans at Oxford, and was soon best of friends with the three Germans on his stair; "the real problem," he wrote a month later, "is to meet the Englishmen." Adding to his father's gray hairs, he considered it an honor when the previous year's rowing crew went on a drunk and piled into his rooms, making enough racket to awaken most of the college. "They . . . were very effusive in their assurances that I was a 'jolly good sort.' To day I dare say the most of them wouldn't recognize me on the street, but that is the usual thing." However, the captain was sober and had asked him to tea the following afternoon.[13] Meanwhile, his compatriots were having some good-natured fun behind his back. Jakob Larsen of Iowa, who was studying classics at Queen's, wrote: "We laughed at his effort to acquire an extreme English pronunciation while the rest of us tried to keep the pronunciation we brought from home. We always claimed that he could not be consistent, so that he might take a băth in a băth tub."[14]

III

Larsen remembered Hubble as being full of astronomy before he came to Oxford, and not terribly interested in jurisprudence. Skirting the real reason why he had opted for the law over science, Edwin explained himself by stating that he would have to make money after his return home. He applied for senior standing in hopes of being exempted from the Law Preliminary Examination, which was slated for the end of term in December, and would require an enormous amount of preparation. But the Warden of Queen's, who deemed his background in physics largely irrelevant to the task at hand, denied his request.

Within weeks he was complaining of too many pressing social obligations and the need to exercise to keep in condition. Though he had never before set foot in a racing shell, he decided to learn to row, a somewhat embarrassing undertaking:

> We freshmen are "tubbed" each afternoon. That is we are put two in a "tub," a two oared half-racing boat and commanded by an old oarsman who teaches us the science and art of rowing. It's a very slow process at first, but if I can learn I think I can make the college crew as a heavy man is much in demand.[15]

It was while he was tubbing that he got his first taste of the town-versus-gown contempt that in an earlier time had triggered riots leading to murder. "The boys do as they please as regards townfolks, and it is a very common sight to see them all going down to the river . . . in their 'shorts'—which are our track-suits. They all go right through the main streets." He hastened to reassure his mother, "You may be sure that I don't." Nor had his self-identity been totally eclipsed by his rampant Anglophilia, he claimed. "[E]veryone carries a cane—another convention which I do not conform with."[16]

He was soon second-guessing his desire for greater companionship. "My rooms are very popular as a 'salon,' as I intended they should be, but now that I must work so hard, it will be a bit annoying."[17] His three gabled windows overlooked the grave-studded churchyard of St. Peter-in-the-East, a melancholy prospect in the dismal autumn rain. The furniture in the larger of the two rooms, whose walls he described as "rambling and delightfully irregular," consisted mostly of odds and ends. A small painted cupboard held tins of loose-leafed tea, which he brewed for guests in the small fireplace with great ceremony. It also contained his growing collection of pipes and tobacco, for he had taken up his father's habit of smoking. A small "bedder" opened on one side, with just enough room for a narrow iron bed, a washstand, and a dresser. There were only two other rooms in the college cheaper than these, and in neither of them could he stand upright.

He sequestered himself amid a litter of books and more books during most afternoons and evenings, studying, reading, and daydreaming by the fireside, the cuckoo clock on the wall, a gift of his German friend Hans-Lothar Gemmingen, ticking off the minutes and crowing over the passing hours. When Jennie wrote that two young women he liked had recently married, he replied that many people believe one can learn to love almost anybody.

> It may be so, but I am still dreaming away & awaiting perhaps a rude awaking, wonderful realization, or more dreams upon dreams until the end of all dreaming. "What's the starlight of her glances, compared with the real stars overhead." The nearest approach to happiness I know is in dreaming, having nought to do with external things.[18]

His tutors were all pleasant men, yet three months of stupefying toil had left him strangely dissatisfied.

> Labour which is labour and nothing else becomes an aversion. . . . Work, to be pleasant, must be toward some great end; an end so great that dreams of it, anticipation of it overcomes all aversion to labour. So until one has an end which he identifies with his whole life, work is hardly satisfactory.

This lament was not intended for John's eyes, to whom Edwin wrote separately, and with less candor. Jennie's heart must have gone out to her troubled son, but she could do little to ease his anguish. Sharing his romantic inclinations, she may have pinned her hopes on another dream he revealed in closing, the dream of a destiny not decided: "I sometimes feel that there is within me, to do what the average man would not do, if only I find some principle, for whose sake I could leave everything else and devote my life."[19]

His spirits had improved markedly by mid-December, with word that he had passed the Law Preliminary Exam and the prospect of spending Christmas vacation in Germany. He had crammed an entire year's work into three months of study and was now in a position to complete his law degree in two years rather than the usual three. He reported the good news to his father in a businesslike letter written from London before crossing the Channel. As further proof that he was hewing to the straight and narrow, he described his first encounter with Parliament. Edwin, William Ziegler of Iowa, and Joseph "Bart" Worthen of Vermont had walked over to the House of Lords and demanded admittance on the strength of being American law students. "The rouse worked and we had the privilege of seeing the House of Lords sitting as the Supreme Court of Appeals of the Realm."[20] The day before, the trio had visited the Court of King's Bench—England's highest criminal tribunal—and thrilled when the Lord Chief Justice made his appearance.

Mr. Hazel, his law tutor, had given him six enormous books to take with him to Germany, and he would be reading history as well next term. The "Divers" loomed in spring, an oral exam on the New Testament in the original Greek. He was still rowing, and it seemed likely that he would make the College Eight. Boxing was also on his agenda, as was trying out for the varsity track team.

Most of his impressions of Germany were too vague to be set down in black and white. Still, the Teutonic preoccupation with order had not gone unnoticed. "Berlin is fairly a model city, clean, well planned, well regulated. London loses much by comparison." He had been able somehow to get a good look at Kaiser Wilhelm II and his wife, Augusta Victoria. Disregarding the Emperor's withered arm, he deemed Queen Victoria's chronically impulsive grandson "a very fine looking man, stern, but scarcely fierce," not at all like his recent caricatures in the British press. "His wife, Die Kaiserin," he added for Jennie's benefit, "is very motherly looking."[21]

Edwin's hopes of garnering a Blue in athletic competition were dealt a setback when he jammed a foot in the jumping pit, dislocating his ankle, which confined him to his rooms for two weeks. His situation was made more rueful by the fact that, for reasons of time, he had given up the place he had recently won on the College Eight, Queen's rowing team, and with it a good chance of making the Varsity Eight, something no American had ever done. German friend and fellow Rhodes Scholar Alfred Eugen Popp, who occupied the same stair, had also withdrawn, depriving Queen's "first togger" of its heaviest men and creating consternation in the rowing circle. While an embarrassed Edwin convalesced, Queen's weakened crew was roundly thrashed by Balliol, Christ Church, and New College. The men of New College then went on a tear, smashing every window in both quads, and reducing the glass conservatory in the Warden's private garden to shards. "New is a 'Blood College,' " he complained in a letter to Jennie, "and all its men are wealthy. The only punishment they will incur, will be severe fines and a couple of weeks 'grating' [confinement] I reckon."[22]

Jennie had been the one to broach the question on both parents' minds: Why hadn't their son written anything about church? When he was sixteen Edwin had taught a Sunday school class of nine- and ten-year-old boys back in Wheaton, a labor he continued to perform after entering the University of Chicago, where chapel was compulsory. Now he reluctantly admitted that he had fallen away from Sunday school, and "perhaps I haven't been to the Baptist Church as much as I ought. At any rate I intend going next Sunday." He preferred the more intellectual sermons delivered at nonconformist Mansfield College Chapel, with its unusual stained glass and elaborately carved stalls. When "a big man" visited Oxford, Edwin sometimes donned the required cap and gown and went to hear him preach at the University Church, St. Mary the Virgin.[23] Helen recalled that her brother once wrote of a brief flirtation with Catholicism during which a friend gave him a rosary.[24]

Rather than deal directly with the question of his wavering faith, he attempted to placate John by describing the many religious shrines he was visiting more out of historical interest than piety.[25] During a long bike ride covering fifty miles, he stopped at Dorchester to see the ruined abbey in whose consecrated precincts Cromwell's cavalry had quartered their horses and knocked the plaster saints from their niches. Bending low at the waist, the absorbed pilgrim made his way slowly across the flagstone floor, reading the inscriptions to the departed and yielding to an overwhelming sense of graveyard melancholy. With pencil and notebook he copied the epitaph of a young woman who had died in 1799, then sent it along to his brother Bill.

If thou has a Heart fam'd for Tenderness and Pity, consider this
spot:
    In which are deposited the Remains of a Young Lady whose artless
Beauty Innocense of Mind and gentle Manners,
    obtained for her the Love and Esteem of all who knew her. But her
nerves were too delicately spun to bear the
    Rude Shakes and Tottlings of this transitory World.
    They gave way. She Sunk and Died a Martyr to Excessive
Sensibility.[26]

He made an extra copy for himself and vowed to return one day. Elsewhere, he philosophized, "I shall be content to adore from a distance,
which, as Dante said, is productive of the very best of human efforts."[27]

IV

Edwin was most frequently seen in the company of Hawaiian-born Herbert Stolz, a Phi Beta Kappa graduate of Stanford who was studying
medicine at Queen's. Handy with his fists, the square-jawed, bull-
necked middleweight won a Varsity Blue in boxing by defeating Cambridge's best. Edwin wrote Jennie that the two had "stuck togather
from the first" and had become best friends. "When I feel lonely or
out of sorts, then I instinctively turn towards Herb, sure of finding that
true comradeship—that sympathy which charges the most commonplace of commonplaces with deep restful meaning."[28] During Eights
Week, the annual rowing competition with Cambridge, the two had
raced to see who would be the first to meet the "very pretty" daughter
of a Queen's don, one Miss Allen. Edwin won and went to lunch, to
the dance, and on a picnic with the family, while Herb, much to his
victorious friend's amusement, was captured by another don's wife—
"a horrid designing woman" with two "impossible daughters."
    It had taken some time, but the Americans were now "of the bunch."
Following the concert marking the end of the competition, Edwin,
Herb, and two other youths drew up chairs and sat under the colon-
nades of the front quad, its gray stones softly illuminated by Japanese
lanterns. They demanded Edwin's "mandoline," which he had purchased in Germany and taught himself to play. "We started singing
and soon had a good bunch of thirty men around—splendid looking
chaps in full dress. We set there until one o'clock before breaking up."[29]
    Other evenings ended on a macabre note. If a student sat at night

in a certain room off a Queen's stairwell, he would hear footsteps ascending the ancient treads. He would listen for a knock on the door or a voice, but nothing happened. When he opened the door, no one was there. Edwin relished a good mystery; together with Herb and a few others, he decided to seek an explanation. They sat in the room for several evenings, and, sure enough, heard steps on the stairs that paused at the door. Yet each time the hallway was empty. The puzzle continued and their studies began to suffer. Finally, in desperation, someone suggested an experiment. One of the group was sent to the bottom floor and told to climb the stairs, enter the room, and close the door. After a long interval came the ghostly follower, and with it the mystery's deceptively simple solution. When the old oak treads were pressed down, one after another, they stayed in place for some time before snapping back in succession up to the last step on the landing.[30]

There was even more whispered talk about the haunted bedroom overlooking the quadrangle on the first floor. It had once been the residence of Jeremy Bentham, and its dark history mocked the political theorist's dictum that the greatest happiness of the greatest number is the fundamental and self-evident principle of morality. Long after the utilitarian's departure, the young don who occupied the room informed the Warden that he was going on a walking tour along the Cornwall coast. He never returned. After some time the quarters were allotted to another man, who had no knowledge of the vanished don. The first night the newcomer stayed in the room, he awoke with a start: crawling toward the bed out of the dimness was someone whose clothes were sodden and whose twisted expression filled him with such terror that he fainted. Those to whom he described the incident the next morning recognized the intruder as the lost don. They said nothing, but the shaken resident was given a different place to stay. The following term the room was allotted to another new man, who experienced the same creeping horror. After a third attempt produced the identical apparition, it was decided to turn the quarters into a reading room, which was padlocked at sundown.[31]

Edwin, who knew this story well, had been in residence less than a year when word arrived from Germany of an accident that sent chills down his spine. A Queen's student and friend named Petrie had gone bathing in a lake with a friend and disappeared while the two were swimming toward an island. After a long search, the body was finally found lying at the bottom of twelve feet of clear water. "I myself can see the picture," Edwin wrote a friend of his brother Bill's. "Queen's

Coll. seems cursed with some sort of Water Spirit—can it be a Lorelei I wonder?[32]

Edwin gradually drew close to Elmer Holmes Davis, Indiana's contribution to the Rhodes class of 1910. The glib, cyclically pessimistic Phi Beta Kappan from Franklin College had undertaken the study of classical literature at Queen's and was a member of the same dining table as Edwin and Herb Stolz. Davis dreamed of becoming a journalist in New York after completing his degree, but the odds against making it in the city seemed dismal to one reared in tiny Aurora, Indiana. He and Edwin often talked of their plans for the future. When Davis once revealed his doubts by remarking that he would rather be first in the provinces than second in Rome, Edwin replied, "Why not be first in Rome?" Davis found this rather astonishing.[33]

The Yanks' home away from home was the American Club on Corn-market Street. The second-floor rooms were open daily in the afternoons, when members would drop in to catch up on events back home by perusing the large selection of American periodicals. Evening meetings consisted of speeches and formal debates modeled after the Oxford Union. Davis, who alternately served as secretary and president, debated Edwin on the question: "Resolved that this House condemns the recent action of the Conference of Governors attempting to revive the doctrine of States' Rights." Edwin described it as "a wild battle," ranging from "the doctrine of the Interstate Commerce Commission to the twilight zone of conflict between State & Federal authority." He was the Honourable Mover of the question, which carried 40 to 32.[34]

Present in the audience was the diminutive Tennessean John Crowe Ransom, who alternated with his friend Davis as club secretary and president. An unpretentious young man of pleasant features and tidy habits, the nascent poet was held in awe by his countrymen for what seemed his suicidal decision to read "The Greats" at Christ Church, generally recognized as the most prestigious of all the programs offered anywhere in the university. It was a rare Rhodes Scholar who spoke anything but English, whereas Ransom not only had been studying classical Latin and Greek for eight years but had actually taught both languages. Yet it was what Ransom did not know and Hubble did that most interested the Tennessean in the tall, self-possessed son of Queen's. "I remember at Oxford when I was reading Kant in translation, he was reading Kant in German, and this impressed me. And that when he borrowed books from me he returned them. He was very handsome."[35] Virginian William Alexander Stuart of Balliol, Ransom's closest friend and the future class historian, never forgot Ed Hubble's "jolly little trick" with a match. After scoring a debating point, he would light his pipe, flip the match into the air so that it described a circle, and catch it, still burning, as it came down.[36]

The cut and thrust of reasoned argument sustained Edwin's tepid interest in the law. He wrote his mother that the very best type of Englishmen seem to be in the Law School, and he had met several of them since becoming a charter member of the new Law Club, which served as a moot court. His first case, a complicated one involving contracts, pitted him against an Australian. As "consel" for the plaintiff he was concerned. His research confirmed that the letter of the law favored his opponent. All he could do was attempt to convince the three-judge tribunal that "my client is really justified in his claims."[37] He never said how he had fared, but he soon volunteered to argue an even more difficult case, one in which "the Law seems to be all on the other side. I shall simply . . . plea for absolute justice."[38]

Meanwhile, he wrote Grandfather Hubble that he was getting on fairly well with every subject except Roman Law, which is "awfully dry. It must be done, however, so I'll face it like a man." His greatest temptation was to "chuck everything and just wander in mind as well as body. Don't see where I'm going but I'm on my way."[39] In a rare moment of candor, he dropped his guard in a letter written to both parents after fourteen months at Oxford:

> Have decided to cut out all diversions and study nothing but Law. Sometimes I wonder whether it was best that I decided on Law. More and more I feel that I am a student rather than a business man and that the theory rather than the practice is the field in which I would best succeed. Not that I by any means dislike the idea of working— on the contrary I am anxious to begin, but I rather doubt the outcome. However I don't care much either way. My ambitions at present are a book, an easy chair and a fireplace. They will probably change some day, perhaps too late.

The sense of a higher calling divulged to his mother a year earlier no longer seemed so compelling: "Wishes have generally been realized in the past, how long this will continue is very questionable."[40]

He did not tell his parents that he was spending time in the company of Yorkshireman Herbert Hall Turner, Oxford's Savilian Professor of Astronomy and director of the University Observatory. A gifted mapper of the skies and a leading figure in celestial photography, Turner was an excellent lecturer and, in addition to scores of scholarly articles, the author of four popular books on astronomy, including the minor children's classic A *Voyage in Space*. The house assigned the Savilian Professor is located at the east end of the east wing of Radcliffe Observatory, one of Oxford's most beautiful structures. From its center rises an octagonal tower four stages high modeled on the Tower of the Winds at Athens. Edwin, accompanied by his host, likely walked round the tower on the balustraded platform, which is surmounted by a wind gauge and globe, supported by the figures of a heaven-wielding Atlas and a prodigious Hercules.

The nature of their discussions is unknown. Indeed, word of this association only came to light after both men were dead. Daisy Turner wrote that she and her husband had known Edwin ever since his undergraduate days. "When he came to lunch for the first time, and said good-bye, he said in his gentle voice, 'It's mighty good of you, ma'am, to have had me to lunch like this.' " The Turners were charmed by his colloquial manner, a vestige of his Missouri upbringing which sometimes surfaced in his letters to his parents and grandfather, an additional reason for affecting the English tongue. Mrs. Turner described another occasion when she had asked him to a dinner she was giving for a close friend. He did his three-card trick again and again without anyone spotting the right card. She could still remember her friend remarking, after he had left, "You said you had asked a Queen's undergraduate to dinner, but you never said he was an Adonis."[41]

V

Edwin, who was approaching his twenty-third birthday, now weighed 190 pounds, twenty more than when he had arrived at Oxford. Remaining narrow through the hips, he had filled out in the legs, chest, neck, and shoulders, giving greater definition to his athlete's lines. His eyelids had a way of straightening that narrowed his gaze to full concentration, and he had learned to use his commanding physique and accompanying demeanor to full advantage. While a certain captain argued with a Moravian count over who was to have the next waltz with a future duchess of England, he bore the charming little bone of contention away in his arms, "like a Young Lochinvar," to the dreamy melody "Night of Love." "It was great fun," he wrote Jennie, "it always is to look a man in the eyes and best him by sheer self possession."[42]

The dislocated ankle had prevented him from facing Oxford's other heavyweights for the right to compete against Cambridge for a Blue.

Based on recent experience, he was confident that he could have out-pointed the only man capable of giving him trouble. "Bill Ziegler and Herb Stolz will be the American Stars while I am left in my chair."[43] Nothing further was said about boxing in the surviving letters, but he later claimed that he had fought George Carpentier, the French heavy-weight champion, to a draw in an exhibition match.[44]

Beautifully proportioned, with handsome features that attracted women to the sport as never before, the 175-pound Carpentier had already claimed the European middleweight title by knocking out Brit-ish champion Gentleman Jim Sullivan in two rounds. Moving up to the heavyweight division, he used his lightning right hand to dispatch Bombardier Billy Wells, England's other great hope, in a mere 73 sec-onds, including the count. While the Frenchman may have sparred a few rounds with the young American, it seems doubtful that they ever boxed.

Oxford's baseball team, of which Edwin was captain, manager, first baseman, and relief pitcher, created something of a stir in the land of cricket when it debuted in the spring of 1911. Their first outing against a squad from London resulted in a lopsided 16–1 victory. "The English folk were very 'bucked' (favorably impressed) with the game, and wanted to know why we didn't play more."[45] He explained to them that the Americans weren't expected to introduce new sports. A rematch took place at the Crystal Palace Exposition on June 21, the day before the coronation of King George V and Queen Mary. Afterward, Edwin and his victorious teammates walked the streets of London in their white flannels until two in the morning, captivated by the unusual sights and sounds of Coronation Eve.

His hopes of winning a Blue were now tied to his performance in track. Herb Stolz wrote his parents in California that Oxford's inter-collegiate meets reminded him of his high school days when one en-tered almost every event. "Hubble of Chicago and I constitute the main part of the Queen's team. He goes in for three events and I try five."[46]

For Edwin these were the same events that had won him his second varsity "C," the high jump, hammer throw, and weight (shot put). In a pinch, he was also available for running the hurdles and the shorter distances. He succeeded in ousting the varsity high jumper during the spring of 1911, his first season of competition. A week later, he tied a Cambridge high jumper for first in a meet sponsored by the London Athletic Club, garnering him the first of many ribbons and medals. A steady flow of photographs and press releases detailing his exploits crossed the Atlantic, where his adoring sisters faithfully mounted them in the family scrapbook. His father also saw to it that news of his son's achievements made the papers: "Edwin Hubble, son of John P. Hubble, state agent in Kentucky of the National of Hartford," one article began, "is a Rhodes scholar at Oxford, and is doing things up brown."[47]

The opportunity Edwin had been waiting for came a year later, during the annual Oxford-Cambridge track meet in London on March 23. Though it rained all week, preventing any records from being set, he described it as "a magnificent meet." The tenacious duel ended in a 5–5 tie, with his friend Bill Ziegler saving the day for Oxford. Edwin did not win either of the two weight events he had entered, but he placed high enough to earn a Half-Blue and informed his parents that he was now wearing his "colors."[48] While it was a proud moment, his priorities had changed since his undergraduate days at Chicago; "sports," he informed his mother, "are not so important as they used to be so they cause me very little worry."[49] In a letter to Lucy dated only a few weeks before his triumph, he calculated that he had read three hundred books since leaving home, an average of five a week, only about a tenth of which were fiction. "[T]he time was, I think, well spent, and the ground wide, covering from Comparative Religion to Russian History."[50]

VI

Edwin had settled into different rooms on the opposite corner of the inner quad at the beginning of his second year. He was also chosen head of his table, making him responsible for the Rhodes men, both old and new, and some fifteen other Queen's scholars besides. He took particular delight in giving the Americans their first lesson in brewing a proper cup of tea. The memories of Queen's traditions and rituals would stay with him for life: the trumpet call that announced dinner; the medieval grace; the Latin song and Boar's Head procession on Christmas Day. Breaking his promise to his father, he partook regularly of the strong, thrice-brewed ale from Queen's brewery, the only college distillery left operating at Oxford, and developed a budding connoisseur's taste for French wines.[51]

When Queen's was selected to host the annual Rhodes dinner, Edwin sat at the head table with Lords Milner and Grey, several M.P.s, a university law don, "and many lesser lights in the educational and political world." He was held in thrall by Sir George Parkin, who had organized the Rhodes Scholarships abroad in hopes of cultivating future presidents and ambassadors. "Parkin . . . is a man of one idea," he wrote his parents, "an idea which permeates his entire life, his actions, his speech. That idea is the British Empire, how it shall be extended, centralized, and made to control the whole world." It is such men, "those who find some rule which they can accept wholly and fully, and live by that rule through thick and thin, who are considered great."[52]

He had traded in his American clothes for baggy plus fours and a Norfolk jacket. Contrary to his previous declaration, he now sported a cane and sometimes donned a black cape, which added to his already commanding presence in the shadowed streets and dark lanes of old Oxford. His social contacts had expanded to the point where he was receiving more invitations to dinner and high tea than he could accept; weekends in the country were common. "If only you could see your son now," he wrote Jennie from Trethill, Worcester, "visiting in a swagger English Home." His host's father was the High Sheriff of the county, a first-rate businessman, a town counselor, and head of half a dozen committees. "We dress for dinner. I come up to my room to find the evening clothes—shirt—collar—tie—everything laid out waiting." It was just like the stories one reads, golf in the mornings, bridge in the evenings. "The guest is given [the] run of the place and treated as one of the family. He looks out for himself, can boss the servants and hunt his own amusements."[53]

Mothers and fathers, as well as their daughters, were succumbing to his natural charm. During Eights Week a Queen's man named Atkinson approached Edwin, informing him that his cousin refused to leave a party without first meeting the American. Edwin had had his eye on the Welsh beauty all week, but he had not approached Atkinson for fear of breaching etiquette. Following the introductions, he talked fashion with her mother, politics with her father, and farming with her brother. "She and I—well I can't remember, but I think it was astronomy. She is really beautiful. I am very sure that I ought to go see the fruit farm on the border of Wales, soon."[54] As before, his spring fever eventually ran its course. "I shall without question return whole heart," he assured Jennie in another letter. "It will be a good many years before I shall even be in a position to think seriously of such things."[55]

With six weeks of vacation at Christmas, another six weeks at Easter, and four months in the summer, Rhodes Scholars had plenty of time to travel, and were expected to. They would often head for the Continent, spend themselves down to a shoestring, then hole up in some English village until term time and the next allotment of funds. Edwin favored the Sign of the Swan in Tetsworth, where, for a guinea a week, he was given his own room and served four meals a day. The village, located twelve miles from Oxford, consisted of one store, a post office, and two dozen cottages with thatched roofs. He studied like a grind during the mornings and dispensed sage advice to the card-playing statesmen of Tetsworth in the afternoons.[56] In the village of Boar's Head nearby, John Crowe Ransom had graduated to reading Kant in German.

Edwin and an English friend spent one of their holidays visiting another Oxford student in the Andalusian city of Cádiz. Their host's father was an exporter of sherry, and the young men toured the cellars, where they were introduced to the Twelve Apostles, a dozen giant casks of amber wine, ranging in taste from very dry to sweet. There was a visitor's book in which each guest signed his name before he met the Apostles and then, after downing generous samples of the twelve sherries, he signed his name again. The signatures were compared for consistency, leading to much teasing and laughter. Saying nothing about his own conduct, Edwin reported seeing many famous names in the book, including that of Alfonso, King of Spain.[57]

Germany claimed more of his vacation time than any other country. Riding his bicycle as many as 90 miles in eight hours, he had covered over 1,000 miles during his second visit to Central Europe in the summer of 1911. He got his first look at the Danube following a hard day's ride. It reminded him of the Ohio in size and color, but it flowed more swiftly than any other river he had seen, except the Niagara. Forgetful of Queen's Lorelei, he informed his parents that he was off for a swim. He was well out from shore when he was hailed by two policemen in a motorboat, who warned him that the river was forbidden to swimmers because of the dangerous current.[58]

It was an outwardly more self-confident and sophisticated Edwin who returned to Germany to spend the summer of 1912. He was drawn to Kiel, capital of Schleswig-Holstein in the northwest, and Germany's chief naval base. "The people are of a much finer stock than in most places—Here one sees the real Teuton—Blond and Strong. Flaxen hair, wonderful complexion, and blue eyes. And they are open hearted."[59] Coming directly from the rigid conventions of England, he found life in the city delightfully free and easy.

His deep love of England notwithstanding, Edwin saw enough of German military power to reconfirm an earlier judgment concerning the outcome of a possible European war. Over tea two rabid young socialists had first libeled the Bishop of London, then got on his nerves by projecting the glorious success of English arms in the near future. "It's just such conceited young fools," he bristled, "who will lead England into war."[60] The scenario was clear—England and France on one side, Germany on the other. If only Russia would stay out of the Balkans. Still, "it's almost impossible to conceive of really big trouble now adays."[61]

He also observed that the one and only thing the Germans go to extremes over is honor, which must be defended with saber and pistol. He had gained entree to the secretive dueling fraternities and had witnessed half a dozen saber fights, earnest affairs in which plenty of blood was let. The blunted but razor-sharp *Schläger* was not so desperate as the saber but required more skill. "One must not move the head a hair's breath—scarcely wink an eyelash during the whole fight—let his whole cheek be laid open, his ear sliced off or his nose divided." To act otherwise was to invite disgrace by dismissal from the corps. He was toying with the idea of taking up the *Schläger* himself; a German who wanted to learn how to box was eager to exchange lessons. All that was holding him back was the fear that the scars might not appear in the right place.[62]

He had recently taken up with a naval officer named Kruger, whom he described as "a wild reckless fellow of 30 who knows every one from the Admirals to the Musical Hall Artists." They swam together in Kiel Bay, and Kruger took his new friend to a private tennis club where Edwin called attention to himself by defeating the best players. "I've met a good many of the Officiers wifes. The women in Kiel are mighty good looking, but I'm beginning to realize how it is there is a pistol duel a month."[63]

Without stopping to think that his son might have been gilding the lily, John, waxing biblical, hastened to reassert the control that had rarely been challenged over the years:

"As a man thinketh, so he is," and when thoughts of bloodshed and conflict are uppermost in the mind . . . the finer and more lofty elements of character must suffer.

In the land and time you will live, the duelist scar is not a badge of honor, and while knowledge is a priceless jewel, the knowledge of the use of the sabre and pistol would have no value. You are equipping yourself for what I hope and believe will be a useful career. At any rate you have been entrusted with the ten talents and will be expected to render a good account of your stewardship.

I am relying upon the promise you made your mother and me, that you would not touch nor taste alcoholic liquor in any form. I believe you will live up to this, but why subject yourself to the temptation. . . .

Character, my son, is the real test and you have had every opportunity to fit yourself for great things. From what I know of Kiel, I cannot but believe that you have not selected the best place for your vacation.

I know you will realize that I am preaching this sermon, because I love you, and not because I want to take away any pleasure and in any wise restrict your action. Be honest, truthful, sober, and use the opportunities given you and you will be sure to accomplish great things for yourself.[64]

Not yet satisfied that he had done justice to his Christian scruples, John continued to sermonize after his son had made his way back to Oxford. He raised the issue of drinking again, then turned to a more urgent matter:

I also want to impress upon you the necessity of staying with your church. Do not get out of the habit of attending. It is the greatest comfort in life, and the man who thinks he can get along without his Creator, will wake up to a sad realization of his error, perhaps too late.[65]

None of John Hubble's other letters to Edwin survive, but Helen characterized some of them as "blistering."[66] Properly chastened, the entitled son replied, skirting the issues raised by his father while pushing the right emotional levers. "I mean business now and shall do real studying."[67] He would be spending the next few months reading German texts in the new field of corporate law and had already asked John to send him information concerning the Kentucky bar examination; some friends in Chicago were doing the same for Illinois. A mollified John pronounced himself "much pleased" with his son's new direction: "Opportunities come in the night, and we should be ready to grasp them at once."[68]

Years later, Edwin would tell a very different, if dubious, story about his adventures in Kiel. He had gone for a swim the day he arrived and responded to a cry for help. Racing over to the floundering victim, he discovered a blond, rather pretty woman of about thirty, whom he dragged ashore. The next day her husband, a titled naval officer of fairly high rank who sounds suspiciously like the "wild reckless" Kruger, called to thank him and extend an invitation for lunch. The two became good friends; Edwin played tennis with the other naval officers to whom he was introduced and swam with the German almost every day. Thus he was stunned when his friend called at his rooms and said, with some embarrassment, that Edwin's conduct toward his wife, to whom he had been nothing more than courteous, made it necessary to ask for satisfaction.

Where a lady was involved, it was customary to use pistols rather than sabers. The two met alone, facing each other from across the officer's oak-paneled dining room. Edwin intentionally shot wide, into the woodwork. His opponent did the same. They bowed and, without saying a word, the American left the house and soon departed Kiel, but with ostensibly matching souvenirs. On either side of his face, between the ear and temple, were small crescent-shaped scars inflicted by the German student he had taught to box in return for lessons in wielding the *Schläger*.[69]

VII

In March 1911, Edwin had written his mother encouraging her not to worry about the family's planned move to Louisville until the time came. By June, the Hubbles were settled in their rented home at 1318 Brook Street, a pleasant three-story brick Victorian not far from John's downtown office. Jennie hired what would be her last "girl," who slept on the third floor in the stifling summer heat.

The two years spent in Shelbyville had been happy ones, and the move would never have been made were it not for the fact that John's health was deteriorating. The malaria persisted, sapping his strength and resulting in the loss of twenty pounds. Jennie, who was cautioned not to alarm her son, couldn't keep herself from asking him to come home for Christmas, although she did not say why. Edwin was determined to live within his budget and reluctantly declined. "I do hope the whole 'Family' can get togather for Xmas sometime," he wrote nostalgically to Lucy after learning that his sister had spent the holidays with their grandparents. "[A]ll the Springfield folks—or most of them—call me 'Ed'n.' Away from there its Mr. Hubble. There's only one other place I run to that whistle—in Marshfield."[70]

An ailing John Hubble returned to Marshfield for the summer apple harvest, which, he informed Edwin, was among the best on record. While there he was examined by his brother-in-law Dr. Edwin James, who discovered what doctors in Chicago and Louisville had missed. Though it was true that John had malaria, his failure to respond to treatment had little, if anything, to do with the mosquito-borne malady. He had contracted Bright's disease, or nephritis. The chronic fever, the headaches, the discoloration of the urine were classic symptoms of the renal damage that would soon result in total kidney failure. No operation or known medicine could reverse the process. At best, John had only a few months to live.

An unsuspecting Edwin was about to complete his B.A. in Jurisprudence, second class, leaving him another nine months to read law on the side while pursuing whatever studies he wished. Impressed by the example of John Crowe Ransom, he flirted with the idea of taking a Bachelor of Literature degree, "one of the swaggerest which Oxford gives."[71] A certain Mrs. Gaynor had sworn that he was to end up a writer, and to that effect gave him a heart-to-heart talk on the danger of missing the wonderful opportunity Oxford afforded those with a literary bent. Even his tutor had gotten into the habit of criticizing his work more favorably, although how Edwin, who seems never to have consulted a dictionary, escaped his censure remains a mystery. "My essays have been steadily improving and I am learning to expand my subject without using superfiless words. This weekly essay writting is a great thing for me."[72] He had also gone to a Spanish lecture, "just to see what its like."[73]

Only once in his letters did he broach the subject of astronomy, but once would have been enough to discomfort a terminally ill father who had every reason to believe that his son's course was firmly set.

> I've lately made one or two excursions into the realms of astronomy—much to the astonishment of some good science men of the University. I as a Law man am supposed to know nothing of the subject. But lately a few of the older fellows have taken it into their hands to speak slightingly of [Forest Ray] Moulton and his work. I soon saw they spoke not from 1st hand knowledge, but from reviews and rumors. So very modestly I started into the arena and, before leaving, had the exquisite pleasure of routing them completely. Moulton with his massive directness is a type unknown in England.[74]

By the fall of 1912, Edwin had learned of his father's illness and imminent death. None of the letters from this period survive, but Helen remembered that her brother wrote home asking for permission to return as soon as possible. John, who by now was permanently bedridden, would not hear of it, fearing that his son would never go back to England. The relatives began to gather as the end neared in January 1913. Martin, who was now confined to a wheelchair, was unable to make the journey from Springfield, but Mary Jane came, accompanied by several of their children. Surrounded by his own offspring, John drifted back and forth across the boundary of consciousness before slipping into a coma. The fifty-two-year-old insurance agent died peacefully in his own bed at four in the afternoon on the nineteenth, a Sunday. The funeral was held at the house on Brook Street two days later, after which John was buried in Louisville's Cave Hill Cemetery rather than the family plot back in Springfield, a reflection of a harsh new economic reality.[75]

Word reached Edwin by cable. He wrote to his mother of dining with a clergyman friend, then passing the evening discussing questions of religion and faith. Before parting, the two had entered a chapel and said some prayers. He promised that the family would all live together after his return, and praised Henry, "a brick," for heading the household in the meantime. "I've a tremendous lot of work on my hands, and am keeping myself too busy to think."[76]

The death of one's father is one of life's crucial events—a deliverance as well as a blow. The surviving son may be filled with dread as he seriously contemplates his own mortality for the first time; all at once, there is no more becoming, everything is being, unwinding. Yet he may also experience a new sense of freedom, especially when the father has been overbearing and censorious. Edwin reacted by entering the social whirl of his last Oxford spring. He assumed the presidency of the Cosmopolitan Club; he dined with a foreign countess; he danced into the small hours during Eights Week; and he engineered a party on the river in honor of a close friend. Lest he seem too cavalier in the circumstances, he sought to reassure Jennie:

> Tomorrow fortnight I leave Oxford, headed for Home. Of course, in a way, I am just a bit sorry to leave the place where I have spent three pleasant comfortable years. But its merely a chapter in life and I am profoundly thankful to finish and turn to the new larger chapter. For the new holds greater and more honourable tasks than the old. I am glad, awfully glad to feel that I am at last going back to help you just as much as I can. It will be little enough at first, but I hope and mean to make it substantial soon.
>
> I think you worry over me and imagine me returning puffed up and discontented. Honestly I come gladly to do all that I can, not so much with a sense of duty as with a sense of my right to help.[77]

The only family member to see Edwin during his three years abroad was his Aunt Louisa May, who toured England and the Continent with the equable Mr. Dickerson in tow. Writing to Aunt Janie, her sister, from Hamburg, Louisa May waxed enthusiastic. "I have just finished a three days visit with the big boy. I have looked him over carefully and he meets my greatest expectations. Not a trace of anything foreign except that he calls men 'chap' and says, 'It's quite all right. . . .' He has put on the social polish he lacked. Only needs a little more travel to finish him."[78]

CHAPTER FIVE

HEAVEN'S GATE

I

Edwin's long-anticipated arrival was a shock to his excited younger sisters, who expected their brother to look just as they remembered him. Eight-year-old Betsy and fourteen-year-old Helen circled the chair occupied by the tall stranger, their eyes uncomprehending. "He had on knickers," Betsy recalled, "and men didn't wear knickers then." A signet ring graced his little finger, and he was wearing a wristwatch he had won for high jumping, one of the first the girls had seen. "That to us was feminism."[1] Their brother pronounced certain words with a strange accent; when he left the house, he was rarely without his Oxford cape and cane. This foppery had not set well with the Warden of Queen's, and was reflected in his final assessment of the young American: "Considerable ability. Manly. Did quite well here. I didn't care v[ery] much for his manner—but he was better than his manner. Will get A."[2]

Edwin's return and the need for more space resulted in a family move to 1287 Everett Avenue, a pleasant frame house on the border of one of Louisville's more elegant neighborhoods. Edwin, rather than the senior Henry, sat at the head of the table. Following her son's instructions, a plump and graying Jennie brought him the kettle and boiling water for brewing tea in the English manner. As at Oxford, the unsliced bread was carried in on a board for cutting at the beginning of the meal. Edwin proudly hung his English pipes on the display rack in the library, where he passed much of his leisure time reading. He had become adept at blowing smoke rings, as though, by recalling his father's presence and reproducing his effects, it were possible to atone for a lifetime of emotional separation and injury.

For all John Hubble's faith in his profession, he had left his family on tenuous footing. His only major asset beyond a $5,000 life insurance policy was a joint interest in the Hubble Land and Fruit Company, valued at $3,382. His once comfortable income had virtually dried up during his long illness, forcing him to take out a personal note for $500. He died owing hundreds more to creditors, ranging from the milkman to his clothiers, and there were doctor bills and funeral expenses to pay. In the interest of economy, the body had not been shipped to Springfield for interment in the family plot, nor had Jennie purchased the lot next to John for herself. By law, the widow was entitled to half of what remained, or $3,451. The other half of the estate was divided into equal shares of $493 and distributed among the couple's seven children.[3]

Henry, who continued living at home, became an inspector with the Kentucky Actuarial Bureau, assuring himself a steady if modest income. Lucy, now twenty-five, her face invariably wreathed in a smile, commuted to Finchville on Mondays and Thursdays to give piano lessons for fifty cents each. These took place in the art building next to the grade school, the only heat seeping in from the adjacent classroom through a hole in the wall. She dressed well, but Elizabeth Winlock, a former student, suspected that Miss Hubble was in need of money or she would not have ridden the train nearly an hour each way from Louisville, sometimes staying overnight with a local family. Lucy's perpetual gaiety masked an inner pathos; fearing spinsterhood, she created an imaginary beau whose make-believe letters were read to a girlfriend who lived across the street.[4]

Bill was a semester away from completing his Bachelor of Science in agriculture at Wisconsin and spent the summer of 1913 at home working on his thesis, "A Study of the Factors Influencing the Churning of Cream." With eight mouths to feed, Edwin became convinced that he could save money on groceries by taking over the shopping. Strict accounts were kept for a month; when his total was compared with Jennie's, he was forced to concede defeat and sheepishly relinquished control of the pantry.

Henry, "the dreamer," and Edwin, whom his sisters considered almost as impractical, brought the family to the brink of financial ruin. Though the details are sketchy, the two talked their mother into investing heavily in a doomed business venture. A faded clipping from a family scrapbook refers to Henry's brief stint in the electrical business but carries no date. Whatever happened, the episode so traumatized the sisters that they were unable to speak of it nearly eighty years later.[5]

Among the notations in Edwin's Rhodes Trust Record is the following: "Heard from Campbell in September 1913 that H[ubble] was back in Louisville, & had passed Bar Exams."[6] The Campbell referred to by the Warden was Walter Stanley Campbell, a 1908 Rhodes Scholar from Oklahoma who had taken a bachelor's in literature in 1911 and then returned to Oxford to complete a master's in 1915. In between, the future biographer, historian, and novelist of the old Southwest, who wrote under the pen name Stanley Vestal, spent a year teaching English at Louisville's Male High School, and was a frequent Sunday-afternoon guest in the Hubble home on Everett Avenue.

The source of Campbell's information was almost certainly Edwin himself. Having started a B.A. in literature during his final year at Queen's, he soon abandoned this halfhearted pursuit in favor of Spanish, which was less demanding.[7] During the summer of his return, he succeeded in obtaining work translating what may have been legal correspondence for a Louisville import company doing business in South America. He would later claim that the firm offered him a highly paid position there but he turned it down. This much is certain: he never passed the Kentucky bar examination, which was not fully instituted until 1923. Nor did he follow the established method of reading up on the law and presenting himself to be examined before a Circuit Court judge in a county other than his own, a bottle of aged bourbon in one hand, a box of fine cigars in the other. Helen, who was living under the same roof as Edwin, reacted in disbelief when asked about her brother's reputed legal practice: "Where did that information come from? He did not practice law."[8]

It might be argued that Campbell, after hearing that his friend was translating legal documents, had simply jumped to the wrong conclusion, especially since he had known Edwin as a law student in England. But it was Edwin himself who later stated that his father's old friends saw to it that plenty of legal work came his way and that he never lost a case during the year he practiced part-time. Moreover, he had supposedly earned $10,000, which, if true, would have ranked him among Kentucky's legal elite and resolved the Hubbles' financial problems.[9] Admitting to friends that his illustrious education had led to nothing more than inconsequential freelance work may have been more than he could face. Proclaiming himself a lawyer would have also helped settle psychological accounts with a father who had refused to be moved by his son's love of astronomy. Whatever his motives, he had created the one falsehood whose threatened discovery would dog him for years to come.

I I

The female high school students of New Albany, Indiana, were "gaga" over their handsome new Spanish teacher, who, wearing knickers and a flowing cape hinting of the exotic, commuted from across the Ohio River by streetcar. Tittering over what were termed his "Oxford mannerisms," they rushed to sign up for his classes, then rationalized their decision on intellectual grounds in *The Blotter*, New Albany High's student paper: "At first we may wonder why a normal High School student would care to learn Spanish. But with the opening of the Panama Canal, our trade with other nations will increase. We will come into closer contact with South America, hence the Spanish language will be used in our business dealings."[10]

In addition to Spanish, Edwin was hired to teach physics and mathematics, as well as coach boys' basketball. He later confided that "teaching amused him," and he set out to prove his theory that any subject could be understood if it was made interesting enough. He took a class of unsuspecting boys who had persistently failed mathematics, and led them gently and guilefully to the level of college juniors. All was well until their older brothers returned home on vacation from various universities and told them what they were doing. The class instantly lost confidence and sank.[11] Earl Hale, a student in Edwin's course in quadratics, told a different story. While most of the boys idolized their worldly pedagogue, he was way over their heads, and couldn't bring himself down to their level.[12]

Hale also had Edwin for study hall, and noticed that he was considerably more interested in his own reading than in what the students were doing. He kept pleading for quiet, to no avail. Finally, he hit upon the idea of making the largest, worst-behaved boys proctors, putting them in the back of each row, and giving them the right to discipline anyone who was talking or passing notes. A troublemaker would be tapped on the shoulder, led to the locker room, and threatened with a beating. From then on peace reigned and he returned to his astronomy books undisturbed.[13]

Edwin's nostalgia for England and the life he had known at Oxford sometimes returned in waves, and others paid the price for his sense of loss. Betsy thought him very selfish when it came to the treatment of their mother. "Momma would have to have tea for his Rhodes scholar friends every Sunday afternoon. And she would make these tiny little cinnamon rolls like they had in England."[14] Much of the conversation passed Jennie by, but, along with the girls, she thoroughly enjoyed the attention of the young men as well as the music played by Lucy, Bill, and Edwin, who had brought his mandolin home with him. The only time Jennie put her foot down was when he tried to teach Betsy Spanish. He refused to let his sister speak any English, an injunction that extended to the dinner table. An uncomprehending Jennie, to whom dining with her family had always bordered on the sacred, brusquely called a halt to the practice.

The family traveled to Springfield for a long-anticipated reunion with the Hubble grandparents during Edwin's first Christmas home. Now in his late seventies, Martin was confined to a wheelchair, while a periodically addled Mary Jane was given to disappearing from home, then showing up on her daughter-in-law's doorstep in Louisville a day or two later. Helen, a high school freshman of considerable beauty, was thrilled when Edwin took her to her first dance during the visit. She enjoyed it so much that he invited her to accompany him whenever it was his turn to chaperon a school dance in New Albany, where the students treated them like royalty. "I thought it was just heaven on earth," she wistfully remarked. "It was," an envious Betsy volunteered, "it was."[15]

Edwin towered over a coterie of young admirers—both male and female—for whom he spun stories of his Oxford days and whom he introduced to the mysteries of English culture. The most intriguing member of this circle was Amy "Ro" Rosbraugh Roberts, a petite and demure brunette, next to whom Edwin was often photographed and whom he was said to have been courting. Ro's younger sister, Lydia, and adolescent brother, Jack, who was in his study hall, were also members of the group. The Robertses lived in a rambling frame house in a New Albany suburb called Silver Hills, with a sweeping vista of the unspoiled Ohio River valley. On nearby Highland Avenue resided the Hale family, two of whose six sons, Earl and Davis, followed Edwin almost everywhere he went. So frequent were his visits to the Hale home that the guest had his own camp bed. It was on the Hales' carriage sweep that he set up a borrowed telescope and waxed eloquent on the constellations deep into the night.

Occasionally joined by others, these six took all-day hikes to such local geographical landmarks as Bald Knob and Spickert Knob. The women wore blouses and ankle-length skirts, the men, shouldering canvas packs containing everything from marmalade to a tea kettle, dressed in long-sleeved shirts, coats, caps, and ties, walking sticks cut from saplings at the ready. Jack Roberts, who had the eye of an artist, chronicled these outings with his Brownie camera, creating an aura of noblesse oblige on the Ohio. Sometimes the women were left behind, as when Edwin and Jack hiked all the way to Wyandotte Cave in Harrison County, rising before dawn and returning well after midnight. The Hale brothers were so enthralled by Edwin's tales of cycling across Europe that they later rode from New Albany to Philadelphia, after which Earl repeated his hero's example by cycling through much of England and Europe, all the way down the Italian peninsula. Despite what were described as "several good offers," Ro was more interested in music and her family than in a husband. She never married and lived out her life in comfortable spinsterhood.[16]

By February 1914, Edwin's popularity among students had broadened to include much of New Albany. He had coached the best basketball team in school history to an undefeated season. On the strength of five victories over teams from across the river, an overzealous sports writer for *The Blotter* claimed the championship of Kentucky for his school, the highlight of the string coming against Lexington, when star forward Julius Joseph hit a basket while occupying a window seat beside the court.[17] The greater test came in mid-March, during the Indiana state championship tournament in Bloomington. The train carrying the coach and his eight-member team, bolstered by forty dollars in community donations, pulled out of New Albany amid "a veritable roar of unrestrained enthusiasm." In its first game, New Albany doubled the score against New Winchester, then won a defensive struggle against Mishawaka, 13–5, the following evening. The scenario was much the same the next night, except that New Albany found itself on the short end of a 14–9 score against a faster, larger Clinton five, thus having to settle for third place in the state.[18] Coach Hubble was considered to be as much a hero as his players. When *The Senior Blotter* became available two months later, his picture appeared on the yearbook's dedication page, along with the following inscription:

To
Edwin P. Hubble
Our beloved teacher of Spanish and Physics, who has been
A loyal friend to us in our senior year.
Ever willing to cheer and help us
Both in school and on
The field,
We
The class of 1914 lovingly dedicate this book.[19]

III

Edwin had kept his promise to Jennie that the family would remain together following his return from England, but a year of teaching high school had convinced him that Louisville was a trap. If he did not leave soon, he might never realize his dream. With the school year about to end in May 1914, he wrote Forest Ray Moulton, his former astronomy professor at Chicago, asking about his prospects for entering graduate school and obtaining some type of financial assistance. He explained that he had spent the past year settling his father's estate and teaching high school in Indiana. He further noted that he had passed the Kentucky bar examination, entitling him to practice law.

Moulton replied that the small number of fellowships in astronomy for the coming year had already been assigned and that it would probably take Hubble three years to complete a Ph.D. The astronomer then wrote Edwin B. Frost, director of the University of Chicago's Yerkes Observatory at Williams Bay, Wisconsin, setting forth Hubble's case and urging Frost to come up with additional funds with which to lure him north: "Personally he is a man of the finest type. Physically he is a splendid specimen. In his work here, altogether, and especially in science, he showed exceptional ability."[20]

At Moulton's urging, Edwin, too, wrote Frost, emphasizing his laboratory experience and the surveying skills he had honed in the Wisconsin woods. During his student days at Chicago he had learned how to "take time" with the three-inch Bamberg transit and had mastered the small equatorial telescope. He was teaching physics and possessed a good knowledge of German, French, and Spanish. In contrast to his letter of inquiry to Moulton, this one said nothing about the bar examination.[21]

As Moulton had suspected, Edwin's letter struck a sympathetic chord. Frost had long complained of the dearth of good assistants at the observatory, a lament echoed by his fellow directors around the country, who despaired of the lack of job opportunities for graduating astronomers. Matters were made worse at Chicago by the fact that undergraduates were hampered by equipment which Moulton characterized as "a joke." "There is no opportunity for me to give students any inspiration."[22]

Frost wrote Edwin that if he could get by on a $120 tuition scholarship plus $30 a month for room and board with Mrs. Sawyer, the local innkeeper, he was welcome to begin his studies at Yerkes on October 1. It was Frost's hope, however, that the two might take one another's measure before then. "I should very much like to have you come and spend a day or two here for the purpose of becoming mutually acquainted and discussing the work."[23] Meanwhile, Edwin would do well to complete as much background reading as possible in astronomy and celestial mechanics.

In another of his unproofed letters, Edwin replied that he was "thouroughly delighted" with Frost's offer, and would be in Williams Bay in the autumn. "This city of Louisville is horribly barbarian in the matter of scientific literature, but I have Mr. Baly's 'Spectroscopy,' and Mr. Moulton's 'Celestial Machanics' at hand, and I can also reveiw my Calculus."[24] Frost was so taken with his enthusiasm that he mailed him two books from his own library and promised him more if needed. Second thoughts had also convinced the director that it would be best if he planned his arrival for the last week of August rather than early October. It would enable him to attend the annual meeting of the Astronomical Society of America on the campus of Northwestern University, after which he could come directly to Williams Bay. Frost was taking the liberty of submitting his name to the council, "so that on your arrival at Evanston, you will regard yourself as a member."[25]

When next photographed, a poised and elegantly attired Edwin Hubble stands in the first row of the Astronomical Society's group photograph, having strategically positioned himself a short distance away from President Edward C. Pickering and Vice Presidents George C. Comstock and Frank Schlesinger. Near the end of the second row from the rear, virtually eclipsed by others, is a wan Vesto Melvin Slipher, a fastidious and mostly self-taught astronomer, whose paper had stolen the show. When Slipher finished reading his findings, every astronomer present rose as one and cheered—a spectacle never before witnessed at a scientific meeting.

Percival Lowell had brought Slipher to his eponymous observatory in Flagstaff, Arizona, in 1901, perhaps as a counterbalance to his own flamboyant and driving personality. The onetime farm boy from Mulberry, Indiana, always dressed in a suit and kept his tie perfectly knotted even when alone in the dark at the 24-inch Lowell refractor. Slipher remained quietly skeptical of his boss's belief in canals on Mars (later found to be an optical illusion), refusing to cut corners or to publish until he was certain of his facts.

In 1900, a year before Slipher came to Flagstaff, Lowell commissioned the gifted instrument designer John A. Brashear to provide him with the most efficient spectrographic equipment that could be constructed. A major breakthrough in understanding spectra had occurred in the nineteenth century when physicists realized that spectral lines could be produced artificially in the laboratory by passing a beam of white light through a bottle of gas. Different gases absorb light from the spectrum at specific wavelengths, producing a distinctive pattern of shaded bars and lines reminiscent of a piano keyboard. It logically followed that this same pattern could be derived from the solar spectrum and those generated by the countless other stars, and with it a chemical map of the universe. It was also known that light, like sound, behaves as if it were a wave; the farther it travels through space, the more it "stretches out." Thus, if a light source is moving toward a viewer the spectral lines are shifted toward the blue, or blueshifted; if the light source is moving away from the observer the spectral lines are shifted toward the red, or redshifted. The amount the lines shift—the so-called Doppler shift—reveals the radial velocity of the light source.

After a decade of refining his skills on the major planets, Slipher was ready to begin his spectrographic observations of the nebulae, those great circles and whirlpools of distant light whose origins, makeup, and location in the universe were as yet undetermined. Following Lowell's instructions, he trained the refractor, mounted with a slitted spectrograph and camera, on the massive Andromeda spiral, employing iron and vanadium as his comparison elements. A spectrum obtained on September 17, 1912, revealed that the nebula was blueshifted, racing toward the sun at the astonishing speed of some 300 kilometers per second, the greatest rate of movement ever recorded for an astronomical body. Slipher next fitted the spectrograph with a lens 200 times faster than usual and proceeded to take additional plates, each of which confirmed his initial measurement. Spectrograms and consequent radial velocities were obtained for forty other nebulae and star clusters, the results of which seemed even more incredible. In contrast to Andromeda, most nebulae were found to be hurtling away from the sun at speeds as high as 1,100 kilometers per second, a figure startling enough to bring any astronomer to his feet.[26]

John A. Miller, Slipher's undergraduate teacher at Indiana University and a member of the audience, could scarcely contain himself: "It looks to me as though you have found a gold mine, and that by working carefully you can make a contribution that is as significant as the one that Kepler made, but in an entirely different way." An ebullient Ejnar Hertzsprung, the great Danish astronomer, wrote Slipher from the Astrophysical Observatory at Potsdam: "My hearty congratulations to your beautiful discovery of the great radial velocity of some spiral nebulae. It seems to me that with this discovery the great question if the spirals belong to the system of the Milky Way or not is answered with great certainty to the end that they do not."[27] To Hertzsprung, the speeds of the spirals appeared altogether too great and their distances too vast for them to be gravitationally bound to the stellar system. They must constitute separate islands in the universe, of which the Milky Way is but one of many yet uncounted.

Slipher had privately come to a similar conclusion. Yet despite a later claim that "no other factor could vitiate the plates' message," he decided to keep his own counsel for the time being.[28] No less an authority than William Wallace Campbell, the director of California's Lick Observatory and the leading American expert on radial velocity, had hoisted a warning flag: "Your high velocity for [the] Andromeda Nebula is surprising in the extreme. I suppose, as the dispersion of your instrument must have been very low, the error of your radial velocity measurement may be pretty large."[29] Nor could one dismiss the photographic work of George W. Ritchey, a cantankerous but deft astronomer working with the 60-inch reflector of the Mount Wilson Observatory, the world's greatest telescope. In 1910 Ritchey had presented photographs of twelve of the larger spirals, which he termed "nebulous stars." It appeared that none of these spirals contained sufficient matter to be classified as galaxies comparable to the Milky Way. In the final analysis, everything hinged on obtaining credible estimates of the distances of the nebulae, and then establishing a correlation with the velocities of recession. Only then would astronomers truly know whether the Milky Way is the universe entire or a minute speck adrift in an infinitely vaster eye of creation.

IV

In September 1892, three days after the University of Chicago opened its doors, Charles Tyson Yerkes, the Chicago streetcar magnate who built the city's elevated train system, offered to purchase what was then the world's largest telescope—a 40-inch equatorial refractor—and donate it to the university. Yerkes also agreed to build a giant observatory to house his gift. The heart of a smoky city, its polluted skies illuminated by expanding waves of electric lights, was no place to conduct delicate observations of faint nebulae and comets. After being deluged with offers of free land from Illinois to California, the university trustees accepted a donation of fifty-three acres on the north shore of Wisconsin's Lake Geneva, a pristine body of water, eight miles long and one mile wide, some seventy-five miles by rail from Chicago. The closest community, the village of Williams Bay, was still illuminated by candles and kerosene lamps. The nearest electric lights were seven miles distant in the small resort town of Lake Geneva, with a year-round population of 3,000.

The home for the huge telescope, which went on exhibit in the Manufacturers Building at the World's Fair, was to be as grand in scale as its prize. University architect Henry Ives Cobb chose the Romanesque style and the configuration of a Latin cross, with three domed telescope towers and a meridian room at the extremities, recalling the church and monastery of Monreale. The interior of the brown-brick structure was to be ornately decorated with intricate terra-cotta carvings, including Apollo in his streaking chariot and caricatures of that lesser god, Charles Yerkes.

When Yerkes Observatory was opened to the public in 1897, following three years of construction, eager visitors made directly for the great tower at the western end of the main building, where the huge 40-inch refractor pointed skyward, like a giant cannon. Its steel tube, 60 feet long and weighing 6 tons, stands on a cast-iron column consisting of four sections, tapering from 5 × 11 feet at the base to 5 × 10 feet at the junction with the head. Masterful examples of Victorian engineering, these four sections are bolted together and rest on a cast-iron base 14 × 18 feet, which in turn is firmly anchored to a massive brick pier supported on a concrete foundation 32 feet long and 5 feet thick. The column and head together rise to a height of 43 feet and weigh 50 tons.

Overhead stretches the 140-ton dome, 90 feet in diameter and 60 feet high. This massive inverted bowl, sheathed by wood and covered with roofing tin, revolves on 26 wheels activated by an electrically driven cable. The observing slit, 13 feet wide, extends from the horizon to a point five feet beyond the zenith. The two shutters which cover this opening are 85 feet long and move simultaneously from the center outward, remaining parallel to each other in all positions. Directly below the dome's base are 32 windows, arranged in three rows, balancing inner and outer weather so that the telescope will not further distort the faint light penetrating Earth's atmosphere.

Dwarfed as the Lilliputian observer is by the dome, his attention is soon drawn to the great expanse of hardwood beneath him, which creaks and shudders like the deck of a heaving ship. To make the telescope accessible for observation at all levels, the entire observing room floor, 75 feet in diameter and weighing 37½ tons, rises like an opera set at the press of a button, the solitary astronomer in apotheosis at heaven's gate.

As keeper of the gate, forty-eight-year-old Edwin Brant Frost possessed the bearing and patrician credentials worthy of his station. Signing letters and documents with an ornate hand resembling holy writ, the square-jawed director gazed imperiously at the world from behind a pair of dark-rimmed glasses, looking more like a piece of granite sculpture than living flesh and blood. Conscious of his intimidating demeanor, the self-deprecating Yankee wrote of "my normal New England chill."[30] A descendant of old Puritan stock by way of Boston and Hanover, New Hampshire, Frost had earned a B.A. and M.A. from Dartmouth College, where his father, Carleton Pennington Frost, was dean of the medical school and later a trustee. In 1897, after serving as professor of astronomy and director of Dartmouth Observatory, Frost was lured west by the opening of Yerkes and the prospect of overseeing research with the then largest telescope in the world.

As it happened, the director was an optimist in disguise as well as a man of good cheer and gallantry. By the time Hubble arrived in August 1914, an uncomplaining Frost was enduring the worst fate that can befall an astronomer. Plagued by cataracts which would ultimately claim his sight, he could no longer use the great telescope and was forced to call on Edwin and his fellow graduate students to read his correspondence and scientific articles aloud.

The director's condition would have mattered less had not the staff of Yerkes been decimated in recent years. Irreplaceable astronomers such as George W. Ritchey, Walter S. Adams, Francis G. Pease, and Ferdinand Ellerman had joined the intellectual migration westward to California, where the new Mount Wilson telescope was about to revolutionize the study of the stars. Of those who remained at Williams Bay, Edward E. Barnard, a mirror image of Justice Oliver Wendell Holmes who had raised eyebrows in his youth by virtue of some spectacular visual discoveries, was the senior astronomer. Barnard had completed an outstanding series of wide-field photographs of the Milky Way, but he had no theoretical training and neither advised nor instructed graduate students. The other faculty members have been less flatteringly characterized as "astronomical lightweights." John A. Parkhurst and Storrs B. Barrett, a B.A. who doubled as Yerkes's librarian and secretary, were left behind when the "first team" departed for Pasadena, while Oliver J. Lee, a longtime Yerkes assistant, had just become an instructor. A once competent spectroscopist himself, Frost was devoid of any creative ideas with which to compensate for his impaired vision. Well before his eyesight began to fail, the director had settled into a routine observational program that varied little over the years.[31] In the span of a single decade, Yerkes had receded from the astronomical frontier into the hinterland.

Between October 1914 and May 1916, Edwin registered for only one three-hour course each quarter: Research at Yerkes Observatory. No regular classes were offered at Williams Bay on the grounds that hands-on work with the telescopes and the other equipment was "of more advantage to the qualified student than set courses of instruction." If the ambitious regimen outlined in the university catalogue had been followed, Edwin would have conducted research in solar physics with the spectroscope, spectroheliograph, and photoheliograph; taken micrometric observations of double stars, planets, satellites, and nebulae; photographed comets, stars, and nebulae; and completed additional work in visual and photographic photometry.[32] In light of his interests and early publications, it appears that he did little work with many of Yerkes's more complicated instruments—especially those unrelated to the nebulae and the stars—preferring instead to develop the observational and photographic skills needed for navigating the night skies.

At Yerkes, as at all major observatories, the amount of "seeing time" on the premier instrument rarely met the demand, leading to the establishment of a rigid pecking order. After resident astronomers, priority was given to visiting scientists with distinguished credentials, followed by lowly graduate students hoping one day to make a name for themselves. Edwin took his turn on the 40-inch refractor, participating in the ongoing radial-velocity program. But most evenings found him alone in the heliostat room, where the 24-inch reflector built by George Ritchey was attached to a temporary mounting on the concrete floor. Largely forgotten, the instrument was just what he needed to pursue his independent photographic studies of faint nebulae, which he commenced to do after first receiving the gatekeeper's benediction.

Inspired, perhaps, by the lingering adrenaline from Slipher's announcement, Edwin trained his telescope and camera on an object known to astronomers as NGC (for *New General Catalogue*) 2261, which he described as "the finest example of a cometary nebula in the northern skies."[33] After taking fifteen plates over the course of six months in 1915 and 1916, he compared these photographs with one taken with the same telescope in 1908 by astronomer Frank Jordan and with another obtained from the widow of Isaac Roberts, who had photographed NGC 2261 in January 1900. At first, Hubble could not believe his eyes and turned to Barnard, who had also photographed the nebula, for confirmation. In less than eight years the following edge of the nebula had bulged to display a larger degree of convexity than before, strong evidence that the object was small and relatively near in astronomical terms, or else this change would have gone undetected. Soon he announced his maiden discovery to the scientific world in the prestigious *Astrophysical Journal*, assuming the mantle of the cautiously conservative researcher he would always remain: "No attempt is here made to explain the phenomenon, save to remark that, unless we are witnessing a phenomenon of illumination, the nebula must be very near."[34] He then singled out twelve faint stars, all of the 15th magnitude, or dimmer, which he had photographed while working on NGC 2261. The plates were individually mounted on an instrument called a blink microscope alongside others taken by Jordan, Ritchey, and Parkhurst. In the alternating flashes of light and dark he was able to detect proper motions, small angular changes in the stellar positions, which he then computed mathematically. "So far as I am aware," he wrote in the resulting article, "these are the faintest stars in which appreciable motion has been found. In view of the small number of fields examined, . . . it is reasonable to suppose that considerable numbers of such faint stars exist in the immediate neighborhood of our *Sun*."[35] Frost was so impressed by his graduate student's initial foray into celestial dynamics that he presented slides of his work on the variable nebula at the annual meeting of the National Academy of Sciences in Washington, D.C., in April 1917. Hoping to conserve his vision as much as possible, Frost had asked Edwin to draft the text and spoke for only three minutes, a brief but auspicious beginning.[36]

V

Edwin often trod the sloping, half-mile path between the observatory and the wooded shoreline of Lake Geneva, a sandwich, swimsuit, and towel in hand. He sometimes tied a long fishing line to his trunks and swam as if trolling from a boat. "If I had caught a fair-sized fish," he once joked, "I think I should have come back without my swimming-suit." He later recounted the story of walking out on a pier, where he passed a vacationing middle-aged professor and his wife. As he turned back, he saw the woman fall into the lake. He quickly shed his coat and dove in after her. She caught hold of him but continued to struggle. "I didn't like to knock her out so, as the water was not deep, I sat her on my shoulders, which just brought her head above the surface, and walked under water until it became shallow and I could put her down." In contrast to the German husband at Kiel, the professor showed no emotion when thanking the aquatic hero, "nor did he seem particularly glad to see her again."[37]

The Hubble sisters visited Williams Bay on several occasions, and enjoyed being paddled around the lake in a canoe by their brother or one of his admiring friends. After pledging not to utter a word or bother anyone, Betsy was taken into the sanctum sanctorum, where she gazed through the great telescope and reached out, attempting to touch the stars with her hand. Together with their mother and Henry, the girls had moved from Louisville to Madison, Wisconsin, in the summer of 1916, assisted by Edwin, who turned the venture into a minor disaster. After struggling with box upon box of books, he gave up and told his literary-minded sisters that most of their cherished novels must go. Only works of nonfiction would be shipped north, tears of protest notwithstanding.[38]

Their standard of living eroded, the Hubbles rented a small, three-bedroom house at 1826 Chadbourne Avenue. It was the time of Henry's rumored breakdown, and he would not return to the insurance business for at least a year. Helen began classes at the University of Wisconsin in the fall, while Lucy continued teaching piano. Although Edwin was only sixty miles distant, it was his younger brother Bill who, in Betsy's words, "had all the responsibility for the family."[39] He had enlisted in the army after graduating from Wisconsin in 1914 and was dispatched to Fort Leavenworth in northeastern Kansas, where he managed the post creamery. Every penny the youth could spare was sent home to support his mother and sisters, a pattern that continued unbroken after he reentered civilian life. While the Hubble sisters admired Edwin, it was Bill, who willingly sacrificed marriage and a family, that they adored. A grateful Helen later wrote: "Bill is really our 'unsung hero.' Betsy and I felt that Bill is really indirectly responsible for Edwin's accomplishments. Bill gave [up] his *dreams* to do the mundane things of necessity."[40]

Having proven himself, Edwin was granted one of two $320 fellowships his second year, and recommended for the top award of $520 his third year. When a financially pinched R. D. Salisbury, dean of science, reduced the amount to the same figure as before, Frost wrote an uncharacteristically sharp letter protesting the dean's decision. "I know that we do not often get the $520, but this was an exceptional case with no other applicants, and [Hubble] is an exceptional man and needs the money." The battle-scarred dean refused to budge: "There are scores of men of one department and another who appeared to have as high [a] claim as Mr. Hubble. This, of course, does not in any way disparage his title to high rank."[41] As if in agreement, the fledgling astronomer stands shoulder to shoulder with Frost and Barnard in the annual photo of Yerkes' faculty, students, and staff.

Edwin passed his French and German exams in May 1916, remained at Yerkes through the summer to gather additional information for his dissertation, and then returned to the university for the fall and winter quarters to complete his final courses in mathematics and celestial mechanics. To make ends meet, he taught a class in introductory astronomy and accepted the post of graduate resident head of Snell Hall, disparagingly known on campus as the Y.M.C.A. of men's dormitories. Consisting of four stories, a basement, and some sixty live wires, the dorm was famous for its "pajama parades" and marathon pillow fights, which quickly escalated into indiscriminate mayhem as masses of bedclothes, books, and personal belongings flooded the corridors. To Edwin's surprise, he was greeted by an atmosphere of peace and decorum. He waited expectantly for the other shoe to drop, but it never did. Not once did he have to so much as call anyone down. Puzzled by this, he subsequently found out that his reputation as a C-man and heavyweight boxer had preceded him.[42]

On July 23, 1916, Walter S. Adams, the assistant director of Mount Wilson Observatory, near Pasadena, wrote his superior, George Ellery Hale, to inform him that data on the mirror tests of the new 100-inch reflecting telescope would soon be ready. Although the instrument itself would not go into regular operation for over a year, it was not too early to begin thinking about the necessary additions to the mountain's already impressive staff. By October, Adams seemed quite certain that he had found at least one of the five astronomers needed, a graduate student named "Hubbell" whom he had recently met during a visit to the University of Chicago campus. Of course, Hale, as director, was free to do his own checking up.

The astronomer did just that. He began by consulting with his good friend and colleague Henry Gale, a University of Chicago physics professor, who had nothing but good things to say about Hubble.[43] Hale next talked to Frost, whom he had preceded as director of Yerkes Observatory and crowned his successor. Writing from the Massachusetts Institute of Technology in Cambridge on November 1, Hale informed Adams that he had offered Hubble a position at $1,200 a year, pending the completion of his dissertation.[44] The only sticking point had been the young man's loyalty to Frost, who had originally hoped to employ him at Yerkes. After learning that the funds would not become available, Frost, who had recently given up riding trains at his occulist's suggestion, told Edwin that he must finish his dissertation and go.[45]

VI

The first two letters Edwin wrote on the new typewriter he had long wanted to purchase were dated April 10, 1917. One was addressed to Frost, the other to Hale. Four days earlier, Congress had declared war on Germany after its government announced that it would engage in unrestricted submarine warfare in an attempt to break British control of the seas. Bread was being rationed in Edwin's beloved England, where the royal family had just renounced its German names and titles. But mostly he was haunted by the images of the gifted young men he had known at Oxford. Few of them, himself included, had acknowledged the terrible forces at work beneath the surface of things, the violence that had begun to stalk the world outside. Of those who did, none possessed the power to alter events. Now their corpses lay scattered along the Western Front like so many leaves in windrows, their names soon to be etched in the stained glass of their college chapels in imitation of fallen knights of old.

Citing preparedness legislation enacted by Congress in June 1916, Edwin informed Frost of his intention to apply for a commission in the Officers Reserve Corps. In addition to his application, he would need letters of recommendation from five citizens. "If you consider me a worthy target, please aid your country by sending such a letter to Major O. W. Bell, U. of Chicago." He hoped to have the first draft of his dissertation ready in two weeks. "When that is in shape, it will be possible to rush the examination a bit, will it not?"[46] Frost replied immediately, enclosing the requested letter and promising to advance the time of his orals. "After I receive your first draft of the thesis, I will go over it promptly so that there may be no delay on that account."[47]

Multiple favors were asked of Hale as well. Would the Mount Wilson director hold the position he had offered Edwin until he completed his military service, and would Hale write Major Bell on his behalf? Since there is no solid evidence that the two men had actually met face to face, these were bold requests and would explain why he closed with an unusually servile caveat: "Personally, if you know of any reasons why we who can rush through our degrees in time, should not follow this course, several of us would be very much concerned in hearing them."[48] Hale, who was already in Washington organizing the nation's wartime scientific effort at the National Research Council, could think of none. "I agree with you that under the circumstances it would be natural for you to apply for a commission." He promised that Hubble would have a place at Mount Wilson after returning from the army, and enclosed a letter recommending him "as one most worthy of receiving a commission."[49]

"Photographic Investigations of Faint Nebulae," as described by the dissertation's author, whose mind was already elsewhere, "seems so scimpy." Yet there was little to be done; it was now the first of May and his orals were scheduled to take place at Yerkes on the morning and afternoon of Saturday, the twelfth, only three days before he was to report for duty. "Of course it was a lot of work getting out the positions," he continued in his cover letter to Frost, "but it does not add appreciably to the sum total of human knowledge. Some day I hope to study the nature of these nebelflecken to some purpose."[50]

Aware of the pressure Edwin was under, the sympathetic but uneasy director tried to put the best face on a difficult situation. "I think that in the typewritten form it does not fully represent the work which you have put into it." It would be easy to beef up the dissertation by including a chapter on variable nebulae, particularly NGC 2261. Indeed, his recently published article could be incorporated almost verbatim. Frost had attempted to contact him about his other concerns during a visit to campus two days earlier, "but the janitor who receives telephone messages at Snell Hall does not appear remarkably intelligent."[51] Had it not been for the war, Frost would have almost certainly insisted that Edwin revise and expand this rather meager offering.

University regulations required that at least one typewritten copy of all dissertations be deposited in the main library. Hubble's either has been lost or, more likely, never found its way there as a result of delayed revisions and the haste with which he departed campus. The published version, consisting of seventeen oversize pages, did not come out until 1920, and only then because of Frost's dogged perseverance and skillful editorial assistance.[52]

By the time Edwin began photographing NGC 2261 and a number of other nebulae, about 17,000 of the dim formations had already been catalogued. An estimated 130,000 more were within existing telescopic range. "Extremely little is known of the nature of nebulae," he wrote, "and no significant classification has yet been suggested; not even a precise definition has been formulated. . . . It may well be that they differ in kind and do not form a unidirectional sequence of evolution." At least some of the great diffuse nebulosities, connected as they are with stars visible to the naked eye, lie within our stellar system. So, too, it would appear, do the planetaries, massive gaseous clouds at even vaster distances from the sun. Others, the giant spirals, with their enormous radial velocities and insensible proper motions, apparently lie outside our system. Beyond these ill-defined classes are the seemingly numberless faint nebulae, revealing themselves as nothing more than scant markings on the photographic plate. "They may give gaseous spectra, or continuous; they may be planetaries or spirals, or they may belong to a different class entirely. They may even be clusters and not nebulae at all. These questions await their answers for instruments more powerful than those we now possess."[53]

For the present, the study of nebulae must remain essentially a photographic problem for cameras of wide angle and reflecting telescopes of large focal ratio. As an extension of his work with the 24-inch, he concentrated on great regions of the sky in which the smaller and fainter nebulae tend to cluster. Within these larger tracts occur accumulation points about which the clustering is concentrated. Selecting seven such points, the astronomer photographed and catalogued 511 previously unknown nebulae, a tedious undertaking that required well over a thousand plates, each needing a minimal exposure time of two hours. He then attempted to classify them by form, brightness, and size, according to a scheme developed by the German observer Max Wolf. Most of the faint nebulae appeared to be not spirals, as had been widely believed, but what are now called ellipticals. At the same time, he was able to confirm that their distribution in the sky avoids the Milky Way, and that most of them do, indeed, occur in clusters. Yet who could be certain of anything until some concrete idea of the nebular distances could be formulated? "Suppose them to be extra-sidereal and perhaps we see clusters of galaxies; suppose them within our system, their nature becomes a mystery."[54]

Though shaky on technical grounds and confused in its theoretical interpretations, "Photographic Investigations of Faint Nebulae" is prophetic of the quest that lay ahead. The mind's eye of its author is already fixed on the 100-inch reflector, nearing completion atop Mount Wilson. "Give me where to stand, and I will move the earth!" Archimedes is reputed to have declared in reference to the lever. With his hands on Mount Wilson's gentle giant, Edwin Hubble would soon be in a position to seek out the very jugular of the cosmos.

VII

Magna cum laude was the judgment of the six-member committee before whom Edwin sat for his final oral examination in astronomy and mathematics. The candidate suspected that Frost, Barnard, Parkhurst, Lee, Moulton, and William D. MacMillan, an assistant professor of astronomy suffering the aftereffects of a severe tonsillitis attack, had treated him rather leniently.[55] It would be late August, more than three months after he departed campus, before the diploma of the newly minted Ph.D. would finally catch up with him.[56]

Private Edwin Powell Hubble had reported for duty on May 15, 1917, at nearby Fort Sheridan, a military reservation on Lake Michigan, north of his brief childhood home of Evanston. There he was assigned to Company 10, joining 150 other Illinois college graduates who began reserve officer training the same week. Three weeks into the accelerated one-month course, he penned a brief letter to Frost. "This military game seems to be a nitch in which I fit. I was fourth man to be made student captain . . . and am an instructor in everything from bayonet work to signalling. This next Sunday I am chosen to represent the Company in a delegation to visit some model trenches." He had already decided not to follow his fellow trainees into the artillery, which would have led to a respectable commission as a first or second lieutenant. "I have been lured into the Infantry by visions of something better than a lieutenancy, so will stay here when the artillery men leave."[57] Captain Graham, his commanding officer, requested that he instruct others in his company in marching by the stars. Any books Frost could lend him on the subject would be appreciated. The director immediately dispatched three volumes and a letter gently chiding him for sending back the proof sheets of his dissertation without making a single correction. Frost volunteered to reconcile certain discrepancies in the data, but Edwin himself must look over these changes very carefully prior to publication.[58]

Hubble was one of ten men awarded his captain's bars on August 15 and ordered to active duty.[59] After furloughing at Lake Geneva, he learned that he had been assigned to Camp Grant, named for the Union general and home of the army's newly organized 86th Division. The training site was a tract of 3,000 acres lying between the Rock and Kishwaukee rivers, five miles south of Rockford, Illinois. The great army Hubble expected to find upon his arrival consisted of workmen rather than soldiers. The post was a din of clanging hammers and squeaking cranes, its long, dust-choked streets carved out of ripening cornfields and filled with straining teams and frantic contractors rushing to complete the new city. A thousand carloads of building materials had been stockpiled since the first delivery of lumber in late June. The blueprints called for 1,400 buildings, 25 miles of sewer line, 22 miles of water pipe, 6 miles of gas mains, and 12 miles of hard roads. On the banks of the Rock River, in the center of camp, were the division headquarters, the Y.M.C.A. and Knights of Columbus auditoriums, the Christian Science Welfare House, the library, the Redpath amusement tent, which would soon yield to the Liberty Theater, the Jewish Welfare Hut, the camp store, the post office, and the telephone and telegraph office. By August 28, when the 1,000 training officers, almost all recent graduates of Fort Sheridan, arrived, the barracks and mess halls, though smelling of freshly sawn pine and new paint, were ready for occupancy.

Dubbed the Black Hawk Division for the Sac chief who had fought both the whites and the Sioux in the region two generations earlier, the 86th was commanded by Major General Thomas H. Barry, a graduate of West Point who won his first promotion battling Indians with the famous 7th Cavalry—Custer's regiment. More recently, the general, a ringer for the heavily mustached French chief of staff, Marshal Ferdinand Foch, had served as commandant of his alma mater and commander of the army's Central Department in Chicago. Revered by his troops, Barry introduced the practice of marching the artillery to a camp at Byron for coaching in Black Hawk's war cry by the town's resident physician, Dr. A. J. "Old Doc" Woodstock—"Kia-kiak! Kia-kiak!"

The train carrying the first 350 of an eventual 115,000 trainees arrived late in the afternoon of Wednesday, September 5. Singing "Hail,

hail, the gang's all here," the men, well-dressed natives of Minnesota, Wisconsin, and Illinois, scrambled out of their cars and assembled, only to be engulfed by a great cloud of dust stirred up by a gathering storm, then drenched while marching to their barracks. The following Monday, Captain Hubble took command of the 2nd Battalion, 343rd Infantry Regiment. "I have twenty-five officers and six hundred men," he proudly wrote his mother. "The battalion is not an administrative unit, hence I do not worry with the responsibility of caring for the personal wants of the men, solely with the duty of training the unit to fight Germans in absolutely the most efficient manner possible. I train the officers, they train the men."[60] Two weeks later, the same day sixty-eight-year-old Mary Jane Hubble died after being paralyzed by a stroke, her grandson was wearing his uniform with a degree of pride bordering on hubris. "They have given me a Battalion to work my will upon and, we hope, to lead to the front," he wrote Frank Aydelotte, a fellow Rhodes Scholar, with whom he remained in close contact. "Stirring times—I can't picture myself missing the gathering, as it were, of the clans."[61]

Reveille sounded at 5:45, nearly an hour before dawn; taps, mournful and evocative, an hour before midnight. To the usual fare of calisthenics, close-order drill, cross-country marching, riflery, bayoneting, and K.P. were added ominous lessons in warfare never before taught the American soldier: camouflage, mines, deep gallery shelter, sniping, cover against shell fire, and mustard gas. In the middle of October work began on a massive system of trenches modeled on those in which the first U.S. troops were already fighting and dying in France and Belgium. When completed, the approaches were named after the streets of Chicago and other cities, the dugouts and firing positions for well-known hotels and resorts. Over and along this system was fought battle after battle, the barbed-wire lines occupied and evacuated, barraged, stormed, counterattacked, until the sandy parapets and firing steps caved in under the strain. Lacking knowledge of a foreign tongue, both enemy and ally invoked the battle cry of Black Hawk, punctuated by the even more familiar "Let's go! Let's go!"[62]

Hubble, who was about to turn twenty-nine, cut a dashing figure in his high-collared khaki uniform, leather boots, and puttees. He quickly gained the attention of his regimental commander, Colonel Charles R. Howland, a tough regular of the old school who possessed little confidence in his untested captain's training and background. One morning, with reporters looking on, Howland strode onto the rifle range and proceeded to fire six bull's-eyes and four 4s for a total of 46. Jumping to his feet, he declared, "I can beat that with my revolver," and fired once with his long-barreled .45, sending a bullet through the target's black center. Hubble calmly stepped to the line and made ten bull's-eyes out of ten for a perfect score of 50. Afterward, it was rumored that the colonel always knew when Hubble was "in the butts" because of the way troops flocked to the range to witness the shooting. After keeping an eye on the former reserve officer for some time, Howland called him aside one afternoon and paid him a high compliment. Many men in the captain's battalion were trying to walk just like him.[63]

So anxious were the men for action that on Saturday afternoons, the one day of the week allowed for recreation, it was not unusual to meet troops voluntarily hiking through the countryside, bent on developing their powers of endurance. Those who had struggled to gain exemption from the service for physical reasons now fought just as hard to remain in it. Patriotic youths occasionally showed up days before their date of induction, betraying themselves as misfits by walking the camp in disheveled suits. But as the weeks turned into months and the constant rumors about shipping out proved unfounded, morale sagged and the monotonous training became drudgery. Distinguished visitors were scheduled in an attempt to bolster esprit de corps. Lieutenant Colonel Fitch, who had served in the Spanish-American War, persuaded his former commander, Theodore Roosevelt, to address the men. Sputtering and flailing, the old Rough Rider spoke of "the wolf rising in the heart" at the onset of battle and of his immortal exploits on San Juan Hill. U.S. District Judge Kenesaw Mountain Landis, who had once attempted to fine Standard Oil of Indiana $29 million and had recently sentenced Victor Berger and six other Socialists for impeding the war effort, held court at Camp Grant, the first time a majority of the men had seen the judicial process at work. Still, the jurist's stern visage and heavy hand proved no deterrent to would-be revelers. The trade with local bootleggers had become so rampant that a special law enforcement unit made up of former Chicago policemen was created to secure the post from the evils of "demon rum."

The winter of 1917–18 settled in early and quickly claimed a place in the record book. Blizzard after blizzard swept down from northern Canada, smothering the camp in an ever-deepening mantle of white. For weeks the seemingly frozen mercury remained near the bottom of the tube, at one time settling at 27 degrees below zero. Ears, noses, fingers, and toes were frozen in the teeth of biting winds and cutting sleet. Yet the training continued, except for two days when drifts reached the barracks' roofs, making it impossible to move about except on skis or snowshoes. Colonel Howland never wore earmuffs. Wishing to appear no less of an example to his men, Hubble also went without them until he froze his ears and reluctantly gave in.[64] With spring the melting snows made rivers of company streets and lakes of drill grounds. The normally placid Kishwaukee thrust its burden of ice over the rifle range, sweeping it clear of bridges, targets, and emplacements. According to the official history of the 86th, only the spirits of the men remained undampened.[65]

VIII

On July 9, 1918, fourteen months after he enlisted, Major Edwin Hubble was examined and found physically fit for overseas duty. The promotion ranking him above an aged and ailing Captain Martin Jones Hubble had come in January, in part as the result of Colonel Howland's "marvelous" and "unequalled" record as regimental commander.[66] Word of the long-awaited order to embark raced through the camp and was greeted with exultation by the officers. It was a different story among the ranks, however, where the tension was palpable. During the previous year, the division had been repeatedly stripped of many of its battle-ready troops, much to the disappointment of the thousands left behind. This was before the growing casualty lists began coming in, each containing the names of hundreds of fallen Black Hawks who, like themselves, had eagerly awaited word from headquarters that it was their turn to go.

From Hoboken, New Jersey, the camouflaged fleet sailed, ship by ship, down the Hudson River into New York harbor and past the Statue of Liberty. It was met by a cruiser, a battleship, a navy blimp, and several sub chasers and escorted out to sea on September 9. Hubble was assigned to the *Walmar Castle*, an overcrowded army transport with hammocks swung in the hold. He was put in charge of the lifeboat assigned to the Red Cross nurses on board and kept his revolver ready should an emergency arise. Much of his time was spent on the bridge conducting submarine watch. The crossing was slow and the weather stormy. Seasickness quickly enveloped the pitching ship. He later confessed that it had taken every ounce of determination he could muster to stay on his feet in order to avoid embarrassing himself before his equally miserable men.[67] The convoy finally rounded the north coast of Ireland and nudged up the Firth of Clyde to Glasgow, where the underweight and wobbly major disembarked on September 19.

Within a week Hubble was back in a much different England from the one he had left four years earlier. At rain-drenched Romsey, ten miles from Southampton, the mud-encrusted division was stricken by the "Spanish flu," part of the worldwide epidemic that would claim more than 20 million lives. Men dropped in the ranks, at drill, and on parade. Hubble's favorite sergeant, a redheaded youngster, stood in line with a temperature of 104 degrees before succumbing a few hours later. Like that of many other Black Hawks, his body never left English soil.[68]

From Southampton the division was ferried across the Channel to Le Havre on blacked-out, zigzagging paddle wheelers. Once in France, the men encamped near Bordeaux while Hubble and many of his fellow officers were assigned to the advanced combat training schools established by General John "Black Jack" Pershing's American Expeditionary Force. During the first week of October, while stationed at Langres not far from Pershing's headquarters, Hubble at last gazed across the ragged chessboard whose pawns were being swept away by the millions. The second battle of the Marne had recently seen the Germans stopped short of Paris. With Foch in command, the Allies launched the Meuse-Argonne offensive, pushing the enemy back beyond the Hindenburg Line, a breakthrough that constituted the beginning of the end.

Hubble, whose military records would be destroyed in a fire many years later, asserted that he had served as both a field officer and a line officer in the trenches and in action. "The hardest thing," he said, "was to see wounded men fall, and go forward without stopping to help them." He also spoke of being trapped in a swaying observation balloon near Pont à Mousson while artillery shells fell around Metz and into the Moselle. Early in November, with the Germans in full retreat, he was knocked senseless by a bursting shell. He awoke, unattended, in a field hospital behind the lines. Besides a mild concussion and some small cuts on his body that left white scars, he had an injured right elbow that he could never quite straighten again. Nevertheless, he got up, found his clothes, dressed, and left without speaking a word to anyone.[69]

Like others treated for battlefield wounds, Hubble was entitled to a wound stripe, an inverted one-and-one-half-inch chevron worn at the end of the left sleeve. In the military record attached to his honorable discharge, the word "none" appears in the blank following each of three categories: "battles, engagements, skirmishes"; "medals awarded"; "wound chevrons authorized." His only decoration was a gold war service chevron.[70] In the confusion of battle, injuries of the type he claimed to have sustained might easily have gone unrecorded. Still, this does not explain the absence of any reference to combat in his discharge papers. In a regretful letter to Frost, he wrote: "I barely got under fire and altogether I am disappointed in the matter of the war."[71]

Hubble and his fellow officers rejoined the 86th at Bordeaux, hoping against hope that they would be called to the front before "the big show" was over. It was not to be. The end was only days away when it was decided that the money each had contributed to a common fund should be spent on a farewell dinner and tribute to the lost battalions. The local wine flowed freely; for the first and only time in his life, Hubble got tight. The next morning, he could not remember what had happened and had to be filled in by his friends. They informed him that he had stood up and made a speech, quite lucid and carefully presented, on the origins of the war. "I wish," he quipped, "I had heard it."[72]

IX

The 86th Division never saw combat, and on November 12, the morning after the Armistice went into effect, the guns of August fell silent at last and the division was dismantled. After eighteen months of dreaming and sacrifice, glory had eluded the handsome major and his fellow officers, most of whom were dispatched to Le Mans to assist in the massive embarkation effort. Still, there was the cachet of the uniform, the vicarious sense of the heroic associated with a great conquering force and a deeply indebted populace, a lingering comradeship among those who had put their lives on the line—or had been willing to do so. After giving the matter much thought, Hubble decided to stay on.

Living the real life of F. Scott Fitzgerald's enigmatic Jay Gatsby, he wound up in Cambridge, England, after serving four months with the occupation forces in France and Germany, where he reputedly dealt with reparations claims and held the position of judge advocate in courts-martial.[73] From Cambridge, in May 1919, he wrote of being in charge of American army students in a dozen British universities. More important, he was renewing his acquaintance with astronomy.[74]

Each morning Hubble walked from his rooms in Herne Lodge at 6 St. Eligius Street to Trinity College, where he passed through the Great Gate on whose massive exterior stands the statue of founder and benefactor, Henry VIII. On the second floor and to the right are the rooms in which Newton formulated the fundamental laws and mathematics of the Scientific Revolution, dreaming all the while of a universe more vast than even his science could encompass. Hubble sat in on a class in spherical astronomy given by Plumian Professor Sir Arthur Eddington and walked the corridors of the elegant Wren Library, where he contemplated Rysbrack's romantic bust of Newton, hefted the great man's walking stick, and gazed on a lock of the icon's silver hair, iridescent in the mullioned light. In the evenings, after completing his duties, he proudly took his place at High Table in the magnificent Dining Hall, with its hammer-and-beam ceiling, minstrels' gallery, and huge portrait of Henry. Then, if the skies were clear, he made his way over to Madingley Rise, the great house near Cambridge Observatory. Its owner, the wealthy English astronomer H. F. Newall, entertained there almost every evening, his guests occasionally strolling over to the observatory to check up on the constellations and nebulae.

Newall himself proposed Hubble for membership in the Royal Astronomical Society at its monthly meeting in London on May 9. Two months later, at a special meeting and dinner honoring visiting American astronomers, Hubble was not only present but seated in a place of honor near the head of the table between astrophysicist Arthur Schuster and Frank Dyson, the Astronomer Royal. It was the first time Mount Wilson astronomers Walter Adams, Charles St. John, and Frederick H. Seares had the opportunity to take his measure. They were doubtless surprised at the prominence accorded the crisply dressed officer, which was hardly commensurate with his scientific achievements. Nor, during this period of inflated nationalism, would his reversion to the accent and Briticisms acquired at Oxford have been much appreciated.[75]

Hubble's self-confidence was all the greater for Hale's pledge that the nearly thirty-year-old astronomer would have a place at Mount Wilson at war's end. During the previous several months, the director had rejected the applications of at least three qualified men, presuming that Hubble would soon be available.[76] When, at last, he had written in June to say that he was almost ready, an impatient Hale replied, "Please come as soon as possible, as we expect to get the 100-inch telescope into commission very soon, and there should be abundant opportunity for work by the time you arrive." He advised him to talk with British technicians about large prisms and short-focus lenses suitable for work on faint nebulae. "It is important to push the light efficiency of your spectrograph to the last possible limit." His salary would be $1,500 a year, $300 more than he had been offered before the war, and Hale assured him of advancement "as rapidly as your work and the funds at our disposal will warrant."[77]

Hubble finally bid farewell to his spiritual homeland and sailed for New York in August 1919 aboard the almost deserted *Imperator*. After landing, he headed straight for California, pausing briefly in Chicago for a one-day reunion with his mother and sisters. He then continued on to San Francisco, where he received his discharge at the Presidio on August 20 and collected the $60 bonus for his service in the European theater.[78] His few mementos consisted of his tin hat, his major's insignia, the crossed rifles of the 343rd Infantry, and a razor-sharp German trench dagger, which he found useful for cutting the pages of French books.

CHAPTER SIX

RECONNAISSANCE

I

While Hubble languished at Camp Grant awaiting word of the fate of the 86th Division, the 100-inch telescope was pointed to the sky for the first time on the night of November 1, 1917. Silhouetted against the blackness of the dome were the figures of nineteen men, ranging from Mount Wilson's director, George Ellery Hale, to Roy Desmond, a janitor's helper. Hale and assistant director Walter S. Adams climbed the long flight of iron steps to the observing platform, not unlike condemned prisoners ascending the scaffold. On the floor below, where only a dim red light marked the location of the control panel, night assistant Wendell P. Hoge pushed a series of buttons, simultaneously instituting three kinds of motion. The observing platform rose and rotated; the dome, its slit already open to the spangled sky, revolved in the opposite direction; the barrel of the telescope itself turned and locked on its target—the planet Jupiter. Hale, not a well man, bent low and looked through the eyepiece. What he saw produced a wave of nausea. Saying nothing, he turned to Adams, an appalled look on his face. The director stepped back and Adams crouched over the eyepiece. In place of the anticipated single image, a half dozen Jupiters—irregularly spaced and partially overlapping—filled much of the lens, as if the surface of the mirror had been distorted into many facets, each of which was contributing its own image.

An irascible George W. Ritchey, who helped mount and polish the mirror, had predicted that air bubbles in the disk might produce the observed effect, a theory he reiterated during the ensuing discussion. As the evening wore on, a second hypothesis surfaced, one seemingly too good to be true. The dome had been left open throughout the day while the workmen were busy with parts of the mounting. It seemed probable that the sun had shone, if not directly on the mirror itself, at least upon the cover above it, thus accounting for the distortion. The men began a tension-filled vigil while the mirror cooled to the temperature of the mountain air. Hale and Adams took turns checking the image, which improved only slightly after three hours. To add to the gloom, word had just arrived of the Italian army's fall at Caporetto, and Adams recalled sitting on the floor speculating on whether Italy was now out of the war.

It was finally decided that everyone should go to bed, but Hale and Adams agreed to meet back at the dome at 3 a.m. Hale lay down without undressing, but sleep eluded him. After an hour of tossing, he switched on the light and began reading a detective story, which did nothing to curb his restlessness. He gave up and returned to the dome at 2:30, where he was shortly joined by a sleepless Adams. They repeated their walk up the echoing steps and issued new instructions. Jupiter was out of reach in the west, so they swung the great instrument northward to Vega, brightest star in the constellation Lyra. With his first glimpse, Hale's depression vanished; the mirror had regained its normal shape during the chill of the small hours, and Vega stood out in the eyepiece as a sharp point of light, almost dazzling in its brilliance.[1]

On the mountain that night was Hale's only invited guest, the English poet Alfred Noyes. While waiting for the mirror to cool, Hale had challenged his friend to write of the "fight for knowledge," of the development of science rather than of war. Noyes responded by composing his epic *Watchers of the Skies.*

> *And now our hundred-inch . . . I hardly dare*
> *To think what this new muzzle of ours may find.*
> *Come up, and spend that night among the stars.*[2]

Hale vividly remembered his first night on the mountain, during the summer of 1903. He lay on a cot watching the stars cross a huge hole in the roof of the "casino," a cabin of cedar logs abandoned a few years after its construction. Seeking a site for a new solar observatory, he had hiked the eight miles to the summit on the old Indian trail leading up the canyon from Sierra Madre. Within a year he had persuaded the recently endowed Carnegie Institution of Washington to provide $150,000 to found the Mount Wilson Solar Observatory, under Carnegie auspices. A year after that the first photograph of a sunspot spectrum was taken atop the mountain with the Snow telescope, while Hale, as always, was dreaming of greater scientific feats to come.

A hint of the boundless and ultimately self-destructive energy that would lead to four mental breakdowns is revealed in childhood photographs of the future astronomer. Restless eyes, pursed lips, and small, clenched fingers betray the deep-seated impatience Hale inherited from his driven father, William Ellery Hale, a Chicago businessman who made his fortune manufacturing the hydraulic elevators essential to the erection of skyscrapers. Chafing against the fixed duties of school life, Hale thought of himself as a freelance, preferring to work on projects of his own, especially ones involving tools and machinery. His small shop in the family home at Kenwood was soon replaced by a building of his own design, complete with a little laboratory where he performed simple chemical experiments, made batteries and induction coils, and observed with a microscope. After constructing a small telescope himself, his father bought him an excellent 4-inch Clark. In 1884, upon turning sixteen, the youth announced that his greatest ambition was to photograph a spectrum, which he succeeded in doing with a small prism affair. Four years later, following his son's blueprints, William Hale built a spectroscopic laboratory on the family property, the nucleus of the Kenwood Observatory at 4545 Drexel Boulevard.

Hale graduated in physics from the Massachusetts Institute of Technology in 1890, where, in contrast to his own laboratory work, he found most of the courses uninspiring. During his spare time he read and abstracted everything he could find on astronomy and spectroscopy at the Boston Public Library. In August 1889, while riding on a trolley car in Chicago, an idea had come to him from "out of the blue." He would plan and build an instrument that would solve the problem of photographing solar eruptions in full daylight and also provide a permanent record of these and other solar phenomena. Hale christened it the spectroheliograph.

Once again William Hale acceded to his son's request by purchasing him a 12-inch reflecting telescope, a better instrument than those owned by many public observatories. An associate professorship in astronomy followed at the new University of Chicago in 1892, shortly after which Hale learned of the availability of two 40-inch lenses at the firm of Alvan Clark in Cambridgeport, Massachusetts. He persuaded the traction magnate Charles Tyson Yerkes to provide for a telescope that would surpass all others in focal length and light-gathering power, and to underwrite the cost of the observatory to house the giant instrument.

The peripatetic Hale, who was soon fighting depression, acute indigestion, and severe pains in the back of his head diagnosed by doctors as brain congestion, was long gone from Williams Bay by the time Hubble began his graduate studies at Yerkes in the autumn of 1914. Having worked his will on a remote lakeshore by constructing the world's most powerful telescope, Hale was engrossed in repeating this exploit on an isolated California mountain.

The casino had been made comfortably habitable in 1904 and a second route to the summit, the so-called Mount Wilson Toll Road, was soon opened. Zigzagging its way up the rugged southern face of the 5,714-foot peak, the "road" averaged two feet in width, sufficient only to permit the passage of the miscellaneous burros, mules, and occasional horses that carried traumatized visitors and supplies to the crest. The limits set by pack trains affected every feature of the design and materials of the early buildings of the observatory. No single structural member of the Snow telescope building exceeded eight feet in length, while the wooden doors of the original dormitory were limited to a width which permitted their transportation on the sides of a burro without dragging on the ground. As it was, several of the doors, when fitted, were found to be slightly rounded at the corners. Wild animals, with the exception of overhunted deer, were plentiful, and the casino housed a small arsenal. The stump of an old pine was the most common target, causing Adams to quip that it might have been mistaken for a lead mine. Communication with the outside world was always problematic, depending on a single iron wire stretched over bushes and an occasional tree branch. Even in the best weather one had to shout so loudly that it sometimes seemed as if the party at the other end could have made out the message without benefit of the instrument.[3]

In 1906, John D. Hooker, an elderly Los Angeles businessman intrigued by astronomy, agreed to Hale's request that he provide the funds for the purchase of a 100-inch disk from the St. Gobain glassworks in Paris. Hooker, after whom the dream telescope would be named, also promised to erect an optical shop and to underwrite the salaries of the skilled technicians needed to figure and grind the surface of the mirror. Half a million dollars was still required for the mounting and observatory building, forcing Hale to turn to the Carnegie Institution once again.

Andrew Carnegie, who had become Hale's friend, decided to conduct a personal inspection of the mountain in March 1910, bringing along Mrs. Carnegie and their daughter Margaret. With the widening of the toll road in 1907, pack trains had been replaced by mule teams and the horse and carriage, although the ascent was still something of an adventure. Fearing that the white-bearded philanthropist would be unable to climb the steps to the floor of the dome, Hale asked Ritchey to design a reinforced box for the purpose of raising the distinguished visitor to the platform of the newly installed 60-inch telescope by means of a hoist. The box was placed on the concrete floor near the foot of the stairs, but when Carnegie arrived he passed it by, charging up the steps like a schoolboy. Sensitive about his stature, he always took the higher position on a slope when photographed with the taller Hale and such other distinguished guests as Hooker and the naturalist John Muir.

While Carnegie seemed interested in the observatory equipment, doubts had been raised about his knowledge of astronomy. These were confirmed on his return to New York, where he proclaimed that 60,000 new stars had been discovered at Mount Wilson, an apparent reference to Ritchey's photographs of a single stellar cluster. The remark elicited a caustic aside from Moulton at Chicago, who stated that one might as well speak of having "discovered 60,000 new gallons of water" in Lake Michigan.[4] However, Carnegie's blunder was soon forgotten when he released the text of a letter to Dr. Robert S. Woodward, president of the Carnegie Institution, pledging an additional $10 million to its endowment and expressing the desire that the work at Mount Wilson be advanced as rapidly as possible.

After many trials, the glassworkers at St. Gobain produced a giant disk large enough to fill a small room. Measuring 101 inches across and 13 inches in thickness, and weighing over 5 tons, it reached Hoboken aboard the steamer *St. Andrew* in November 1908. The New York press dubbed it the most valuable single piece of merchandise ever to cross the Atlantic, and patiently followed its progress to New Orleans, and from there to the optical shop of the Mount Wilson Solar Observatory on Pasadena's sleepy Santa Barbara Street, where it arrived on December 7. After the disk's uncrating the next day, a disgusted Ritchey, who had been assigned the task of grinding it, declared the object useless, plunging Hale into the doldrums. The casting had required three melts to fill the mold, each poured in quick succession. This procedure allowed countless tiny air bubbles to become trapped in the interior, while the prolonged annealing process weakened the glass in several places. St. Gobain volunteered to try again, this time at the firm's expense. A second disk was poured and buried in a huge mound of horse manure, where it proceeded to crack while cooling. Additional attempts also failed miserably.

The original disk lay beneath canvas and cloth for two years while Hale's frustration mounted, leading to a nervous breakdown. Finally, in 1910, Arthur L. Day, a vice president of Corning Glass Company, was asked to examine the rejected specimen. After conducting several tests, Day concluded that the layer of bubbles strengthened rather than weakened it, nor were they close enough to the surface to interfere with the formation of a smooth optical face.

The painfully monumental task of grinding 7,800 square inches of glass took nearly five years to complete, during which a reluctant Hale was forced to replace an unbalanced and increasingly belligerent Ritchey with the optician W. L. Kinney. In the first stage the disk was brought to a spherical shape; in the second the spherical form was changed to a paraboloid. To keep from scratching the surface, all of the optical work, with the exception of the first rough shaping, was carried out with wooden tools of various sizes and forms; jeweler's rouge combined with distilled water constituted the abrasive. The glacial process resulted in the removal of a ton of glass, accounting for the depth of the curve at the center of 1¼ inches. So fine was the craftsmanship that every portion of the surface produced the same focal length to within one part in about 90,000.[5]

Construction of the observatory dome, which Adams ranked close to the erection of the pyramids, began during the summer of 1913. The ground was blasted and graded, after which forms were set for the concrete pier on which the telescope would rest. Thirty-three feet high and 40 feet long with a circular concrete floor at the top, the structure called for more than 100 tons of cement, heavily reinforced with twisted steel rods and I beams. Located within the pier was a room for resilvering the mirror and housing the largely ineffectual apparatus for circulating cold water around the disk at night through coils of pipe.

The road, which had been widened to facilitate the construction of the 60-inch dome, now had to be widened again. A Mack truck, whose solid tires and unyielding springs gave it the riding qualities of a tractor in a plowed field, was purchased in 1912, followed by the acquisition of a more exotic vehicle the following year. One of only a few units ever built for mountain hauling, the cableless truck was powered by a gasoline engine that operated a generator which in turn furnished electric current to the four motors installed in the wheels, whose independent action proved useful on the sharp turns. However, it was discovered that when fully loaded the vehicle could not negotiate the steep grades under its own power and had to be assisted by a team of mules, lengthening the ascent to two days. At the wheel was Merritt "Jerry" Dowd, the only person who understood the truck's idiosyncrasies and who later became chief electrical engineer on the mountain.

In 1914, with war on the horizon, that part of the steel building up to and including the track that supports the revolving dome was completed. Three circular rows of columns were called for: 22 in the inner row to help support the second floor, 25 in the second row to serve as the main support of the dome, and 45 lighter columns along the outside row to carry the exterior sheet-metal wall. After a cement floor was laid at ground level, an inner wall of corrugated galvanized iron was installed and painted black. As winter set in, a lone workman began the long task of grinding the circular rails on which would run the 24 four-wheeled trucks supporting the dome.

As a consequence of the telescope's size and tremendous weight of more than 100 tons, the "closed fork" mounting was chosen, whereby the tube is hung in the center of a rectangular frame of massive steel girders. The entire rectangle is then mounted on bearings at top and bottom, which furnish the east and west motion of the instrument. To relieve the tremendous friction, two large steel floats are submerged in tanks of mercury. The floats bear some 98 percent of the telescope's weight, allowing the celestial cannon to be moved with the touch of a hand.

The Quincy, Massachusetts, firm of Fore River was awarded the contract for the mounting. When finished, it was deemed too large to be shipped by rail and had to be routed around Cape Horn on the *Alaska*. The fear of storms and marauding German submarines added appreciably to Hale's mental anguish. From San Pedro harbor, where it arrived in February 1916, the tube was trucked up the mountain in four increments without incident. By this time the sheet-metal dome, 100 feet high, 95 feet in diameter, and weighing 600 tons, was all but completed, as was the observing platform. This phase of construction had seen the only serious accident of the precarious undertaking. On September 30, 1915, the night assistant wrote the following in the 60-inch logbook: "Mr. Moore ironworker fell from top of 100-in. dome to cement pier, 70 feet, and instantly killed."[6]

The construction of the switchboards, electrical controls, and most other accessories took place in the Pasadena shops or in similar facilities on the mountain. The driving clock, which moves the telescope at a uniform rate corresponding to the earth's rotation, required more than a ton of bronze castings and nearly 1½ tons of iron castings in addition to a 2-ton driving weight. Housed within the concrete pier near the south end, the shaft meshes with a worm wheel 17 feet in diameter attached to the telescope axis. Its movement and the motions of the telescope, dome, and shutters are controlled by over 30 electric motors tied into a single control panel.

With all nearing completion on the mountain, the time had come to silver and burnish the giant mirror upon whose surface would fall not 60,000 but an estimated 3 billion points of light. The windows of the polishing shop were doubled and sealed while the painted cement floor was hosed to keep down the dust. Only filtered air, maintained at a constant temperature, reached the interior, whose ceiling was hung with a canvas screen to shield the glass from any falling particles. Dressed in surgical caps and gowns, the optician and his two assistants were the only ones allowed past the door.

A waxed metal band, lined with fabric, was clamped around the mirror, transforming it into a great shallow dish. Then a solution of caustic potash and water was poured over the glass, whose surface was scrubbed with cotton swabs attached to T-shaped sticks. Made up a month in advance, a reducing solution of distilled water, alcohol, nitric acid, and 5 ounces of rock candy was next added to 25 gallons of water which had been poured on the glass to fill the concavity. Concentrated ammonia in combination with silver nitrate was then added slowly, along with a reserve of silver. After more swabbing the solution was poured off and the mirror rinsed with a hose. The band was removed and the dark silver surface was dried by means of chamois skins and a blast of air from fans. A final hand swabbing with an assist from a motor-driven polisher 3 feet in diameter completed the delicate process. So thin was the burnished coating that the amount of silver remaining on the glass would have made a dish no larger than a 25-cent piece.[7]

The 100-inch mirror was moved safely to the mountaintop on July 1; 1917, eleven years after Hooker's initial bequest. Its installment was delayed another two months when it was discovered that the observing platform was too heavy, requiring that it be taken down and redesigned.

Not until September 22 was the mirror hoisted into its massive mounting. The polishing of the giant teeth of the worm gear was completed October 26 and work on setting up the driving clock was well under way. According to the logbook: "Preliminary observations and some photographs will probably be taken within the next few days."[8] Hale's exact thoughts are unknown as he sped west by train for the crucial trial from Washington, D.C., where he was serving as wartime chairman of the National Research Council. Surely the poet Noyes could not have been far off the mark when he wrote:

> Where was the gambler that would stake so much,—
> Time, patience, treasure, on a single throw?[9]

III

On a late August morning in 1919, C. Donald Shane, a Ph.D. candidate in astronomy at the University of California at Berkeley, was in San Jose waiting for the autostage to the Lick Observatory on Mount Hamilton. His attention was drawn to a tall gentleman in uniform who soon walked over and introduced himself as Major Hubble. Shane recognized the name from papers on nebular work done at Yerkes Observatory, but he was somewhat thrown by the stranger's affected accent. Hubble explained that he was headed for Mount Wilson and had decided to visit Lick along the way. Shane remembered that the handsome guest remained on Mount Hamilton no more than a day, but it was sufficient time to make a lasting impression. Lick astronomers whose careers paralleled Hubble's would address him as "the Major" the rest of their lives.[10]

Hubble joined the staff of the Mount Wilson Solar Observatory on September 3, presumably still in uniform when he arrived at the Santa Barbara Street offices. Though it was poised for a boom of unprecedented duration, Southern California remained a place of quiet rusticity. Pasadena's 20,000 citizens were scattered over a wide area; open country with vineyards and orange groves intersected by rambling dirt roads dominated the landscape. The occasional farmhouse and the mostly small dwellings of this semidesert environment were serviced by private water companies whose fortunes, like those of their patrons, depended on the vagaries of the winter rains and the amount of water they could buy, borrow, or steal. The business district was essentially limited to Colorado Street, at the end of which stood the elevated remnants of an abandoned dream. Had not the automobile come of age, the never completed cycleway would have allowed residents, for the price of a small toll, to ride their bicycles into Los Angeles over an enclosed wooden track free from the pitfalls of roads deep in sand and the offensive residue of horse-drawn traffic.

Hubble had to wait nearly six weeks for his first turn on the mountain, which came in mid-October on the 10-inch Cooke. A three-day run on the 60-inch followed a week later. Inside the logbook's front cover were jotted simple words of advice that to jaundiced eyes of a later time appear homiletic: "The seeing is better than you think it is"; "When tired, cold, and sleepy never make any movement of telescope or dome without pausing and thinking"; "Do not give up in despair and go home, the conditions will improve after midnight."[11] It so happened that the seeing remained poor all three nights because of cloud cover, limiting his yield to a meager nine photographic plates, including a 30-minute exposure of his favorite object, the cometary nebula NGC 2261. A quick perusal of the negative vindicated his initial findings at Yerkes; the triangular mass had undergone striking changes since he had last looked at it in 1916.

The observer was also being observed. Word of Hubble's military service and formidable bearing had preceded him up the mountain, leading the support staff to expect the worst. Milton Humason, a night assistant whose life would intersect with Hubble's in ways neither could have dreamt at the moment, was secretly taking the new astronomer's measure. Thirty-four years later, Humason would write:

> He was photographing at the 60-inch, standing while he did his guiding. His tall, vigorous figure, pipe in mouth, was clearly outlined against the sky. A brisk wind whipped his military trench coat around his body and occasionally blew sparks from his pipe into the darkness of the dome. "Seeing" that night was rated extremely poor on our Mount Wilson scale, but when Hubble came back from developing his plate in the dark room he was jubilant. "If this is a sample of poor seeing conditions," he said, "I shall always be able to get usable photographs with the Mount Wilson instruments." He was sure of himself—of what he wanted to do, and of how to do it.[12]

While the moment was romanticized in the afterglow of a long association, it had taken an unusual man to create so vivid an impression.

After passing its first and most crucial test with flying colors, the 100-inch had sat idle while Hale returned to Washington and the rest of the observatory staff became involved in the war effort. Not until September 11, 1919, some three weeks before Hubble's arrival, was the telescope finally cleared for general use. Then, on September 22, shortly after Francis Pease succeeded in taking the best photographs of the moon ever made with the instrument, it was shut down again because of what was termed "Pittsburgh seeing," a reference to a nearby forest fire whose ashes were drifting onto the sensitive mirror.[13] As it was, Hubble could not have arrived at a more opportune time.

That the mountain could be a lonely place during the holidays is witnessed by Wendell Hoge's poignant entry in the 1912 log of the 60-inch: "Merry Christmas to all the Universe."[14] Yet for Hubble, who was far removed from his family and friends, the isolation hardly mattered. Christmas Eve 1919 found him on top of the world detached from this small planet circling its little star, his fingers and thumbs poised on the buttons of the 100-inch Hooker, the greatest gift any astronomer could desire.

The seeing was next to perfect, as evidenced by the number of 5s appearing after most plates, the second-highest ranking on a viewing scale beginning with 1. His first photograph, H1H (Hooker 100-inch, plate one, Hubble), was of a nebular star, followed by three "good" plates of NGC 2261 and various other objects, including the Great Orion nebula, the elliptical nebula M60, and T Tauri, a type of star displaying rapid and erratic variations in brightness. Like a proper scientist, he attempted to mask his excitement in the brief space allowed for remarks: "A fine open spiral," "striking negative," "good detail."[15]

It took the night assistants some time to learn the correct spelling of the newcomer's name: "Hubble," "Hubbel," and "Hubbell" all appear in the logbook, sometimes on the same page. Yet, according to Humason, they were quickly won over by his direct, no-nonsense approach. When the time came for his run on the mountain, his campaign had been planned well ahead. If the seeing was bad, or the sky was overcast, he sat up longer than his fellow astronomers, waiting until 2:30 a.m. before turning in. Once he dismissed a night assistant, he breached convention by letting him sleep, even when the skies subsequently parted. Speaking little, he delegated responsibility and then allowed subordinates a free hand. In return, he expected results and could be quite stern if things weren't done properly. In Humason's words: "You knew where you stood with him."[16]

Hubble was also respected for never letting a night assistant do for him what he could do for himself. A stickler when it came to the equipment, he was always present when the telescope was changed from a Newtonian or reflecting focus to the Cassegrain, which increases the focal length by causing light to traverse a longer path, thus producing a larger image. This operation called for the delicate manipulation of a giant steel cage weighing several tons. The assistants were also responsible for cleaning up after a midnight lunch served in an unheated concrete bunker beneath the 60-inch. Hardtack and cocoa made up the usual fare, coffee having been banned on the mountain by an abstemious Hale, who considered it unwholesome. Hubble surprised everyone by washing his own dishes before returning to the telescope and making certain that his assistant got a share of the bland offering.

In his mind's eye, Hubble was still a major, an image he cultivated from his first day on the mountain. His usual attire was a shirt, tie, Norfolk jacket, jodhpurs, and high-topped military boots. As time passed, the less confining plus fours of his Oxford days were substituted for riding breeches. The former commander knew how to gain the respect of his subordinates while creating the illusion that he was "one of the boys." On a stormy night shortly after his arrival, Humason walked into the photographic lab about one in the morning and was taken aback when the astronomer inquired about the night assistant's whereabouts earlier in the evening. Humason, who was loath to lie, confessed that he had been playing poker with the janitor and members of the work crew. Expecting to be reported to "the higher-ups" and possibly dismissed, the abashed conspirator could hardly believe it when Hubble, a skilled bridge player, asked if he could play a few hands. The others were startled when the major walked in, but everyone was soon "getting along just fine."[17]

During his off-hours, Hubble relished life in the Monastery, so named by Hale, who delighted in such old-time thrillers as *The Monasteries of the Levant.* Except for the cook and housekeeper, all women, including the astronomers' wives, were banned from the living quarters on the theory that their presence could only distract from the great work at hand. Hale himself had selected the site. Assisted by Adams, he hacked his way downward a quarter of a mile from the Snow telescope, coming out in a small clearing at the very end of the ridge. The slopes fell away into almost sheer precipices on three sides, affording a remarkable view of the San Gabriel Valley and the purple-rimmed mountains beyond. The first structure was ready in six months, but it had burned well before Hubble's arrival, as had the first Mount Wilson Hotel, whose passing in March 1913 was recorded by the astronomer Adriaan van Maanen in the log of the 60-inch: "Closed dome 1 hr. 10

m. account of fire. Searching out radial velocities via spectrographs and burned one good hotel."[18]

The second Monastery proved more ambitious, consisting of some fifteen small rooms, each furnished with a single bed, chair, table, and lamp. At the head of the dormitory corridor was the library, whose shelves held duplicate copies of journals and books found at Santa Barbara Street, enabling staff members to keep up with recent developments in their fields. Rocking chairs of dark varnished oak faced a row of windows overlooking the valley. It was here, next to a carefully built fire consisting of exactly three logs, that Hubble often read and mused before dinner, a pipe firmly clamped in his boxer's jaw.

A bell sounded at 5 p.m. in winter and at 6 p.m. the rest of the year. The diners, each wearing a coat and tie, made their way from the library into the adjoining room to participate in a formal rite reminiscent of High Table. The man scheduled to observe on the 100-inch was seated in the place of honor, his status confirmed by a personalized napkin ring. To his right sat the 100-inch assistant, followed by the 60-inch observer. To the head observer's left was seated the chief solar observer, his assistant, and so forth down both sides until the wooden rings ran out and clothespins, the badges of the callow and the lowly, took their place.

Not until dawn would the astronomers return, stiff from the cold and mentally exhausted from hours of concentration at the instruments. Even the finest telescope in the world had its idiosyncrasies. As the great tube of the 100-inch tilted in pursuit of a nebula or an individual star, it would either forge ahead or lag behind its target, the result of minor imperfections in the gearing. Equally annoying was its tendency to sway. During the war, mercury became very expensive, so the float tanks were lined to halve the required amount. When the temperature changed, the linings buckled, creating a permanent defect. Moreover, light falling on the mirror through the mountain air was subject to differential refraction, and produced a shift in the apparent location of the stars, much like a straw when viewed sideways through a glass of water. To compensate, the astronomer had to stand on a migrating platform with an eye to the separately moving eyepiece while punching the buttons of a handheld control paddle to speed up the telescope or slow it down. The dome also had to be advanced simultaneously, and one had to make certain that its slit remained open in front of the mirror. Next to sleeplessness, cold was the greatest adversary. On the most bitter nights fingers and toes grew numb while tears literally froze the observer's lashes to the eyepiece.

In contrast, the walk back to the Monastery through the silhouetted pines assumed a spiritual quality as the valley below gradually took form in the emergent light. The serenity of the fading night, the knowledge that a fragment of the yawning cosmos had been captured on a photographic plate marked it as a moment of reflection and reverie. The Buddha himself would have been at peace atop the mountain.

I V

In a letter to an all but blind Edwin Frost, Hubble described the housing situation in Pasadena as "an exceedingly unstable structure. Rooms can be found from day to day, but it is rare that a desirable room is vacant for more than a couple of days."[19] After bouncing from one boardinghouse to another, he finally moved into a small home at 55 Euclid Avenue, where he shared the first floor with Clinton Judy, head of the Humanities Department at the recently christened California Institute of Technology. The second floor was occupied by Harry Wood, a seismologist on the staff of the Carnegie Institution. The small, book-filled rooms and little open fireplace reminded Hubble of his Oxford digs, as did Judy, who had spent time at Oxford as an independent scholar. Rather than cook, the bachelor trio dined at the Peacock restaurant nearby.

Hubble did not own an automobile; when not on the mountain he walked to and from the observatory offices in the white neoclassical building at 813 Santa Barbara Street. His door was about halfway down the left wing on the first floor. It opened into a spacious room facing south, to the street, its tall, oak-trimmed windows shaded from the morning sun by venetian blinds. Like the others, it was furnished with a large flat-topped desk, a revolving chair, and a single straight chair, a subtle reminder to the visitor not to overstay his welcome. Opposite the desk was a tall metal cabinet that held his latest collection of plates; the rest were stored in the quakeproof vault below. The walls were barren of pictures. The only object that could be called personal was a rock impregnated with fossils, which sat on a corner of the desk. The one feature that distinguished Hubble's office from the others was its private washroom, a reminder that, until recently, it had been occupied by the ailing "sun worshipper," George Ellery Hale.

Along the carpeted and paneled halls of the first and second stories were the offices of other suited gentleman, who, from outward appearances, might have been bankers, stockbrokers, or business executives.

Because of Hale's frequent and prolonged absences, most administrative duties were assumed by the indefatigable Walter Adams, who had been appointed assistant director in 1917. The son of New England Congregational missionaries, Adams grew up in Kassab, Syria, a village near Antioch, from which he hiked with his father along roads once trodden by the armies of Persia, Greece, Rome, Arabia, and, later, the knights of the great Crusades. The family returned to Derry, New Hampshire, in 1885, which enabled eight-year-old Walter to enter regular school for the first time. His preparatory education eventually took him through Phillips Andover Academy, where he excelled in mathematics, physics, and chemistry. "Very early in my education," he later wrote, "a strong preference developed for exact subjects with concrete and definite answers as compared with those involving alternatives and the exercise of considerable judgment."[20] As a Dartmouth College undergraduate, Adams enrolled in the only course offered in astronomy. His teacher was a rising young professor named Edwin B. Frost. When the latter was called to Yerkes by Hale, Adams followed him and undertook graduate work leading to an A.M. at the University of Chicago. After a year of studying celestial mechanics at the University of Munich, Adams, who by this time had come under the thrall of Hale, returned to Yerkes as an assistant astronomer. In 1904, the year Hale established his observatory atop Mount Wilson, Adams was present at the creation.

The granite-faced, pipe-smoking New Englander walked to the observatory offices reading *The New York Times*. As punctual as he was precise, he was so regular in comings and goings that people in the neighborhood set their clocks by his passing. Yet his was the patience of a Yankee reared in a land where the ephemeral brush strokes of time meant nothing. Drawn to statistics and spectroscopy, he willingly spent decades gathering data on the positions, brightness, temperatures, and motions of the stars. He expected others to toe the mark in the same fashion and let it be known through the grapevine when he heard a report that someone wasn't at the telescope full-time on a particularly cold night. Self-deprecating and easily embarrassed, he once blushed when a colleague complimented him on his design of a series of powerful and tricky spectrographs, replying, "It is a very low form of cunning."[21] The assistant director's natural reserve disguised both an unexpected wit and the skill of a master storyteller. Hale liked nothing better than to lounge by the fire in the Monastery while his best friend spun visions of an exotic childhood for the man who doted on all things Oriental. Adams also played bridge with Pasadena's most accomplished devotees, and he excelled at tennis, hiking, golf, and other sports. The "mathematical shark" was the best billiard shooter at Santa Barbara Street, a skill acquired over the course of countless matches on the mountain.

A frequent opponent in such contests was the Dutch astronomer Adriaan van Maanen, a Woodrow Wilson look-alike who, though descended from a distinguished line of jurists, cabinet ministers, and naval commanders, submitted without protest to the American nickname of "Van." He had come to Pasadena in September 1912, after spending a year as a volunteer assistant at Yerkes, where he slept in the attic or under the floor of the 24-inch reflector in winter, and in a tent on the grounds in the summer until a fierce thunderstorm sent him scampering for permanent cover. While studying for his Ph.D. at the University of Utrecht, the young van Maanen had come under the influence of the distinguished Dutch astronomer and Hale intimate Jacobus Cornelis Kapteyn. It was Kapteyn, following in the eighteenth-century footsteps of Sir William Herschel, who concluded that the sun is located at the center of a great flattened disk of stars—the Milky Way. Accordingly, there could be no other galaxies, a proposition to which van Maanen wholeheartedly subscribed. From 1908 until 1913, Kapteyn, at Hale's invitation, made an annual pilgrimage to Mount Wilson. He resided in the cottage which still bears his name, and whose guest book is filled with the signatures of the famous who spent a night or two roughing it beside a wood-burning stove, while the mountain dropped sharply away just a few feet from the front porch.

It was natural that Kapteyn's protégé should be accepted by Hale, who immediately set him to work making solar observations by day and photographing spiral nebulae by night in an attempt to measure their internal motions. By the time Hubble joined the staff, van Maanen had already expended years taking hundreds, if not thousands, of plates for comparative purposes. Not surprisingly, he was able to show that the bright centers of the farthest nebulae were rotating once every 100,000 years, slow, to be sure, yet much too rapid for them to be independent galaxies moving away from the sun at the incredible speeds calculated by Vesto Melvin Slipher.

Van Maanen, a gregarious, alert-minded bachelor, full of playful nonsense, loved the company of young people and lived at the Y.M.C.A. for several years prior to purchasing a small house, from which he took the streetcar to his office, almost always arriving earlier than his colleagues. A raconteur and something of a playboy, he joined the Valley Hunt Club, whose fellow members were often guests at his intimate dinner parties, the food for which he prepared and served with élan. Unyielding when he thought himself in the right, he could be as stubborn as he was engaging. He displayed extreme jealousy over the division of time on the 100-inch and never for a moment considered relinquishing his Dutch citizenship. Though Hubble had enjoyed the company of many Europeans, he "scorned" van Maanen from the outset, refusing to address his fellow astronomer by his nickname or to associate with him more than was absolutely necessary.[22] The feeling was mutual, as if both sensed that they were already on a collision course and did not wish to confuse the issue with pleasantries.

To further complicate matters, van Maanen had a friend. "Some time ago," wrote Harold J. Ryan, Horticultural Commissioner for the county of Los Angeles, in a letter to Walter Adams, "I heard from a source that one of the men at or connected with the Mt. Wilson Observatory had made some observations of the variation in the activity of ants at different temperatures." The commissioner wanted to know more about this curious phenomenon and requested that Adams put him in contact with the observatory's amateur entomologist.[23] Written in November 1924, the letter had arrived three years too late. Harlow Shapley, the person in question, had left Mount Wilson in 1921 to become director of Harvard College Observatory.

By his own admission, Shapley, a plainspoken, hot-blooded Missourian, hated the chill of the mountain. "I suffered quite a bit those long, cold nights. I suppose I didn't get as much sleep in the daytime as I needed, for I was running around observing ants in the bushes."[24] The insect detective had begun his surveillance of Formicidae quite by accident. While pausing to rest during a climb from a canyon where he had gone to collect plants, Shapley noticed a stream of ants running along a concrete wall—some going one way and some another. One branch flowed into the shade of some manzanita bushes, slowing perceptibly as it did so. This puzzling behavior caused the astronomer to arm himself with a thermometer, barometer, hydrometer, and stopwatch. A flashlight was added to his kit for nocturnal vigils. Employing "speed traps" 30 centimeters long, he timed the insects' movements at all hours. In the noontime heat the ants ran faster than in late afternoon, and slower still at night. The higher the temperature, the faster they ran, like snakes and lizards, unaffected by humidity, barometric pressure, or the season of the year. Shapley had discovered the thermokinetics of ants. He wrote up his findings and hurried over to the observatory's publication office, where he handed the article to astronomer and editor Frederick H. Seares. The normally craggy Seares, who had once been Shapley's professor, glanced at the title, smiled, and pushed the paper back. Shapley had a reputation as a nonconformist and joker, and Seares assumed that this was just the most recent of his little eccentricities. Besides, Mount Wilson wasn't in the insect business. "On the Thermokinetics of Dolichoderine Ants" and related monographs would have to await a more appreciative audience in such distinguished journals as the *Proceedings of the National Academy of Sciences* and the *Bulletin of the Ecological Society of America*.[25]

The son of a hardscrabble hay farmer and country schoolteacher, Harlow Shapley was born in 1885 on a farm about five miles from the little town of Nashville, Missouri, on the edge of the Ozarks and little more than a day's ride on horseback from tiny Lamar, where the toddler Harry Truman had entered the world the previous year. Harlow and his brother Horace, fraternal twins, were four years old when Hubble was born some seventy miles to the east. The two attended a one-room school on the edge of the Shapley farm, presided over by their older sister Lillian. Because money was scarce there were few books in the Shapley home. The Sunday edition of the *St. Louis Globe-Democrat* was the family's chief contact with the outside world. Willis Shapley, whom his son described as a "paragon of virtue," was without religion. The youth's mother, Sarah Stowell, was ostensibly a "hard-shell Baptist," but to Harlow she seemed more interested in showing off her twin boys at Sunday school than in rearing them in the fundamentalist tradition of her ancestors. The family was Republican on both sides. From very early on, the twins were warned that all Democrats chewed tobacco, and "we were not tobacco-spittin' people."[26]

Though encouraged by his sister to attend high school and college, fifteen-year-old Harlow had completed only the equivalent of the fifth grade when he left home to enter business school in Pittsburg, Kansas, a universe away. He finished the course of instruction within months and signed on as a crime reporter with the *Daily Sun* in nearby Chanute, a rough-and-tumble town in the midst of an oil boom. The community had constructed a new public library courtesy of Andrew Carnegie, the first Shapley had seen. He began reading history, literature, and poetry, committing much of Tennyson to memory, discussing Dostoyevsky with anyone who would listen. After a year in Chanute he departed for Joplin, Missouri, to become a police reporter for the *Joplin Times*. Shapley called it "a miserable little daily," while Joplin, a lead and zinc town, was worse than Chanute. The time had come to fulfill his sister's higher ambitions.

Together with his younger brother John, who was destined for a

career as an eminent art historian, Shapley applied for admission to
Carthage High School, twenty miles from home. Both were denied
entrance due to a lack of preparation. The undaunted brothers walked
the three blocks to Carthage Collegiate Institute, "a Presbyterian out-
fit" quite willing to take what little they could afford. Combining special examinations with only two semesters of residence, Shapley
graduated in 1907, the twenty-one-year-old valedictorian of his class of
three. His senior essay, "The Romantic Values in Elizabethan Poetry,"
while admittedly pretentious, helped gain him admission to the University of Missouri, in Columbia. "From then on," he reflected, "I never
stopped; the vanity probably helped."[27]

Shapley could have rightly called himself the accidental astronomer;
he had undergone no childhood apotheosis while being mesmerized by
a planet, had experienced no hunger to know the stuff of stars. Encouraged by their father to observe a meteor shower, the twins lay down
on a blanket one beautiful August night and promptly went soundly to
sleep, never seeing the streaking Perseids.

Shapley's plan was to make his way as a newspaperman. On arriving
in Columbia, he was stunned to learn that the opening of the much
advertised school of journalism had been postponed a year. He opened
the university catalogue to the alphabetical listing of courses and was
further humiliated when he couldn't pronounce "archaeology." However, the entry on the next page gave him no trouble; "astronomy" was
a familiar word, but one fraught with exotic overtones.

He was set to work in the Laws Observatory by Professor Frederick
H. Seares, a prim, demanding pedagogue who paid the fledgling astronomer thirty-five cents an hour. Seares would occasionally encourage
his student by saying, "That's well done," the most praise anyone could
expect from him. A teaching assistantship followed two years later,
enabling the university's only astronomy major to complete his B.A.
with highest honors in 1910, three years after his arrival on campus.
Not long before Shapley took his degree, he was offered a one-third
interest in Chanute's *Daily Sun* by his former editor, Fred Cone. When
he refused, Cone bristled: "All right. You're going to grow up here and
sit around in fat chairs and eat bonbons."[28] The remark cut deeply.
When the time came to recount his life in print the erstwhile crime
reporter titled his autobiography *Through Rugged Ways to the Stars*.

Shapley remained in Columbia another year, and in 1911 he received
his M.A. In the autumn he boarded a train bound for New Jersey, where
he had been awarded the Thaw Fellowship at Princeton University
Observatory. The Astronomy Department was headed by Henry Norris

Russell, one of the most prominent experts in stellar evolution, part of the emerging field of astrophysics. Amazed by Shapley's knowledge and capacity for work, the aristocratic Presbyterian entered into what a fellow astronomer described as "a beautiful collaborative effort" with the Missouri farm boy.[29]

The objects of their research were double stars or binaries, which revolve around each other, bound by gravity across the void. Of the various binary systems, which encompass roughly half of all the visible stars, Russell and Shapley zeroed in on the eclipsing binaries, so called because Earth lies very nearly in their orbital plane, causing one star to pass in front of the other. By night, Shapley obtained nearly 10,000 measurements with Princeton's 23-inch refractor, which was mounted with an instrument called a polarizing photometer. By day, employing new computing methods pioneered by the mathematically gifted Russell, he sat at a desk armed with two slide rules and a sheaf of complex tables, calculating light curves and rough stellar masses. Scarcely ten binary orbits had been plotted before he entered the hunt; spurred on by Russell, the young astronomer added another ninety. The results of two years' feverish activity proved little short of revolutionary. A hundred stars were found to have densities only one-millionth that of the sun—enormous gas bags, Shapley called them. But what most interested him was the distance factor. Though Kapteyn had placed the sun near the pupil of the Milky Way's cyclopean eye, Shapley was undergoing his first intimations of doubt. Compared with earlier calculations, his orbital work suggested that the distances to the binaries "were pretty darned big." Perhaps the size of the universe itself had been grossly underestimated. Yet when he gave exuberant expression to these private musings during a physics colloquium without first checking with Russell, he was gently chided by his mentor for having gone too far out on a limb.[30]

While never a tobacco spitter, Shapley, whose round face, chubby cheeks, and twinkling eyes gave him the appearance of perpetual youth, had taken to smoking a pipe and forsook his Republican roots by becoming a liberal Democrat. His numerous triumphs over adversity resulted in the development of an inflated ego, yet he took a childish delight in bringing down idols, whom he called "high hats." Always witty, always sparring, he rarely failed to have the last word, whatever the subject. He considered his Missouri upbringing an asset, and took no pains to pattern his speech after that of the highly educated. Only in technical journals were the shooting stars of his youth termed meteors. He treated everyone the same and was naturally interested in what his colleagues were up to. Yet, like the moon, he had his permanently dark side, and those who crossed him were in danger of provoking his dictatorial streak. Beneath it all there lurked the insecurity of the plebeian. "I always worry," Shapley admitted.[31]

At almost twenty-nine, Shapley was virtually the same age when he came to the mountain in 1914 as was Hubble five years later. He had graduated from Princeton with highest honors, had authored a dozen scientific papers, and was about to publish an expanded version of his dissertation on eclipsing binaries that would remain the standard in its field for decades to come. By 1920 the papers to his credit totaled nearly a hundred, the most important of them documenting his astounding conclusion that the Milky Way—still the only galaxy—is some ten times larger than the universe of Jacobus Cornelis Kapteyn. Though Shapley's cosmology had its detractors, one of whom was about to engage its author in the most highly touted astronomical debate of the twentieth century, there was no denying the fact that the vain, impulsive Missourian was looking down on the scientific world from the catbird seat.

Like his colleague and occasionally nonsensical friend van Maanen, Shapley cared nothing for Hubble, whose aristocratic air seemed a cover for the barely tested astronomer's suspect credentials. "He was a Rhodes scholar," Shapley reminisced, "and he didn't live it down," neither seeking nor accepting the hand of friendship on the mountain. Especially rankling were the newcomer's semimilitary garb and disdain for the "Missourian tongue." Hubble spoke "Oxford," employing phrases such as "Bah Jove!" and "to come a cropper." Yet Shapley, who took pains to avoid the patrician, was forced to admit that the ladies Hubble associated with enjoyed "that Oxford touch very much."[32]

V

To one such admirer Hubble seemed as cool as the other side of the pillow and capable of walking between the raindrops. In June 1920, Lick Observatory astronomer William H. Wright received an invitation from Hale to visit Mount Wilson for the purpose of employing his ultraviolet spectrograph on the 100-inch. An inveterate camper and backpacker, Wright often retreated to the High Sierras with a group of young friends who looked to him as their chief and nicknamed him "the Captain." On his way to Mount Wilson, he and his wife, Elna, stopped at the family home of Grace Leib, the daughter of a prominent

Los Angeles banker, to ask that their frequent hiking companion accompany them to the observatory. The petite, athletic brunette did not want to go, but she finally relented to please Elna, who feared staying by herself in the Kapteyn cottage. It was Grace's first visit to the mountain, and the accomplished horsewoman exchanged her normally stylish attire for an old pair of patched riding breeches and a shirt in anticipation of exploring the many trails. As the trio drove slowly up the steep toll road, she attempted to identify the flowers blooming on the slopes, only half hearing the Captain's enthusiastic description of a new astronomer who had visited with him on Mount Hamilton and who showed great promise. "He is a hard worker, he wants to find out about the universe, that shows how young he is."[33] His name was Edwin Hubble, but Wright, along with Shane and most others at Lick, called him "the Major."

That afternoon the women decided to walk the short distance to the laboratory, where they hoped to borrow some books. As they entered the room, Grace's attention was drawn to the commanding presence at the window, studying a photographic plate of Orion, the constellation named for the giant hunter and lover of Eos, goddess of the dawn.

> This should not have seemed unusual, an astronomer examining a plate against the light. But if the astronomer looked an Olympian, tall, strong, and beautiful, with the shoulders of the Hermes of Praxiteles, and the benign serenity, it became unusual. There was a sense of power, channeled and directed in an adventure that had nothing to do with personal ambition and its anxieties and lack of peace. There was hard concentrated effort and yet detachment. The power was controlled.[34]

Hubble turned around and Elna introduced him to Grace. In the afternoons that followed, the Captain and the Major came over to the cottage together and sat on the porch, the San Gabriel Valley serving as their backdrop. They talked nontechnical astronomy, telling tales of star lore and great discoveries for the benefit of the ladies. Hubble spoke of his research as a "dream" and an "adventure," which Grace would later learn were two of his favorite words. After the run was over, he asked if he might drive down with the others, and they left him at his observatory office. When she later put her memory of this charged encounter on paper, Grace omitted the fact that she was a married woman at the time and that Elna Wright was her husband's older sister.

The Captain's visit likely stirred bittersweet echoes of childhood, and of a more recent sense of loss. Martin Jones Hubble, a true captain, had passed away in February from what was diagnosed as paralysis of the heart.[35] An invalid in his final years, the grizzled Civil War veteran, whom his grandson had not seen since their Christmas reunion following his return from Oxford, proudly wore his medals until the last. At nearly eighty-five, Martin had outlived all but one or two of the Springfield pioneers whose reminiscences of bygone days over hog's jowls and turnip greens he had recorded and published through the years. Edwin had not returned for the funeral of his boyhood idol, nor would he travel east to see the other members of his family again.[36]

Dogged by Frost for not having fully settled accounts with the University of Chicago, Hubble at long last returned the corrected dissertation proofs he had neglected throughout the war. "I didn't realize," he wrote sheepishly, "the wretched state in which the manuscript was handed in."[37] Soon afterward "Photographic Investigations of Faint Nebulae" appeared in the *Publications of the Yerkes Observatory*, one of five papers he published in 1920, his first full year on the mountain.

On April 8, together with his fellow astronomers, Hubble participated in a conference on the 100-inch program chaired by Hale. According to the minutes, Adams, Shapley, van Maanen, and Seares dominated the discussions, while the newest member of the Mount Wilson fraternity looked on in silence. It was difficult to argue against Shapley's pronouncement that the distant nebulae presented the greatest challenge of all the unexplored fields, demanding the highest mechanical power and an indefinite amount of observational time. The attack could be mounted from so many different points that none of the several observers need infringe on the territory of his colleagues. The conference ended with the participants formalizing the division of labor. Led by Adams, the spectroscopists or "light men" would have charge of the 100-inch when the moon was at its brightest. As the satellite waned, the "dark men," headed by Seares and Shapley, would lay claim to the mountain. Hubble, together with Seares and Francis Pease—the mechanical genius responsible for the design of the 100-inch—was allocated the darkest time of all.[38]

By his third or fourth run on the mountain, he was moving about the 100-inch dome as easily as if it was daylight, avoiding obstacles, climbing ladders and stairs, leading the occasional groping visitor to the place he wished to go. His familiarity with the night sky amazed Humason, as it would others. The astronomer seemed to carry the whole celestial map in his head. He recognized the approach of the field he was planning to photograph just as it was coming into focus, and would often shout the command to clamp the telescope before the night assistant expected it. Equally impressive was his memory, which may have owed something to his legal training at Oxford. He could immediately put his hand on any plate he wanted, though his cabinet and desk always seemed the quintessence of confusion. He also possessed the uncanny ability to discern a new feature, such as a nova, before a plate was put on the blink machine for comparison with others. His colleagues were surprised when he glanced at their plates and casually pointed out crucial details they had overlooked.[39]

Judging by his early publications, Hubble himself did not yet know exactly what he was after. Shapley had virtually claimed the Milky Way as his own, while a group of astronomers centered at Lick Observatory favored the theory that the universe is composed of not one but countless galaxies scattered through what the poet Milton had called "this wild abyss." Still, these were but theories, neither of which had received a cautious Hubble's imprimatur. Whichever one was correct would depend on a combination of luck and hard data gathered during many long, unglamorous nights on the chill mountain.

Ever since Yerkes, Hubble believed the nebulae to be the key. He thought of them as vast, scattered beacons holding major clues to distances and the distribution of celestial matter. His plan called for photographing the more conspicuous ones first in an attempt to understand their structure and content, as well as their common features. With this information in hand, one could begin forging outward into deeper waters for the purpose of deriving a distance scale based on luminosities. Finally, the question of nebular distribution in space could be intelligently addressed in the hope of defining the structure and limits—if limits there are—of the cosmos itself.

Settling in for a voyage of undetermined time and distance, Hubble, like the first seagoing mariners, began his quest with a galactic reconnaissance. His personal logs reveal that he was making extensive use of the Cooke telescope, whose wide-angle lens was ideal for photographing large swatches of the sky. These plates enabled him to fashion a rude patchwork quilt of the firmament, which he then scoured for traces of individual nebulae. His first year on the mountain found him at the 100-inch a total of forty-one nights, during which he garnered 115 plates. Depending on the nebula being photographed, exposure times ranged from 2 minutes to 9 hours, with most falling between 45 and 120 minutes. Ironically, the seeing was rarely as good as that experienced on his first run.[40] His other nights on the mountain were spent at the 60-inch, his days taken up with developing and assessing his many plates. Added to this were the burdens of the historian. The plates of others figured almost as prominently in his research as his own. As he had demonstrated in his dissertation, even nebulae undergo changes, both marked and subtle, over the years, transformations critical to an understanding of their structure and ultimate classification.

He had no sooner begun making headway than the attention of the astronomical world was distracted by two important developments. Harlow Shapley was about to engage Lick astronomer Heber D. Curtis in a debate on the "island universe" theory before the National Academy of Sciences. Adding to the drama was the widely circulating rumor that the feisty Missourian had decided to leave Mount Wilson for seemingly greener pastures.

CHAPTER SEVEN

THE COSMIC

ARCHIPELAGO

I

The Romans called the luminous celestial band the *via lactea*, or Milky Way. But it was the Greeks who first compared it to milk—what they called *gala*—from which the modern word "galaxy" derives. The Greeks, too, were latecomers in their attempts to fathom the incandescent river that flows only at night, banked with dark shoals, eddied in shimmering pools. The minds of more ancient civilizations had long since named its planets and its constellations, built ziggurats and perfected mathematics to better track their circlings across the millennia, invented religions to curry the favor of resident gods and goddesses whose barges plied the river of light. To the less imaginative Anglo-Saxons it was simply a road; to the Norse with their Christless chivalry, the way to Valhalla; to the medieval knight, a luminous escutcheon unfurled against the heart of darkness.

Then came the arrogant and erudite Galileo, a Renaissance man with a deep sense of history and a cosmic flair for the dramatic. Locked in a race for immortality, the savant turned his homemade telescope on Orion and the Pleiades, two of the oldest inventions of the human mind. What he observed was beyond even his capacity for exaggeration:

> With the aid of the telescope [the material of the Milky Way] has been scrutinized so directly and with such ocular certainty that all the disputes which have vexed philosophers through so many ages have been resolved, and we are at last freed from wordy debates about it. The galaxy is, in fact, nothing but a congeries of innumerable stars grouped together in clusters. Upon whatever part of it the telescope is directed, a vast crowd of stars is immediately presented to view. Many of them are rather large and quite bright, while the number of smaller ones is quite beyond calculation.[1]

Hale had his Noyes, but Galileo possessed a more profound poet advocate in the blind Milton, who met Europe's most vaunted astronomer and returned home to describe the Milky Way in *Paradise Lost* as "The Galaxy . . . thou seest Powder'd with stars."

Right as Galileo was about the Milky Way's composition, he was too optimistic by far in his assertion that "all the disputes . . . have been resolved." Vexed natural philosophers would soon begin asking themselves the apparently insoluble question: "Does the Milky Way constitute the universe entire?"

Among the first to speculate in this regard was Sir Christopher Wren, the gifted polymath of architectural fame. In his inaugural address as professor of astronomy at Gresham College in 1657, Wren imagined that future astronomers might find "every nebulous star appearing as if it were the firmament of some other world, at an incomprehensible distance, buried in the vast abyss of intermundious vacuum."[2] Though this passage is inconclusive, it suggests that its author was thinking of the "nebulae" as extragalactic stellar systems.

A century later, in a series of bold intuitive strokes, the German metaphysician Immanuel Kant anticipated astronomical discoveries that would be confirmed only by the most powerful observational techniques. "The universe," Kant wrote in his genial early work, *Cosmogony*, "by its immeasurable greatness and the infinite variety and beauty that shine from it on all sides, fills us with silent wonder."[3] The harness maker's son imagined a system of stars gathered together in a common plane or thin disk, like those of the Milky Way, but so far removed from Earth that its individual members are indistinguishable with a telescope. Instead, they would show up as a little spot feebly illuminated—circular, if its plane is perpendicular to the line of sight; elliptical, if it is viewed obliquely. Such are the nebulae, countless systems of countless suns so distant in space that they cluster together, so limited in light that they are the most feeble candles in the heavens, yet sister galaxies to our own. In assuming that any large sample of the universe is much like any other, Kant had embraced the principle of the uniformity of nature and is credited with formulating the theory of island universes, a term whose origins derive from the explorer Alexander von Humboldt's multivolume *Kosmos*, published in 1845–62.

With few exceptions, astronomers themselves wanted nothing to do with such speculation, little noting Kant's injunction that the field is wide open for discovery, for observation itself "must give the key." Their instruments were limited and could be better applied to solving problems associated with nearby members of the solar system.

Among the exceptions was the maverick Hannoverian William Herschel, an oboist who joined his father's regimental band in 1753. Because of his youth, Herschel had not formally enlisted in the army and, together with his brother Jacob, later fled to England from Germany, ultimately settling in Bath. Though he had not known an astronomer or constructed a telescope until he was thirty-five, the magnitude of the expatriate's accomplishments won him the title of Prince of Astronomy and rank him as perhaps the greatest celestial observer of all time.

Herschel designed and constructed the largest and finest reflecting telescopes the world had yet seen, and then proceeded to tackle astronomy's most difficult questions. Sweeping the whole of the heavens several times during the thirty most active years of an obsessed existence, he discovered more than 800 double stars and added 2,400 nebulae to the hundred or so previously known. To those who had asked, "Where are we in the universe?" Herschel answered by unfurling his great star map of the heavens. Confirming Kant's speculations, he argued that the Milky Way galaxy is indeed a stellar disk or "grindstone" which bulges somewhat at the center, like a lens. The sun and its attendant planets are located in the middle of the grindstone, a seeming restoration of the preeminence lost when Copernicus plucked Earth from the heart of creation. Neither a theologian nor an egocentrist, Herschel based his honest but flawed conclusions on massive amounts of data garnered through the attempt to formulate accurate distances, or what he called "star gauging."

The nebulae proved an even greater conundrum. With his gifted sister Caroline seated next to him, the astronomer dictated a description of each chalk-colored web and pinwheel that his stationary telescope isolated in the sky, determining its position by noting the elevation of the instrument and the time at which the object passed the field of view. Star gauging had led him to accept Galileo's conclusion that faintness means distance. If, like the Milky Way, the nebulae were massive aggregates of stars, they must be outsiders, some of which "may well outvie our milky-way in grandeur."[4] Then, on November 13, 1790, Herschel observed a nebula consisting of a central star surrounded by a luminous shell which could not be resolved into stars. In a paper published the following year, he admitted the existence of "true nebulosity." His construction of the universe had collapsed, for an unresolved nebula might be small, near, and hazy or, just as likely, a vast star system in the distance. Such was the astronomy of dreams that it was possible the universe contains both.

Hubble was well acquainted with the details of this unfinished story, for he had long nurtured a passion for the history of science, which excited his imagination more than other forms of literature. Never able to pass up a bookshop or stall, he had indulged his bibliomania to the limit of his financial resources while a Rhodes Scholar. Copernicus, Kepler, Galileo, Newton, and Herschel were his icons, and he read what they themselves had written before turning to works about them. With the history of his discipline as prologue and his chosen field of nebular research in equipoise between finitude and the infinite, no one was more conscious of the stakes involved than Hubble when Shapley departed for Washington, D.C., in April 1920 to uphold the honor of the Milky Way.

II

On the evening of what has passed into the literature as "The Great Debate," a fidgeting Shapley found himself seated at one of the National Academy's banquet tables feeling "just horrible." His discomfort had nothing to do with his upcoming presentation or the man he was about to face. In fact, he was rather looking forward to it. Shortly after he had boarded the Southern Pacific for Washington, he discovered that his congenial adversary, the fastidious, bespectacled Heber Curtis, was on the same train. The two astronomers politely avoided talking shop and passed the hours discussing nature and the classics; when the train broke down in Alabama, they walked the tracks in search of ants for Shapley's collection.

The cause of Shapley's distress was the awards ceremony. Some "noble human antique" from one of the government bureaus was being honored for his fight to eradicate hookworm. The endless testimonials were making everyone uneasy, including Johns Hopkins professor Raymond Pearl, who was seated across from Shapley. Pearl jotted a note on one of the menu cards and discreetly passed it over to W.J.V. Osterhout, the Harvard botanist. Shapley glimpsed the message: "Jesus H. Christ!" Osterhout scribbled a reply and again Shapley peeked: "Jesus H. Christ, and the H. stands for hookworm!" Seated at the head table was a visitor from Europe named Albert Einstein, his face beaming. Feeling embarrassed about the presentation, Shapley wondered what Einstein had whispered in the ear of an official from the Dutch embassy, who was there to receive a prize on behalf of the physicist Pieter Zeeman. When later queried on the matter, the official replied, smiling, "I have just got a new theory of Eternity."[5]

As for the debate itself, Shapley later claimed to have forgotten "the whole thing," arguing, with some justification, that historians had turned it into a romance. The affair was largely the accidental brainchild of Hale, who a year earlier had established a fund in memory of his father for the purpose of underwriting annual lectures before the Academy. The Mount Wilson director suggested two possible topics to C. G. Abbot, secretary of the Academy—the island universe theory and relativity. Hale favored relativity, but Abbot replied that only a handful of the Academy members would understand even a few words of what the speakers were saying. "I pray to God that the progress of science will send relativity to some region of space beyond the fourth dimension, from whence it may never return to plague us."[6] The secretary feared the island universe theory for similar reasons and asked Hale whether there were other subjects, such as glaciers or animals, which might make for a more interesting evening. When Hale demurred, Abbot reluctantly wired both Mount Wilson and Lick, and the debate was on.

It was not long, however, before the normally combative Shapley began having second thoughts. He wrote his Princeton mentor Henry Norris Russell that not only were the men at Lick and Mount Wilson taking the upcoming debate very seriously but astronomers from around the country "seem to regard [it] as a crisis for the newer astrophysical theories."[7] Russell, a speaker of legendary note, would certainly want to know this in the event his own views should come under attack in Washington.

Shapley went on to reveal a second, more compelling cause for anxiety. Edward C. Pickering's long reign as director of Harvard College Observatory had ended with his death the previous year. That very day Shapley admitted to pausing at the corner of two streets on his way home for lunch and pondering whether to reach for the choicest academic plum of all: "Should I curb my ambition? Finally, I said to myself, 'All right. I'll take a shot at it.' "[8] Having signaled his intentions, he had recently learned that he would be met in the capital by George Russell Agassiz of the Harvard College Observatory Visiting Committee.

With the sure knowledge that Agassiz would be in the audience "hovering," the Missourian was taking no chances. Fearing that Curtis, a skilled orator who had devoted his early life to mastering Latin, Greek, and other ancient tongues, would go after him "hammer and tongs . . . with his shillelagh," he turned to Hale for help.[9] The latter, over the strenuous objections of Curtis, who was spoiling for "a good friendly 'scrap,' " demoted the debate to "a discussion" with no time for rebuttal, although questions and comments from the audience would be allowed. Shapley further convinced Hale that the forty-five minutes originally allotted each speaker was too long, so the time limit was reduced by five minutes each. Anything else the astronomers wished to say on the assigned topic, "The Scale of the Universe," would be published by the National Research Council at a later date.

The final element of Shapley's strategy concerned his method of presentation. He would treat the two hundred or so members of the audience, only a small number of whom would be astronomers, to an address so elementary that his remarks would stir a minimum of controversy. The definition of a light-year does not appear until page 7 of his nineteen-page typescript, the last three of which are devoted to the intensifier he had developed to facilitate the photography of very faint stars—irrelevant to the theoretical argument, but doubtless directed to a certain member of the audience entrusted with the future of Harvard College Observatory. Curtis, on the other hand, came armed with slides and a technical manuscript and was truly shocked by Shapley's pedestrian offering.[10]

The Mount Wilson astronomer was the first to take the floor. Though his paper constituted little more than a gloss on long years of intensive labor, Shapley's belief that the Milky Way galaxy was the only such entity in the universe was underpinned by sound scientific reasoning. Shortly before completing his Ph.D., he had visited Harvard College Observatory. While in the dome, he had entered into a discussion with the astronomer Solon I. Bailey. Bailey had heard of Shapley's recent appointment at Mount Wilson, which was largely the result of Frederick Seares's influence with Hale. Bailey encouraged Shapley to use the "big telescope"—then the 60-inch—to measure certain stars in globular clusters.[11]

So named because of their resemblance to large spheres, globular clusters are composed of "swarming" stars which sometimes number in the millions. Of primary interest to Bailey were the Cepheid variables, known to astronomers since the late 1700s. Scarce in number, Cepheids are distinguished from other variable stars in the way they change in luminosity. Like telltale clocks, they brighten rapidly and dim slowly in regular fashion, passing through one cycle, or period, in about one day to three or more months, depending on the star. With the aid of Harvard's telescope in Peru, Bailey had already detected a number of Cepheids in globular clusters, but the key to Shapley's future success lay in the largely overlooked work of Miss Henrietta Leavitt, a research assistant at Harvard College Observatory.

Leavitt had focused her attention on the discovery of variable stars. Of particular interest to her were those located in the Small Magellanic Cloud, a mass of stars, clusters, and nebulosities in the Southern Hemisphere. In 1908 she published a long list of variables which included a small number of Cepheids whose cycles ranged from 1.25 to 127 days. She was able to determine that the longer the period, the brighter the star on the photographic plate, a discovery of profound implications. Cepheids displaying similar periods must be virtual twins, no matter their apparent brightness or location in the sky. As "standard candles" they could become beacons for calculating distances across the void.

Like Gregor Mendel's article on the hybridization of peas, Leavitt's paper was ignored. Undaunted, she resumed her intellectual hybernation and subsequently determined periods for an additional nine variables. Four years later, she published a second paper, this time graphing her results for the benefit of hoped-for readers. As before, the plot showed that the brightest Cepheids in the Small Magellanic Cloud had the longest periods.

This time Leavitt's now famous period-luminosity relation did not go unnoticed. Ejnar Hertzsprung, in Potsdam, recognized the implications of what the unsung Leavitt had done. Working with a different sample of thirteen variable stars, he possessed just enough data to calculate an average of the absolute magnitude, or true brightness, of Cepheids in various reaches of the Milky Way, though the brightness of individual stars was as yet beyond his grasp. He then compared the absolute magnitudes of his variables with the apparent magnitudes (the brightness of a star viewed from Earth) of Leavitt's variables of the same period in the Small Magellanic Cloud. The first to employ Cepheids as indicators of distance, Hertzsprung determined the Cloud to be 30,000 light-years from Earth, or what he thought were the fringes of the Milky Way. This was the greatest astronomical distance at the time, but Hertzsprung's colleagues were unaware of it. His article in the *Astronomische Nachrichten* contained a typographical error. The 30,000 light-years were inadvertently shrunk to a mere 3,000, resulting in a distance scale too small by a factor of ten.[12]

As often happens in science, another creative mind was drawn to the same problem almost simultaneously. At Princeton, Russell, working independently of Hertzsprung, employed the same baker's dozen Cepheids and a similar set of premises to arrive at absolute magnitudes that differed little from those formulated by his Danish colleague. In a classic example of understatement, he later wrote Hertzsprung that the analogy between their work and Miss Leavitt's can be pushed "a little farther." He calculated Leavitt's Cepheids to be of the 15th photographic magnitude while the Milky Way variables appeared to be of the 5th photographic magnitude. If the two sets of stars are truly similar and little of their light is absorbed during its ancient journey through space, the distance of the Small Magellanic Cloud is a hundred times that of the Cepheid variables, or about 80,000 light-years. Russell conceded: "This is an enormous distance, but it is not intrinsically incredible."[13]

These developments had taken place during the final year of Shapley's doctoral studies. Russell's astonishing result must have given both men a thrill, coming as it did on the heels of Shapley's confirmation that the eclipsing binaries are located much deeper in space than previously imagined. The chance conversation with Bailey at Harvard had taken place a few months later. Shortly thereafter, Shapley, together with his new bride and collaborator, Martha Betz, moved into a rented house on Pasadena's Villa Street, only a block and a half from the observatory office and shops. The onetime farm boy was soon climbing a mountain, heaving against the backside of a recalcitrant burro. But as Hale had promised, he was given his freedom with the 60-inch telescope, which he promptly focused on globular clusters.

Shapley began by formulating the hypothesis that Cepheids are pulsating stars whose surfaces ebb and flow in great cauldronlike waves. He scoured the mottled globulars for the glimmer of variables yet unknown, a daunting, monotonous undertaking destined to consume years. As the Cepheids appeared one by one, he noticed that those in a given cluster always seemed to show the same period-luminosity relationship that Leavitt had found. And while apparent magnitudes shifted from cluster to cluster, he felt certain this effect was the result of their varying distances. Still, he could never hope to exploit this breakthrough until he devised a trustworthy method of measuring the distances to the swarming stars.

Unfortunately, no Cepheid is located close enough to the sun to be measured by trigonometric methods. Thus, Shapley was forced to rely on the Milky Way Cepheids as a basis for establishing an absolute magnitude scale. He began with the same thirteen variables employed by Hertzsprung and Russell, reducing the already statistically anemic sample to a feeble eleven when two of the light curves appeared suspect. From this smattering of stars he calibrated a new standard candle for the measurement of stellar distances.

Skating on thin ice was Shapley's forte, yet he was well aware of the precariousness of his position. His dogged pursuit of Cepheids continued, until by 1918 he had plotted the period-luminosity relation of 230 stars with cycles ranging from 5 hours to 100 days. All looked well for his argument: Leavitt's variable stars in the Small Magellanic Cloud, those analyzed by Hertzsprung and Russell in the Milky Way, and the variables Shapley himself had identified in the globular clusters appeared identical. These calibrations were supported by the equally meticulous tracking of scores of the less luminous but more common RR Lyrae variables, whose average period is less than a single day. The ice was holding. When the astronomer decided to put pen to paper in a virtual storm of publication after four grueling years on the mountain, he was in a position to do for the Milky Way galaxy what Copernicus had done for the solar system.

Shapley's galactic model was unexpected and audacious—and to many astronomers such as Curtis, inherently unreasonable. Shapley himself wrote Arthur Eddington at Cambridge in January 1918 that the globular clusters had elucidated the "whole sidereal picture" with "startling suddenness and definiteness."[14]

Prior to this, it was thought that the Milky Way was at most 20,000 to 30,000 light-years in diameter and that the sun lay near its center amid a flat lens-shaped disk of stars. In the course of his investigations Shapley found that over 40 percent of the hundred or so globular clusters then known to astronomers are concentrated in less than 3 percent of the sky's area. These in turn combine to form a huge spherical cloud in the general direction of the constellation Sagittarius, an estimated 60,000 light-years in the distance. To Shapley, this massive gathering of stars marked the center of the galaxy itself and could mean only one thing. The planet Earth is part of a solar system located on the outskirts of the Milky Way, whose diameter had swelled to 300,000 light-years, dwarfing the puny Kapteyn universe and knocking conventional wisdom into a cocked hat.

And what of the cosmic archipelago, the theory that the sea of stars is populated by island universes resembling the *via lactea*? A chimera, Shapley argued. The Magellanic Clouds, which some conceived of as satellites of the Milky Way, were suddenly engulfed by the galaxy. The faint spiral nebulae, true galaxies in the minds of others, were to him insubstantial gaseous bodies being driven off by radiation pressure from the Milky Way, which would also explain Slipher's exotic Doppler shifts. That other galaxies could exist was next to unthinkable. The "Big Galaxy," as Shapley dubbed it, and the universe were one and the same.

While eschewing "dreary technicalities" for the benefit of his audience, Shapley briefly set forth three additional arguments before yielding the floor to Curtis, all pertaining to the nettlesome problem of the spiral nebulae. Astronomers had long known that the as yet undefined bodies were dense around the poles of the Milky Way and very few in number in the galactic plane—the so-called zone of avoidance. This was even more true of the globular clusters Shapley had dealt with: 46 of them lay above the plane of the galaxy; the other 47 lay below, in almost perfect symmetry. Surely this must mean that the spirals, like the globulars, were somehow related to our galaxy or their distribution throughout space would be relatively uniform.

Shapley next cited one of the most noted events in recent astronomical history. In 1885 a new star, or nova, had flared near the center of the Andromeda nebula, M31, becoming as bright as the very body with which it was associated. That a single star could rival an entire galaxy composed of millions of stars seemed preposterous in the extreme, for no process then known to physics could achieve such a result. Reasoning by analogy, Shapley concluded that this nova, dubbed S Andromeda, had to be like others within the Milky Way system, several of whose distances were known. Whatever the spirals might be, they are far smaller than our galactic home.

Finally, there was the photographic analysis of Adriaan van Maanen, Shapley's respected colleague, for whom a star would one day be named. When van Maanen compared his series of photographs of spiral nebulae with similar plates taken years earlier, his measurements disclosed transverse motions of the bright knots at their centers. Subsequent observations of all the famous spirals, including M33, M51, M81, and M101, revealed similar rotational motions, or so van Maanen sincerely believed. Again and again, in a score of papers, he calculated the rate of motion at about one hundred-thousandth of a revolution per year. In human terms this figure seems small, even infinitesimal, but when set on a cosmic scale it deals a death blow to the island universe theory. The stellar rim of a spiral only 500,000 light-years from Earth would have to turn at an incredible 30,000 miles per second. Those much farther removed would not only approach but exceed the speed of light! In the words of Yeats, "Things fall apart; the center cannot hold."

It was Shapley the apostate who relinquished the floor. As late as October 1917, only three months before he described the sudden parting of the mists in his letter to Eddington, the astronomer was tilting in favor of the island universe theory championed by his opponent.[15] Now he was even more strongly committed to a radically altered vision of the cosmos, the distance to the center of which he had reached out and measured himself. Immortality seemed an attainable dream, an appointment at Harvard a just reward.

Though Hale had acceded to Shapley's request and modified the rules of engagement, the Mount Wilson director was appalled by his overweening ambition and rush to judgment, and told him so. Russell, who also wrote from Princeton, was equally censorious: "I am sorry to see you join the company of those who advance theories that are 'startling if true.' There has been a great deal too much of this done in the last few years."[16]

Curtis nursed strong misgivings about the reliability of distances obtained by means of Cepheid variables. He thus began his attack with the clever ploy of quoting Shapley against Shapley. In another of his many incarnations he had written: "The maximum radius of the Milky Way is probably not greater than ten thousand light-years and may be somewhat less."[17] Eddington and others had settled on an even smaller galaxy, while Curtis himself was thinking of a galactic diameter no greater than 30,000 light-years, if that.

Rather than debate the dimensions of the galaxy in detail, Curtis was more interested in establishing a direct relationship between the Milky Way and the spiral nebulae. He opened this front by challenging Shapley's view that the spirals are composed of gas rather than stars. The known spectra of the spirals were indistinguishable from those of great clusters of bluish-white and yellowish-white F- and G-type stars, and in no way resembled the spectra of gaseous clouds. As avid a studént of Galileo's writing as was Hubble, Curtis noted, "It is such a spectrum as would be expected from a vast congeries of stars."[18]

Curtis turned next to Shapley's argument concerning the distribution of the spirals. One did not have to resort to an exotic theory of repulsion to account for the fact that so few are observed in the galactic plane. The Lick astronomer's analysis of many spirals photographed edgewise had revealed dark bands of occultating matter. Whether these bands are composed of gas, interstellar dust, or some unknown material was impossible to say. However, he postulated that such rings are the rule rather than the exception. If so, the Milky Way, as one of many island universes, must be similarly corseted. The zone of avoidance is no such thing; the spiral nebulae are present in all directions, albeit hidden from the prying eyes of fogbound astronomers.

The nova of 1885 was subject to another interpretation as well. For at least two years preceding the debate, Curtis had been analyzing photographs of many novae in spiral nebulae, only one of which, Z Centauri, came close to matching the fireworks of S Andromeda. As luck would have it, George Ritchey of Mount Wilson discovered a nova that was still visible, causing astronomers to conduct a reexamination of their old photographic plates. Shapley was among them, and he breathed a sigh of relief when no further examples of a spectacular nature came to light. Still, Curtis refused to give ground. He theorized that another class of nova, more powerful than anything astronomers had previously imagined, had flared far beyond the galactic rim. As was realized a few years later, such phenomena—called supernovae—do in fact exist.

Neither, in the eyes of Curtis, could van Maanen's alleged rotations bear scrutiny. Many of the older plates employed by the Dutch astronomer were poor, indeed useless for determining proper motions. And van Maanen's own images of the spirals were necessarily hazy, requiring much longer intervals of time before one could speak of galactic motion with any assurance—twenty-five years if the proper motions are small, centuries for spirals 100,000 or more light-years from Earth. Curtis would yield on this point and abandon the island universe theory only if, during "the next quarter-century," close agreement among different observers detected annual rotations of spirals equaling or exceeding 0".01 in average value.[19] A youthful forty-seven, the former classics student was willing to bide his time.

Confident of his convictions, Curtis left Washington believing he had carried the day. "I have been assured," he wrote his family on May 15, "that I came out considerably in front."[20] No contemporary account of Shapley's feelings exists, but in old age he reflected that he went away thinking he had triumphed from the standpoint of the assigned subject—the scale of the universe.[21] In truth, the duel between the champions of the two great California observatories had settled nothing. Curtis and his proponents continued to believe in a small Milky Way galaxy, a spiral adrift among a great ocean sea of similar island universes. To Shapley and his followers the Milky Way remained the ocean sea, its farthest reaches marked by faint pinwheel clouds of little substance. Captives of their compelling visions, neither protagonist paused to contemplate the possibility that the truth might lie somewhere in between.

III

Harlow Shapley's decision to seek the post at Harvard was fraught with risk. Only new data could resolve his fundamental differences with Curtis. In this the 100-inch telescope remained the key, for nowhere in the world was there another instrument to rival its tremendous light-gathering powers. Even if he was wrong about the "Big Galaxy," the Missourian still had an excellent chance of making the breakthrough discovery that would place him in the Valhalla of scientific accomplishment. To leave the instrument behind could well open the way for some other Columbus to claim the mantle of "admiral of the ocean sea."

A master at hiding his feelings, Hubble was privately anxious for Shapley to go. He was not accustomed to standing in anyone else's shadow, especially one whose conduct he considered obnoxious and whose ambition matched his own. Finding it difficult to relinquish the trappings of his military service, he little appreciated his colleague's hatred of war and reluctance to fight it out with the Germans in 1917. Shapley's hit-and-run approach to great scientific problems also clashed with his inherent conservatism. With a little luck, another Missourian would soon become king of the mountain.

Shapley had come away from Washington believing all was well on the Harvard front. The interview with Agassiz and a second member of the search committee, Theodore Lyman, had gone smoothly. In a separate interview, Mrs. Shapley had gained the approval of Wellesley College professor J. C. Duncan, whom Shapley dismissed as "another big shot who wasn't so big."[22] Apparently both husband and wife were operating under a false impression, little realizing that Shapley was being considered for a lesser post than the one he was after.

In the eyes of Agassiz and his fellow committee members, Russell was first. The Princeton professor had spoken eloquently from the floor at the debate and had been no less impressive during private talks afterward. In contrast, Shapley, for all his advanced preparation, came off as immature and lacking in force; he confirmed background information gathered from others. Agassiz's assessment was both blunt and to the point: "[Shapley] does not give the impression of being a big enough personality for the position."[23]

Still, Harvard wanted the brilliant, if erratic, astronomer, but as its number two man. The directorship was offered to Russell, who wrote Hale that Shapley would make a "bully second, and would be sure to grow—I mean in knowledge of the world and of affairs; if he grew intellectually he would be a prodigy!" Together, mentor and student could cover the field of sidereal astrophysics, and do some theory as well. "I might keep Shapley from too riotous an imagination—in print."[24]

Hale must have cracked a smile, for he knew whereof Russell spoke. Shapley's most recent brainstorm concerned thalophide cells, whose marvelous potential he described in an exuberant letter to the director:

> Do you know that one day we shall use these . . . to listen to celestial music? Hearing a sunspot, or the revolution of the moon (reproducing them with a phonograph to a gaping audience) will be very easy—more easy than worthy; and I think we can hear a star transit; and possibly the eclipse of Algol. I have been computing on candle powers of stars for this and worthier purposes.

This nonsense was followed by an intemperate claim. "In a moment of bravery I tackled radiation pressure yesterday and astonished myself by finding it easy, and also exciting as to the motion of spiral nebulae. There will be no difficulty at all in driving off spirals."[25]

In the end, Russell turned Harvard down, forcing the selection committee to consider the risks of appointing Shapley director, with all his limitations. Unwilling to grasp the nettle as yet, President Abbott Lawrence Lowell offered Shapley an appointment as the leading "technical man," with the title of Assistant Professor and Astronomer, but not the directorship. When Shapley promptly declined the offer, Hale felt it necessary to intervene. Despite his reservations concerning Shapley's maturity, Hale wrote Lowell to propose that Shapley spend a probationary year at Harvard. If all went well, the directorship would be his; if not, he could return to Mount Wilson and resume his former duties. The compromise quickly won Lowell's support and Shapley's assent. On the evening of March 15, 1921, the night assistant entered the following in the logbook of the 60-inch telescope: "Dr. Harlow Shapley leaves the observatory today to become identified with Harvard College observatory."[26]

There had been a curious development before Shapley's departure for the East. According to Milton Humason, the astronomer had given him plates of M31, the great Andromeda nebula, for examination on the stereocomparator. During the process of blinking the plates, the night assistant discerned images never before seen. He marked their locations in ink and sought out Shapley for confirmation. If he was not mistaken, the plates contained Cepheid variables from beyond the Milky Way. Shapley, who was certain of himself, was having none of this. He launched into a shortened version of the same arguments he employed during the Great Debate, then calmly took out his handkerchief, turned the plates over, and wiped them clean of Humason's marks.[27]

CHAPTER EIGHT

UNCHARTED WATERS

I

In May 1921, William A. Stuart, who had recently succeeded Elmer Davis as secretary of the Brotherhood of Rhodes Scholars of 1910, wrote his former classmates asking them to update their individual "achievements, difficulties, and failures" from the armistice to the present. Jotting his reply at the bottom of Stuart's query, Hubble reverted to the affected tone of his Oxford days.

> My one distinction is that of being the only Astronomer amongst the Brotherhood. Whilst toying with such minor matters as the structure of the universe, however, I sometimes tackle the serious problems of life, liberty, etc., and strive ernestly to circumvent these damnable prohibition laws. The little gods have blessed me with mild adventures but so far I have escaped the alter and the courts. I haven't seen any of the Brothers since those regretful days of uniforms and endorsements. If any of you pass thru Pasadena, please come around.

While the class secretary was willing to overlook Hubble's pretension, the temptation to snipe at him from another angle was irresistible: "You may infer from his contribution that he did not study spelling at Oxford."[1]

His shortcomings with the pen aside, Hubble was more than toying with the structure of the universe. In his recently published dissertation, the astronomer lamented the fact that no reliable system of classifying the nebulae had been formulated to date. Writing in the spring of 1920 to astronomer Edward E. Barnard, a frequent visitor at Mount Wilson, he noted, "My program, in short, is an attempt to find all I possibly can concerning the nature, form, and location of the galactic nebulae, and their relation to the stars involved."[2] One year later, Hubble sent William H. Wright a long letter in which he included a tentative classification of the various galactic nebulae and the types of stars with which they are associated. "There are a few anomalies," he confessed, "but on the whole, the progression is surprisingly definite." He asked "the Captain," an authority on the subject, to point out significant criticisms of his scheme, while making it clear that he had no desire to trespass on previously claimed territory. Nor was he proposing a theory of galactic evolution; "it is merely a matter of systematising data until all theories can be put forward with some degree of confidence."[3]

Wright's reply could not have been more encouraging. "Your progression from the planetaries to the diffuse nebulae is extremely interesting, and I hope you will make something definite out of it." With respect to the question of intruding: "[P]lease forget it. I never owned the nebulae and am not observing them now."[4]

Hubble spent another year photographing with the 60- and 100-inch reflectors before he felt confident enough to submit his findings to a wider audience. In 1922 he gained a seat on the International Astronomical Union's fourteen-member Commission on Nebulae and Star Clusters, presided over by France's aging dinosaur G. Bigourdan, who was still dwelling in the age of visual observations, with no knowledge of recent photographic work. In addition to Hubble, the commission's American members were Heber Curtis, Solon I. Bailey, William H. Wright, and president-elect Vesto Melvin Slipher, then acting director of Lowell Observatory. Among the other committee notables were the venerable Irish astronomer, historian of science, and cataloguer of nebulae J. L. E. Dreyer, H. Knox-Shaw, who would soon become director of England's Radcliffe Observatory, and G. Horn D'Arturo, director of the University of Bologna's observatory.

It was to Slipher that Hubble had written prior to the July 1922 meeting in Rome, setting forth his elegant classification scheme. But Curtis, Bailey, Bigourdan, and J. H. Reynolds, another commission member, who maintained a private observatory at Birmingham, had also submitted models for classifying nebulae, and Bigourdan, hardly an impartial observer, seemed incapable of seeing the forest for the trees. Nothing was accomplished during the session, and Hubble's scheme, which had impressed Slipher, was relegated to an unpublished report.

Faced with the fact that others were attempting to chart the same territory, Hubble decided to play his hand and hope for the best. He redrafted his paper, "A General Study of Diffuse Galactic Nebulae," and submitted it to the *Astrophysical Journal*, which wasted no time bringing it to press.

Hubble first traced the history of the classification of nebulae from visual observation, beginning with Sir William Herschel and ending with the system set forth by Heber Curtis. Each was found wanting in one way or another. In their place, he boldly divided all nebulae into two broad categories, galactic and nongalactic. The galactic nebulae were in turn separated into two groups—the planetaries, so named by Herschel because of their resemblance to Venus, Mars, and Jupiter, and the diffuse nebulae, both luminous and dark. All galactic nebulae, he asserted, are notable for their association with stars. Even the dark nebulosities are detected by their obscuration of starlight. Conversely, the nongalactic nebulae—spirals, spindles, ovates, globulars, and ir-regulars—have no stars definitely associated with them save for the rare phenomena of novae, such as S Andromeda and Z Centauri, which have flared in spirals.

In a rare instance of speculation, Hubble had written in his disser-tation that the giant spirals, "with their enormous radial velocities and insensible proper motions," apparently lay beyond our galactic system. He had grown even more cautious of late, however, to avoid being associated with either camp in the great debate. Readers were warned that the term "nongalactic" should not be taken to mean that these nebulae are "outside" the Milky Way, although they tend to concen-trate in the high galactic latitudes. As to their true nature, only time would tell, perhaps soon.[5]

The new classification scheme was bolstered by a second, equally seminal article published later the same year. In it Hubble homed in on the source of luminosity in galactic nebulae, providing a detailed physical explanation of why certain nebulae glow like miners' lanterns while others seem little more than dim silhouettes against the twilight firmament. In many of the planetaries, which are round and well defined, a faint blue star is located at the center, providing the energy with which the nebula shines. The amount of matter in the nebula, composed of expanding gas supplied by the central star, is thus a minute fraction of the mass of an ordinary star. What is more, the expanding gas is escaping into space, almost certainly the result of a mild eruption on the stellar surface.

More insightful still was his research on luminous diffuse nebulae, which are identified by wisps, filaments, or formless clouds. Hubble had theorized that they must be composed of material highly sensitive to stellar radiation. Light emanating from neighboring stars is "re-emitted" in the exact amounts it is received, accounting for the continuous spectra. In other diffuse nebulae light is supplied by a combination of internal and external stars. After obtaining data on eighty-two such objects with the use of a spectrograph, he was able to confirm his hypothesis. Not only does nebulosity fade with increasing distance from the stars; it obeys the inverse-square law, the very concept Newton had used to compass gravity.

Hubble concluded his analysis of galactics with an explanation of the dark nebulae, to which no mathematical principle applied. Appearing as "holes" against a background of thick but faint stars, these amorphous occulting clouds contain no internal source of light and are too far removed from stellar radiation to glow. Dreadnoughts of the dark, they were of only passing interest to the celestial mariner.[6]

11

While Hubble continued to refine his classification scheme, Shapley both flourished and fretted at Harvard. The astronomer had so impressed the powers that be that on October 31, 1921, only six months into his trial year, he was able to write Hale a single-sentence letter: "You will be interested to hear that I was today appointed Director of the Harvard College Observatory."[7]

Although Curtis remained the champion of the proponents of the island universe theory, his own research on the spirals had ended when he moved from Lick to Allegheny Observatory in 1920. His place on the ramparts was taken by Knut Emil Lundmark, a Swedish astronomer of generous girth and saturnine disposition. Regarded as brilliant by some, Lundmark had come to the United States in 1921 and spent two years conducting nebular research at Lick and Mount Wilson. His spectrograms of the spiral galaxy M33 convinced him that the starlike points in the nebula were indeed stars, and he published this conclusion within months of his arrival.[8]

Shapley saw red. He immediately fired off a heated letter to the author, asking Lundmark why he had ignored van Maanen's rotation measurements of the spiral in question. He also wrote van Maanen describing what he had done: "I thought that perhaps he knew of some reason why your results should be suspected of systematic error. Apparently, he has none, but is still more or less lured by his own hypothesis."[9] Startled and seemingly cowed by Shapley's biting criticism, the younger Lundmark, who hoped to conduct research at Harvard, wrote two conciliatory replies, apparently settling the matter. Thus Shapley was stunned on receiving the April 1922 issue of the *Publications of the Astronomical Society of the Pacific*, whose pages he associated with the enemy camp. The journal contained a paper by Lundmark—"On the Motions of Spirals"—in which he openly challenged van Maanen's measurements.[10]

His Big Galaxy under assault, an enraged Shapley, who believed Lundmark had stabbed him in the back, again vented his feelings to van Maanen:

> Isn't he clever and wide spreading, and sweeping in his statements? I could talk to you (through this dictating machine) for an hour or so concerning the frailties of his discussion. But I will not in this letter, which I wish you would answer as soon as you can, and is merely to ask if you care to reply to him in any way? . . . I very much desire not to enter any controversy in these matters, and certainly dislike to run the risk of alienating the affection of a young man like Lundmark. He seems to be very touchy, judging from my correspondence with him last winter.

Shapley's anger masked the even deeper feelings of a haunted man. He had risked much by relying on his friend's measurements of proper motions and was now beginning to wonder if van Maanen was too slender a reed on which to balance a universe. In what was tantamount to a cri de coeur, he begged to know, "How seriously do people take Lundmark, do you think?"[11]

Finally, in April 1923, Shapley met Lundmark face to face at Harvard College Observatory. As he reported to van Maanen, he asked him pointedly what he thought of van Maanen's proper motions in spiral nebulae. Lundmark's reply stunned his host. "To my surprise he said he thought the best thing to do was to accept your measures at present. Apparently his work on Messier 33 convinced him that something was there. Hence, congratulations." Lundmark had also suggested that he did not believe very seriously in the island universe theory, "but one would never suspect that from his printed papers." The chance to take Lundmark's measure had made his adversary appear much less threatening than before. "From all that I had heard and read, I was surprised to find that he was not a more brilliant genius than Russell."[12]

Tensions eased, but it was not long before Lundmark recanted his confession after repeating his measurements of M31 and failing to detect any systematic motion of the spiral. Slipher, who believed more fervently than ever that the spiral nebulae are hurtling away from Earth at breakneck speeds, also refused to be swayed by van Maanen's findings. In a New York Times article guaranteed to give Shapley a headache, he proclaimed Dreyer nebula No. 584 the "Celestial Speed Champion," located millions of light-years away.[13]

III

On October 1, 1923, six months after the Shapley-Lundmark tête-à-tête, the Wisconsin astronomer Joel Stebbins wrote to Hale describing his recent visit to Mount Wilson. He had sat up long after midnight with Hubble and Francis Pease, talking shop. "Those of us who come back to our small places feel how modest our opportunities are, and yet after all it is not the material equipment but the intense activity and the spirit of research which impresses one at Mt. Wilson."[14] Though the wistful Stebbins did not go into the details of these conversations, they doubtless concerned Hubble's recent attempts to photograph novae in spiral nebulae. If the flaring stars could be located in sufficient numbers, their mean apparent brightness could be calculated, thus enabling astronomers to estimate the distances to spirals à la Cepheids.

While Stebbins was writing Hale, Hubble was on a run with the 100-inch reflector, his ninth of the year. On the night of October 4, with the seeing rated at less than 1, the poorest possible without closing the dome, he locked the instrument on a spiral arm of M31 and took a 40-minute exposure. Despite the adverse conditions, the plate of Andromeda yielded a "suspected" nova. His curiosity aroused, he repeated the procedure the next night, increasing the exposure time by five minutes.

The seeing was much improved, and plate H335H, destined to become the most famous ever taken, confirmed the suspected nova. Further examination of the plate revealed the existence of two more stars, both of which Hubble concluded were novae as well.[15] It was the last night of his run, and he went down the mountain, still conspicuously clad in knickers, weary but deeply satisfied at having given birth to triplets.

Once back in his office, he undertook a more detailed study of the plate. He rifled the observatory files for earlier photographic examples taken by Shapley, Humason, Ritchey, and himself, and soon discovered that one of his three "novae" was subject to cycles; it alternately brightened and dimmed, something an exploding star should not do. Plotting the object's light curve, he determined that it had a period of 31.415 days. Then, by exploiting Shapley's distance-measuring techniques, he found it to be at least 300,000 parsecs from Earth—the equivalent of a million light-years—more than three times the diameter of Shapley's entire universe. There seemed little doubt: Andromeda's spiral arm was bejeweled by a Cepheid variable, making the giant nebula an external galaxy composed of stars by the millions! Hubble opened his personal logbook to page 156 and made the following correction: "On this plate (H335H), three stars were found, 2 of which were novae, and 1 proved to be a variable, later identified as a Cepheid—the first to be recognized in M31."[16] He then took out his pen and in capital letters printed the abbreviation "VAR!" for variable on the plate.

In late February 1924, Shapley was rocked by a bolt from the blue. "You will be interested," Hubble began, "to hear that I have found a Cepheid variable in the Andromeda Nebula (M31). I have followed the nebula this season as closely as the weather permitted and in the last five months have netted nine novae and two variables." He enclosed a diagram of the first variable's light curve, which, "rough as it is, shows the Cepheid characteristics in an unmistakable fashion." It was well within the borders of the spiral arms, situated against a background of faint mottled nebulosity resembling grains of porridge. The second variable, about half a magnitude fainter than the first, was located just at the edge of the arms, and was too dim for a reliable determination of its period from the data at hand, although twenty-one days seemed a reasonable estimate. He concluded his incendiary missive on a gleeful note, as though consciously rubbing salt in the wound. Shapley, who hated Hubble's spoken "Oxford," found the written form no easier to swallow: "I have a feeling that more variables will be found by careful examination of long exposures. Altogether next season should be a merry one and will be met with due form and ceremony."[17]

English-born Cecilia H. Payne, who would soon become Harvard Observatory's first Ph.D. graduate in astronomy, happened to be in Shapley's office when the letter arrived. After scanning its contents he held the document out to her, remarking, "Here is the letter that has destroyed my universe."[18]

Shapley replied to Hubble within a week: "Your letter telling of the crop of novae and of the two variable stars in the direction of the Andromeda nebula is the most entertaining piece of literature I have seen in a long time." He then added a series of numbered comments which "may be of interest or service to you." Shapley reminded Hubble that Cepheid variables with periods of more than twenty days "are generally not dependable." Hubble's first variable exceeded this limit by more than a half, while the second, though more promising because of its shorter period, was still under investigation. Hubble's experience should also have shown him that in the denser part of the nebulosity every star is a variable, depending upon the exposure time and the manner in which the plate is developed. Harvard owned a number of plates containing these false variables in magnitudes comparable to the ones he purported to have found.[19]

If Hubble was hoping for unconditional surrender, which seems unlikely, he was disappointed. Using the 100-inch like a celestial battering ram, he launched an assault on NGC 6822, a nebula located near the star-rich clouds of Sagittarius. Again, Shapley was among the first to hear of his discoveries. "I have normal curves for nine variables," he wrote on August 25, 1924. "The periods run from 64 to 12 days and the curves all appear to be of the normal Cepheid type." There was more. He had confirmed another dozen variables in Andromeda, at least three of which were Cepheids. Still another fifteen or so variables had turned up in the other great spiral, M33, and variables "have been suspected in Messier 81 and 101, but I cannot call them established as yet." His hauteur in check, he gently edged Shapley in the direction of the inevitable: "I feel it is still premature to base conclusions on these variables in spirals, but the straws are all pointing in one direction and it will do no harm to begin considering the various possibilities involved."[20] In plain terms, the Great Debate was over.

There was little a crushed Shapley could do but swallow his pride and surrender the standard. Writing from Woods Hole, where he was vacationing, he admitted that Hubble's "exciting letter" had stirred ambivalent feelings: "I do not know whether I am sorry or glad to see this break in the nebular problem. Perhaps both."[21] It was a good time to be dredging for starfish. That morning Shapley and his companions had snagged a globular cluster of about one thousand in twenty fathoms. Somehow, it was as if he had never been at Mount Wilson, never wielded the giant telescope that could have made his name as immortal as Hubble's. A year earlier, sensing that the end was approaching, he had exclaimed in a letter to Hubble, "What a powerful instrument the 100-inch is in bringing out those desperately faint nebulae."[22]

The rest of the astronomical community was agog. Russell, who first heard of the discovery from James Jeans, a recent visitor at Mount Wilson, wrote Hubble from Princeton in December, "It is a beautiful piece of work, and you deserve all the credit that it will bring you, which will undoubtedly be great. When are you going to announce the thing in detail?" The American Association for the Advancement of Science was due to meet shortly in Washington, D.C., and Russell urged Hubble to submit a paper. It would be a splendid forum for a major scientific announcement. "[Y]ou ought, incidentally, to bag that $1000 prize."[23]

Three weeks later, on New Year's Eve, Joel Stebbins, who held the position of secretary of the Council of the American Astronomical Society, was having dinner with Russell at Washington's Hotel Powhattan. Both were in town for the joint meeting with the A.A.A.S. In the course of the conversation, Russell asked, "Has a young man named Hubble sent in a paper?" When Stebbins replied in the negative, Russell exclaimed, "Well, he is an ass!" Later, the two joined a group of astronomers in the hotel parlor, where the discussion led naturally to the topic of Hubble's Cepheids. As they warmed to their subject, someone suggested they draft a joint telegram to Hubble, urging him to send a wire containing an abstract of a paper that could be presented to the society and placed on the program. This message in hand, Stebbins, together with Russell, headed for the front desk. Stebbins was just reaching for a Western Union blank when Russell noticed a large envelope with his name on it resting against the wall. Stebbins spied Hubble's name in the upper left corner, and the two hurried back to the parlor and their astonished colleagues. "There it was," Stebbins recalled, "complete with diagrams, just what we wanted. It was no breach of confidence to read and discuss the paper there and then; the title was 'Cepheids in Spiral Nebulae.' "[24]

The next day Russell, with Shapley looking on, read the paper from the convention floor. Afterward, the Council of the Society selected it as the one to be recommended for the A.A.A.S. prize. Russell and Stebbins were chosen to draft the cover letter to the Committee on Awards. Stebbins vividly remembered Russell pulling out his ever present slide rule to compute how much the universe had expanded as the result of Hubble's work. "This paper," they concluded,

> is the product of a young man of conspicuous and recognized ability in a field which he has made peculiarly his own. It opens up depths of space previously inaccessible to investigation and gives promise of still greater advances in the near future. Meanwhile, it has already expanded one hundred fold the known volume of the material universe and has apparently settled the long-mooted question of the nature of the spirals, showing them to be gigantic agglomerations of stars almost comparable in extent with our own galaxy.[25]

On February 7, Hubble received the following telegram from Burton E. Livingston of the Committee on Awards: "American Association Prize this year divided and five hundred dollars awarded to you for paper given at Washington meeting."[26] Not until he received a follow-up letter did Hubble learn the name of his fellow prizewinner. Dr. L. R. Cleveland of Johns Hopkins was recognized for illuminating a very different universe of minuscule proportions—the physiology of protozoa in the digestive tracts of termites and other insects.

Though Hubble had tied for first place among 1,700 papers, Stebbins, in a letter to Adams, expressed disappointment that his fellow astronomer had been forced to share the glory with another. At the same time, "there are several of us who are smiling on how the period-luminosity curve has come around to work against some of Shapley's ideas."[27] He was frankly puzzled, however, by Hubble's delay in publishing his discovery. So, too, was Russell, to whom Hubble wrote on February 19, 1925, misspelling his champion's name: "The award came as a joyous surprise. . . . We realize that the business was about 99% Russel and 1% Hubble."

> The real reason for my reluctance in hurrying to press was, as you may have guessed, the flat contradiction to van Maanen's rotations. The problem of reconciling the two sets of data has a certain fascination, but in spite of this I believe that the measured rotations must be abandoned. I have been examining the measures for the first time and the indications point steadily to a magnitude error as a plausible explaination. Rotation appears to be a forced interpretation.[28]

He wrote Stebbins much the same a few weeks later, adding, "You can realize how delicate a business such a discussion will be."[29]

The day before Hubble drafted his explanatory letter to Russell, a beleaguered van Maanen, on whom the wolves were fast closing, wrote Shapley, "After Hubble's discovery of Cepheids I have been playing again with my motions and how I look at the measures. I cannot find a flaw in M33, for which I have the best material. They seem to be as consistent as possibly can be."[30] Shapley, his patience nearly spent, shot back, "I am completely at a loss to know what to believe concerning those angular motions; but there seems to be no way of doubting the Cepheids, providing Hubble's period-luminosity curves are as definite as we hear they are."[31] When, years later, the bitter astronomer was asked about his reasons for sticking with van Maanen's measurements for so long, he replied, "They wonder why Shapley made this blunder. The point . . . is that van Maanen was his friend and he believed in friends!"[32]

IV

The astronomer Margaret Harwood first met Edwin Hubble while visiting the Frosts at Yerkes in 1915. The then graduate student, whom she remembered as polite but aloof, gave her a tour of the observatory and grounds. It was her impression that Hubble was attracted to the Frosts' daughter, Katherine. The astronomers crossed paths eight years later, in 1923, at Mount Wilson, after Adams, during one of Hale's many absences, temporarily suspended the rule barring women from using the instruments. Hubble remembered Harwood and was more friendly on this occasion, but still somewhat reserved. Harwood, who had five brothers, thought his conduct a natural defense mechanism, for he was considered "a very eligible bachelor."[33]

Much of the observatory staff had long been preparing for the total solar eclipse of September 10, normally a time of warm temperatures and clear skies. Two observation sites had been chosen, one at Point Loma, near San Diego, well within the path of totality, and the other at Lakeside, close to the edge of the path. Harwood was invited to join the Point Loma team, which included a reluctant Hubble, whom Adams had dragooned into serving as timekeeper. He was to watch the clock, counting the seconds aloud to keep the observers on schedule. Not only did the eclipse have nothing to do with his astronomical interests; he was preoccupied by other matters.

Lick Observatory also mounted an expeditionary team, funded in part by the Los Angeles banker John Patrick Burke.[34] Burke's daughter, the recently widowed Grace Leib, was planning to join the Lick observers, including her former brother-in-law, William H. Wright, at a remote location south of the Mexican border. Hubble had hoped to meet Grace there until Adams scuttled this plan by requesting that he be present at Point Loma. Instead, the smitten pair agreed to meet in San Diego after fulfilling their respective obligations.

Unfortunately, the normally pristine weather turned foul only minutes before totality, a sudden thunderstorm blotting out the sky. Harwood watched as the astronomers emerged, one by one, from the makeshift darkroom with nothing to show for their efforts. She sensed they were trying to look sad but were thinking, "Now I can get back to my work."[35] With three weeks still remaining before his fateful run with the 100-inch, Hubble collected Grace and her mother, then headed up the coast to the Burke summer home at Pebble Beach. There the couple swam in the chill ocean, walked the misty shore, and made plans for a simple and very private wedding.

A daughter of privilege, Grace had spent her childhood in San Jose, located in the heart of western California's rich fruit-growing area. Her father and mother, Luella "Lulu" Kepford, both natives of the Midwest, came to California in 1891, where the Irish Catholic Burke became vice president and director of the San Jose and Santa Clara Railway Company. He was admitted to the bar in 1901 and practiced law three years before assuming the position of vice president and manager of the Bank of San Jose. Grace Lillian, the first of two daughters, was born in 1889. She was described by her childhood friend the novelist Susan Ertz as a small girl with hazel eyes and a face of charming vivacity. Friendly, clever, and athletic, Grace rode to school every day on a pony, trailed by the family pet, a fox terrier named Montmorency.[36]

San Jose, like Walnut, Iowa, from which he had come, proved too small for the ambitious banker. Burke moved his family to Los Angeles, and swiftly climbed the financial ladder to become vice president of the First National Bank, as well as a member of its board of directors. The Republican millionaire took an active role in civic affairs, serving multiple terms on the boards of the Los Angeles Chamber of Commerce, the local chapter of the American Red Cross, and the Tuberculosis Association. The Burkes and their live-in maid inhabited an elegant but not ostentatious ten-room home at 505 Lafayette Park Place. When Lulu and the girls went shopping, they were driven about town by a chauffeur in one of the family's two Cadillacs.

After the move south, Grace and her sister, Helen, nicknamed

"Max," were enrolled at Marlborough School, a prestigious private institution for girls in West Los Angeles. An excellent student, Grace had the misfortune of running afoul of a mentally unbalanced Latin teacher, who hectored her the semester long before giving her a failing grade.[37] Ultimately vindicated, she received her high school diploma in June 1908 and enrolled in Stanford that fall.

A romantic who loved literature, Grace chose to major in English. Her grades, which were straight A's, did not prevent her from leading an active, if selective, social life. She pledged Alpha Phi, joined the English Club, and represented her sorority on various prom committees. With her horse stabled nearby, the coed was able to spend many hours riding through the still undeveloped countryside, where she further polished her skills as an amateur naturalist by observing trees, flowers, birds, and wildlife.

Victorian dress still dominated women's fashion. When not attired in riding coat, breeches, and high-topped boots, Grace donned long dresses or flowing skirts complemented by intricately stitched blouses with collars secured at the throat by cloth flowers. Her thick, dark brown hair was worn in a generous upsweep and knotted at the back after the popular French style known as chignon. Small-boned and of light complexion like her Irish ancestors, she was neither beautiful nor pretty, but something in between.

By her senior year the soon to be crowned Phi Beta Kappan had pledged herself to Earl Warren Russell Leib, a member of Stanford's class of 1911. A year older than his fiancée, the tall, angular geology and mining major was from one of California's best-connected families. Earl's father, Judge Samuel Franklin Leib, was arguably San Jose's most prominent citizen, and certainly one of its wealthiest. A graduate of the University of Michigan, Leib made his fortune practicing law before undertaking a brief but distinguished career on the bench. His personal friendship with the railroad magnate and Republican governor Leland Stanford gained him a seat on Stanford University's board of trustees, which he later headed as president. The judge, an avid horticulturist whose correspondents included the New England plant-breeding genius Luther Burbank, maintained 110 acres of French plum trees and substantially increased his wealth by perfecting an innovative process for the drying of prunes. After the harvest, the gracious Leib home on Alameda Street, considered the city's most beautiful, was filled with lively music, fine wine, and excellent food.

As up-and-coming attorneys in a small city, Burke and Leib had doubtless gotten to know one another well before John moved his family to Los Angeles. Their wives may have socialized, while their children attended the same schools. The only possible drawback to the match was religious, the Leibs being Episcopalian, the Burkes Catholic. The bride's wishes prevailed; Grace and Earl were married by a priest in Los Angeles on December 23, 1912, six months after her graduation from Stanford. Rather than buy or rent a house of their own, the newlyweds moved into the Burke home on Lafayette Park Place.

It remained their residence eight years later, when Grace first set eyes on Hubble in the laboratory on Mount Wilson. The couple had no children, and thirty-two-year-old Earl, who had recently taken a new job as a geologist for the Southern Pacific Company, was away in the field much of the time, assaying coal deposits and mines. In June 1921 he traveled to Amador County southeast of Sacramento, site of the famous strike that sparked the "Rush of '49." At 1:30 on the afternoon of Wednesday the fifteenth, the geologist entered the vertical shaft of an idle coal mine near the village of Ione for the purpose of obtaining samples. According to his death certificate, he had descended the ladder to a point fifty feet below the surface when he was overcome by gas, causing him to fall the remaining forty feet to the mine floor.[38] The body, its legs and back broken, was not recovered until seven o'clock in the evening, and only then by means of ropes and grappling hooks because the accumulation of gas made it impossible to venture underground.

A newspaper account of the tragedy made note of the curious fact that the victim, who had ten years of mining experience, had not worn a gas mask. This, in turn, sparked an investigation by the State Industrial Accident Commission, together with a coroner's inquest.[39] Whether Leib died of asphyxiation or of internal injuries from the fall was never determined. The body was returned to San Jose by special train two days later. Grace and her parents came north for the funeral at the Leib home at 10:30 Saturday morning, with the Reverend Noel Porter, rector of the Trinity Episcopal Church, officiating. Earl's brothers and other members of the family served as pallbearers. Following the service, the body was taken to the local cemetery for burial in the family plot.[40]

Grace and Edwin were romantically involved by the following year, when he gave her the first of many books, his worn Oxford copy of the *Symposium* translated by Shelley.[41] Although she destroyed his early letters along with much else, an excerpt from one penned in 1922 survives in her account of his life as an astronomer:

These are hectic days on the mountain. The 60-inch is laid up for repairs on the dome and the occasion has been seized to reroof the monastery. I slept the first morning in the Kapteyn cottage and awakened to the rattle of compressed air riveters. This morning I tried the monastery, but hammers on galvanized tin were no better. The clouds that are piling up in the south offer a temporary solution that is really welcome.[42]

Several years later, on the anniversary of their first meeting on the mountain, they spoke wistfully of June 1920. Edwin remarked, "Time means nothing when I think of you."[43]

The courtship was discreet. Following a run, Hubble would head for the Burke home and a waiting Grace, who had cut her long hair and adopted the popular marcel wave of the twenties, along with shorter skirts. Staying the night, he read to her by the fireplace in the library, often in the company of her parents. As Grace recalled: "The only really satisfactory times I could remember were when Edwin used to call on me, and sometimes read aloud."[44] She confessed to a yearning for a balancing spirit, which "I . . . recognized for the first time when I saw it, in Edwin."[45]

With his family two thousand miles away, Hubble completed the long process of reinventing himself. His stories of heroism in the Wisconsin woods, of boxing the greats, of saving young women from drowning, of practicing law, and of leading frightened men into battle were dutifully recorded by Grace, seemingly without the slightest question as to their validity. It was an impressive display which helped win her over, and her parents as well. In a grand gesture that further stirred sympathetic feelings, he offered to give up astronomy for the law so that Grace could continue living a life of material ease. She demurred, as would have been expected, telling Edwin she would not marry him if it meant forsaking his love for the stars.[46]

The wedding took place at 7:30 on the morning of Tuesday, February 26, 1924. Bowing to John Burke's wishes, the couple agreed to have the service performed by Michael O'Halloran, the family priest. After breakfasting with Grace's parents and sister, the couple drove north to the Burke cottage at Pebble Beach. Isolated on six acres of untouched woods consisting of Monterey pines and white-barked oaks, it stood on a rolling plateau that looked down across the meadows to Stillwater Cove and south to Point Lobos. Edwin built a fire, and, after it had burned down, Grace grilled chops over the coals. A cat they named Black Abbott suddenly appeared out of the darkness. After eating his share of dinner he curled up on Edwin's lap before the fire. When the *Los Angeles Times* arrived two days later, Grace was deeply relieved to see no notice of personal events. "If any reporters appear," she cautioned her mother by letter, "please discourage them violently from printing anything."[47]

V

The Hubbles returned within a week and immediately began preparations for the second stage of their extended honeymoon. In mid-March they boarded the California Limited for New York City, where a normally composed Grace, who had never been to the East Coast, was "simply speechless" viewing a brontosaur and a dinosaur egg from Mongolia at the Museum of Natural History. In a letter to Lulu she described them as "the finest things in New York." They then took the train to Boston and Harvard to confer with Shapley, who, only a month earlier, had received the first devastating letter from Hubble concerning Cepheids in Andromeda. From Boston they traveled north into a frigid Canada and settled into their comfortable cabin aboard the rechristened steamship *Montlaurier*, part of the war booty captured from the Germans. Bound for Liverpool, the aging liner sailed into an ominous North Atlantic and was soon overtaken by giant swells that, in Grace's words, tossed the passengers about "like so many bugs . . . on a chip." Most, including Edwin, became seasick and spent the voyage in bed belowdecks. Grace, her adrenaline pumping, was much too excited to succumb to the foul weather. Her new husband wrote her parents that she was the only passenger who survived. "She mothered me, lying low in my bunk, and still found time to captivate everyone on board from the Captain to the bugler."[48]

The enchanted land of which Grace had read so much was just returning to life when they arrived early in April. She described the 200-mile train ride from Liverpool to London as a whirl of gorgeous green countryside, trees, pastures, and streams sprinkled with lovely villages, brick houses, and thatched roofs set against a vivid background of flowers, both wild and domestic. Standing in Westminster Abbey, once the church of a Benedictine monastery, the nonpracticing Catholic experienced her most moving moments in London. Time, like a slow and ancient glacier, had passed through the sanctuary, its flow wearing the pavement stones into soft hollows. "I had not realized," she wrote home, "that the work of men could be invested with such awe. And I have never realized before the sacred permanence of the dead." She paused to pay silent homage to her "old friend" Samuel Johnson, with

David Garrick lying beside him. The stilled bards of Poets' Corner—Chaucer, Browning, Tennyson—were accorded equal tribute, as were Isaac Newton and his neighbor, Sir John Herschel.[49]

At Oxford the couple stayed at the Mitre, a former coaching inn six centuries old. Word of Edwin's recent discovery had preceded the couple's arrival, and both were deeply touched when H. H. Turner, Oxford's Savilian Professor of Astronomy, came riding up on his bicycle to pay a call. Following a cordial visit, Turner sent the same telegram to many of his colleagues: "Hubble is here with a lovely bride," and, as Grace noted, "they all asked us to come and stay."[50]

From Oxford they proceeded to Cambridge, where they had been invited to stay with the H. F. Newalls at Madingley Rise. It was Newall who had proposed Hubble for membership in the Royal Astronomical Society while he was stationed at Cambridge after the war. Arthur Eddington, who was among the first physicists to grasp and embrace relativity theory, was invited to lunch, an even more prized dinner invitation having been withheld because the Newalls did not approve of those who stood for radical innovation. There were no bathrooms in the great house, and the Hubbles, like their hosts, took hip baths by an open fire. When they went for rides it was by carriage rather than automobile. Grace was impressed, finding the Newalls "delightfully civilized and critical; they judged and they chose. And it was, while the going was good, much pleasanter than the way of life that replaced it."[51]

Back in London, Hubble was guest of honor at a Royal Astronomical Society dinner, after which he spoke of his breakthrough discovery and classification of nebulae. The assembly drank a toast "to Hubble and Mrs. Hubble," and called his attention to a remark made about him in the venerated British humor magazine *Punch*, a measure of his newly won fame.

They crossed over to the Continent by way of Calais. Edwin lunched with the director of the Paris Observatory and purchased several rare books, including a very good Pliny, bound in vellum. From Paris the couple headed for Basel, Lucerne, and Interlaken in central Switzerland. The dedicated mystery buffs made the pilgrimage to Reichenbach Falls, where Sherlock Holmes and Professor Moriarty had danced with death above the maelstrom 300 feet below.

After England, it was Italy that laid claim to their hearts. Strolling the vast corridors of Florence's battlemented Palazzo Vecchio, once the home of Medici patriarch Cosimo I, the Hubbles came upon a suite of small, beautifully proportioned rooms. "When we build our house,"

Edwin remarked, "let's have it something like these rooms."[52] They visited the tomb of Galileo, and then made their way to Arcetri in the Tuscan countryside, the town to which, in 1633, the nearly blind and disgraced astronomer had been sentenced to perpetual house arrest by the Inquisition.

They returned to Oxford in time for Eights Week, then embarked from Liverpool on the voyage back to Montreal via the St. Lawrence. They reached California in May, after an absence of nearly three months. The evening of their arrival found the astronomer back on Mount Wilson for a week's run. When he came down, it was to find that Grace had settled them in an apartment at the northwest corner of Wilson and San Pasqual streets, across from the Caltech campus in Pasadena. Although the quarters were small, consisting of a living room and tiny kitchen on the first floor, a single bedroom and bath on the second, it suited the longtime bachelor very well. "I never thought," he confessed to his bride, "that I should have such a nice home."[53]

VI

In 1922, George Ellery Hale had suffered the fourth in a series of what he termed his "nervous breaks." He soon departed for Europe and the Middle East, where he remained more than a year, taking the waters and consulting with numerous physicians in hopes of finding a cure. When his health failed to improve, Hale, now fifty-five, decided it was time to take stock of his life. In March 1923 he drafted an extraordinary letter to John C. Merriam, president of the Carnegie Institution. Its three single-spaced pages read like a clinician's journal. In addition to multiple nervous breakdowns, Hale had survived an acute appendicitis, a diseased gallbladder, and a severe kidney infection, all of which had required surgery.

Add to this list daily evidences of congestion of the head, with frequent acute phases; many attacks of hemorrhoids, several of which have kept me in bed for weeks, while one involved a severe operation; and repeated cases of lumbago or similar trouble, and you have a catalogue fit to rejoice the soul of a pathologist! He would be still more delighted if it were extended back to include the typhoid fever, repeated dysentery, colitis, and other difficulties of earlier days. . . .

Congestion of my head is caused chiefly by worry, excitement, responsibility, discussion of any scientific subject, attendance at scientific meetings, lecturing (mainly because of the defective memory of faces and inability to recall names—even when well-known—when needed), and continued mental work.

This and more had convinced him that it was time to step down as director of Mount Wilson. He recommended that his assistant Walter Adams, who had been in charge of the observatory most of the time in recent years, take his place. If the Carnegie Institution should see fit, he would accept the title of Honorary Director, for much was yet to be done during the fifteen years he hoped he had left. "I have never been skillful in carrying on steady . . . work. In fact, I am a born adventurer, with a roving disposition that constantly urges me toward new long chances."[54]

Adams, whose affection for Hale was genuine and deep, wired Merriam in hopes of maintaining the status quo. When Merriam replied that Hale was adamant, Adams agreed to become director at a salary of $8,000 a year, the same amount appropriated for Hale's new position.[55]

Hubble, who had seen little of Hale since coming to Mount Wilson, probably viewed the change in leadership as part of the natural order of things, perhaps even welcoming it. No friend of Shapley's, he had doubtless followed with interest the running conflict between Adams and the other Missourian before he departed for Harvard. Adams had questioned Shapley's distance measurements, which he believed rested on too limited data and flawed methods. This disagreement soon invaded the realm of the personal, coloring Adams's perception of the man. A puritan by way of the Middle East, he disliked prima donnas, and even Shapley's closest friends could hardly acquit him of that charge. As Shapley's ally, van Maanen, who had developed the reputation of a playboy, was to Adams another example of a reprobate masquerading as a serious scientist. Early in 1918, Shapley wrote a friend, "I feel very sure that if I should go away from here no opportunity would be given me to return so long as Adams had the deciding voice. . . . [V]an Maanen and I are in ill-favor because we try to do too much."[56]

With Shapley long gone and Hubble's fame about to mushroom, the latter had suddenly become his own lightning rod. According to Grace's journal, her husband had spent three months abroad in 1922, visiting England, France, and Portugal.[57] Now he had taken another three months at full pay to escort his bride across much of Europe while his Mount Wilson colleagues, Adams included, made do with a month's vacation. What would become an annual rite was bound to rankle. Known for keeping his own counsel, the new director began compiling a mental list of this and Hubble's other transgressions.

VII

Hubble liked to bring distinguished visitors home to dinner, especially those from England. Grace relished her role as hostess and was soon showing some of the world's great scientists around Los Angeles. The first of her charges was the bachelor Quaker Arthur Eddington, who arrived the September following their marriage. The modern Archimedes, with whom they had lunched at Madingley Rise, had recently published his magnum opus, *Mathematical Theory of Relativity*, which Einstein later spoke of as the finest presentation of the subject in any language.[58] Grace collected Eddington at his hotel nearly every afternoon and drove him to the Flintridge Country Club. She sat with a book on the grass while her guest dove from the springboard, swam across the pool and back, and dove again. He told the Hubbles that he went for a swim in the Cam every morning, summer and winter, and revealed his secret for judging the temperature of the water, which Grace promptly forgot. She only remembered Eddington as silent and far away in his thoughts.[59]

More vivid were her recollections of James Jeans. Appointed research associate at Mount Wilson in 1923, the president-elect of the Royal Astronomical Society returned to Pasadena in 1925 to continue his pioneering work on stellar dynamics and the evolution of nebulae, a subject of more than passing interest to Hubble. By this time Jeans had been awarded most of the major scientific prizes of the day, and, along with his sometimes rival Eddington, was soon to be knighted by George V. Jeans's aloof bearing was partly a shield for a shy and sensitive spirit, which had suffered permanent damage during a childhood dominated by a passion for organ music and clocks rather than human companionship. While in England, Grace was told by Rose de Marcellus, the scientist's neighbor in Dorking, that he terrified her. Whenever she said anything, Jeans raised a censorious eyebrow and drawled, "Rawther obvious."[60]

The Hubbles were among the few who came to know Jeans in simple unconstrained friendship, prizing his challenging wit and boyish sense of the ridiculous. He told them of his distasteful undergraduate experience at Cambridge with Bertrand Russell, "who was maliciously personal and spiteful in argument." (Of Jeans, Russell later told the Hubbles, "[He] is a wicked man; Eddington is a saint.") Jeans was accompanied on this visit by his wife, Charlotte, and their daughter, Olivia, with whom he had made up a game in a language only they understood, Jeans's watch serving as the third character in this conspiracy against the outside world. The Hubbles drove the family to a beach house near Santa Monica, where Edwin and Olivia, in defiance of a chill wind, put on their bathing suits and entered the rough surf. He picked up the child, lifting her above the big swells, much to the delight of a father who preferred the organ to the sea.[61]

Jeans's visit afforded Hubble the opportunity to update his English colleague on the progress of his nebular classification scheme, in which the physicist-astronomer had displayed a keen interest. Hubble knew what he knew, and the lack of attention given his work by the International Astronomical Union continued to chafe. The fact that he had retaliated by publishing two articles on the subject was well and good, but he was playing for much higher stakes. The Commission on Nebulae and Star Clusters had decided to compile a new catalogue of the nebulae, based, as far as possible, on photographic rather than visual observations. Yet it was difficult to envision the project moving forward until an acceptable method of classification could be arrived at. In July 1923, with Slipher now chairing the commission, of which he was also a member, Hubble typed up his extensive notes on his system of nebular classification and mailed them, together with the appropriate photographic data, to Slipher in Flagstaff. "Instead of working them up into an article for publication," he wrote, "I thought it would be better to offer them to the committee on nebulae as a basis for discussion, the outcome of which might be a system of classification approved by the committee and sanctioned by the I.A.U." The men at Mount Wilson, including a visiting Henry Norris Russell, "have looked over the notes and have expressed their approval."[62]

This scheme differed considerably from the one Hubble had published in 1922. While the galactic nebulae were allowed to stand as before, he had begun the process of separating nongalactic nebulae into only two classes. Spindles, ovates, and globulars were subsumed under the categories of ellipticals and spirals.

The elliptical nebulae are amorphous in appearance, resembling nothing so much as blobs. Hubble further subdivided them according to how flattened they look. Those that are perfectly circular he called E0 nebulae; several steps removed are the E7 ellipticals, the flattest of their kind, reminiscent of a lens or a somewhat elongated football.

The more elegant double-armed spirals, as their name suggests, have a characteristic pinwheel structure. These Hubble divided into logarithmic (later termed "normal") spirals and a smaller but distinctive subgroup called barred spirals, so named because their curved arms originate from the end of a "bar" running through their nuclei rather than from the nuclei themselves. Just as the ellipticals are classified according to their degree of flatness, the spirals, both normal and barred, are divided into subclasses based on the tightness of their arms. An Sa designation denotes a normal spiral with tightly wound arms; SBa is the symbol for its counterpart within the barred category. Sc and SBc spirals are the most loosely wound of all.[63]

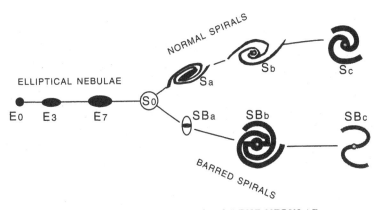

HUBBLE'S CLASSIFICATION OF THE NEBULAE

Finally, Hubble was left with the anomalous—nebulae that did not fall readily into any of the three previous categories. To the nongalactic ellipticals and spirals he was forced to add a catchall designation, "the very few" he termed irregular nebulae. "Irregular," he wrote, "calls particular attention to the lack of rotational symmetry, the chief obstacle to the normal classification of these objects."[64] Warped, dented, or twisted, some of the irregulars seemed in the process of exploding, whereas others had apparently been bent out of shape by primordial collisions or close encounters with trespassing neighbors. Both the Small and Large Magellanic Clouds fit within this category.

In a lecture delivered at the University of Toronto in February 1924, Russell remarked that "Jeans' theory [of nebular evolution] seems to be matched by things we have in the sky."[65] This perceived relationship was equally apparent to Hubble. Jeans, in his 1919 Adams Prize essay,

Problems of Cosmogony and Stellar Dynamics, had speculated that the elliptical nebulae are spirals in the making. Beginning as globular masses of gas devoid of structural form, they pass through a progressive series of changes, becoming more and more flattened until the long axis is three times the short one. At this point the elliptical nebula begins the transformation into a spiral, its arms coiled closely to its center or nucleus. In later types the arms are more and more splayed, building up at the expense of the dimming center. Finally, the arms begin to break up into flagelliform condensations composed of individual stars.

The alluring prospect of fitting observation to theory worked its magic on the normally cautious astronomer. Applying Jeans's evolutionary morphology to both categories of spirals, Hubble termed their respective members "early" (tightly wound), "middle" (intermediate), and "late" (loosely wound). He would later design and publish his famous cosmic "tuning fork" diagram, in which the handle is formed by ellipticals evolving into dual tines, one composed of normal, the other of barred spirals.[66]

Slipher did nothing with the notes Hubble had submitted to him. In February 1924, after a silence lasting seven months, Hubble wrote to advise him that "Mr. Hale thinks I should publish the system of classification. I would prefer that it go through the committee if that is feasible within a reasonably short time."[67] Slipher duly had copies made, and these reached the hands of commission members by the end of May.

Comments on Hubble's notes, most of which were quite encouraging, began to trickle in the following month. Curtis pronounced the scheme "very thorough and complete," but expressed a preference for what he termed the astronomer's "simpler classification" system quoted in Slipher's 1922 report.[68] Knox-Shaw found the work "especially interesting," yet chose to reserve his opinion until a later date.[69] Dreyer, writing in August, also thought the work of "great interest." The commission "can hardly do better than to take it as a basis when deciding on the form of the 'Description' column of the General Catalogue."[70] Although he waited until October to respond, Harvard's Solon Bailey was the most positive of all: "In general I do not see how it can be much improved at this time."[71]

Committees being what they are meant that Hubble was not yet safely out of the woods. His enthusiasm notwithstanding, Bailey balked at the terms "early," "middle," and "late" as applied to the spiral nebulae "since the order of evolution does not appear to be sufficiently well established at the present time, and it is undesirable to introduce theories into a classification, if it can be avoided."[72] Slipher was of a similar mind. These same terms were singled out in his marginal notes as holding an "evolutional significance that the present state of our knowledge hardly warrants."[73]

The other fly in the ointment was Shapley. The Harvard Observatory director's preeminence as an authority on star clusters had prompted Slipher to add him to the commission in early 1925. In October of the previous year he had written to Hubble protesting the latter's use of the term "nongalactic nebulae" when referring to the spirals. After all, it was Hubble himself who had lately proved them to be independent star systems. Would not "galactic nebulae" or simply "galaxies"—Shapley's choice of nomenclature—be more appropriate?[74] Going over Hubble's head, he wrote Slipher in the same vein.

There was no denying the fact that Shapley had scored a telling point, but Hubble was in no mood to be dictated to by the man whose universe he had demolished. When he finally published his classification scheme, the nongalactic nebulae became the "extra-galactic nebulae," and would remain so on Mount Wilson until his death in 1953, when the term "galaxies" was universally adopted.

Disappointment followed. The commission next convened at Cambridge in July 1925, a meeting Hubble did not attend. The need for a new classification system was discussed at length, but it was finally decided not to adopt Hubble's model, which was thought too deeply encumbered by theoretical bias. Nor was the time "yet ripe" for the compilation of a new general catalogue of the nebulae.[75] If Hubble chose to go ahead and publish, it would be without the long-sought sanction of the I.A.U.

He had made up his mind to proceed by the spring of 1926 when, suddenly, the astronomer became embroiled in a major controversy. Knut Lundmark, with whom he had jointly authored a paper on the nova Z Centauri in 1922, published an article bearing the all too suggestive title "A Preliminary Classification of Nebulae."[76] As a recently appointed member of the commission, he had been present at the Cambridge meeting, and what had gone on afterward was obvious to Hubble. He immediately wrote Slipher, crying foul:

> I see Lundmark has published a "Preliminary Classification of Nebulae," which is practically identical to my own, except for the nomenclature. He calmly ignored my existence and claims it as his own exclusive idea. I am calling this to your official attention because I do not propose to let him borrow the results of hard labor in this casual manner.[77]

His anger continued to build during the coming months and finally boiled over in a broadside to Lundmark reminiscent of Newton's indictment of the presumed plagiarist Leibniz:

> This is a very mild expression of my personal opinion of your conduct and unless you can explain in some unexpected manner, I shall take considerable pleasure in calling constant and emphatic attention, wherever occasion is given, to your curious ideas of ethics. Can you suppose that colleagues will welcome your presence when they realize that it is necessary to publish before they discuss their work?[78]

Although they agreed on little, Hubble was aware of Shapley's animus toward the enigmatic Swede and wanted him to know that they were of one mind on the subject. Lundmark "has mixed the good with the bad, facts with fancies, in such a manner that the general significance of his results are entirely misleading."[79]

Of one mind on the question of character, but not on the charge of plagiarism. While Hubble fumed and Lundmark protested his innocence, Shapley confided to Russell his belief that Lundmark's scheme "is considerably better than Hubble's."[80] He also wrote to Adams, Hubble's superior, hoping to undermine confidence in the revised classification scheme, "which places a tremendous majority of those [spirals] now photographed in the waste-basket class Q—that is, unclassified, and for general statistical studies essentially useless."[81] Shapley's solution was to devise and publish his own little-remembered scheme, combining what he saw as the virtues of Hubble's and Lundmark's systems while endearing himself to neither antagonist.[82]

Hubble's seminal paper, simply titled "Extra-Galactic Nebulae," appeared in the pages of the *Astrophysical Journal* a few months after Lundmark's article went to press. Employing the strongest language allowable in a professional publication, he attempted to settle accounts with the interloper by dressing him down before his peers: "K. Lundmark . . . has recently published a classification, which, except for nomenclature, is practically identical with that submitted by me." Moreover, "Dr. Lundmark makes no acknowledgments or references to the discussion of the Commission other than those for the use of the term 'galactic.' "[83]

Keeping faith with Jeans, Hubble published both his classification scheme and his evolutionary hypothesis as contained in his 1923 notes to the Commission on Nebulae and Star Clusters. Yet there is little question that he had been sensitized by the commission's criticism of his case favoring nebular evolution. He thus drew a clear distinction between what he had photographed with the 100-inch Hooker and questions of a largely theoretical nature:

> Although deliberate effort was made to find a descriptive classifi-
> cation which should be entirely independent of theoretical consider-
> ations, the results are almost identical with the path of development
> derived by Jeans from purely theoretical investigations. The agree-
> ment is very suggestive in view of the wide field covered by the data,
> and Jeans's theory might have been used both to interpret the obser-
> vations and to guide research. It should be borne in mind, however,
> that the basis of the classification is descriptive and entirely inde-
> pendent of any theory.[84]

The passing years would cast doubt on the premise of nebular evolution outlined by Hubble. Jeans, whose champion Hubble had become, was eventually forced to admit that all attempts to explain
dynamically the origin and development of the spiral arms were
failures—and remain so to this day. Conversely, Hubble's somewhat
modified classification scheme has weathered the test of time and continues to serve as the accepted standard for the vast majority of galaxies. The great astronomer Walter Baade, Hubble's colleague at
Mount Wilson for more than two decades, spoke eloquently of the
genius of its formulator during a lecture given at Harvard in 1958:

> I have used [the classification scheme] for thirty years, and, al-
> though I have searched obstinately for systems that do not fit it, the
> number of such systems that I finally found—systems that really pres-
> ent difficulties to Hubble's classification—is so small that I can count
> it on the fingers of my hand.[85]

CHAPTER NINE

MARINER OF THE

NEBULAE

I

In the spring of 1926, construction of the house the Hubbles had dreamed about while visiting the Palazzo Vecchio on their honeymoon was nearing completion. Edwin had come upon the heavily wooded site during one of his long Sunday walks and had taken Grace to see it just before they were married. At the time, San Marino's Woodstock Road was little more than a circular path winding its way among giant oaks bordered by heavy undergrowth. The spur of land dropped down to a shallow streambed that marked the boundary of the Patton Ranch, birthplace and sometime residence of George Smith Patton, Jr., a brash and fearless battlefield officer, who was wounded and decorated for bravery in 1918 while commanding a tank brigade in France.

From the stream the view lifted above the dark line of trees marking the Huntington Gardens to Mount Wilson, with its shining towers far beyond. Farther still to the east were higher clusters of mountains, snow-covered during much of the winter. To the south, one looked down across the San Gabriel Valley to Mount San Jacinto, and to the southwest to Palos Verdes. It was obvious from the nearby escarpment that the property was on an earthquake fault, which made it that much more interesting to Hubble, who took pleasure in pointing out its characteristics to visitors. He was even more pleased when two geologist friends at Caltech named the rift the Hubble-Huntington Fault. Less pleasant was the $500 bill he received from the man hired to prune the many trees and clear the lot of dead wood. The work coincided with the newspaper accounts of his being awarded the A.A.A.S. prize, and he remarked to Grace that it seemed strange the two amounts were equal down to the last penny.[1]

Still, money was suddenly the least of Hubble's worries. He had recently been promoted from assistant astronomer to astronomer, and saw his annual salary nearly triple to $4,300 after seven years at Mount Wilson.[2] The new house, along with many of its furnishings, was a wedding gift from the Burkes.[3] Beyond this, Grace had considerable assets of her own. Earl Leib had been the son of a wealthy man; in addition to her inheritance, his widow had almost certainly collected substantial life insurance and other benefits from the Southern Pacific Company. An occasional diary entry hints at interest-bearing funds set aside for her by her father. Never again would Hubble have to second-guess himself for choosing astronomy over the practice of law.

Their architect was the nearly forgotten Joseph Kuchera, one generation removed from Europe and known for the style called Spanish Revival. His first sketch evoked what Hubble termed "the sense of recognition" of the great Renaissance palace in Florence. "We knew," Grace wrote, "that what we saw was our house."[4]

Most evocative in this regard was Hubble's study, located at one end of the sunken living room. Its polished oak floor, thickly plastered walls, and barreled ceiling, which would soon become discolored from the rising smoke of his many pipes, gave it the feeling of a monastery. He worked at a simple desk beneath a high stained-glass window set within an arching scallop shell of ribbed plaster, like a luminiscent pearl. Recessed bookcases laden with leather-bound volumes lined the study walls. In the circular niche above one of these was a small marble copy of the seated Galileo at Ferrara, his gaze fixed on the globe in his hands.

The beehive fireplace in the study corner was little used. Preferring a generous fire to a small one, Hubble piled oak logs high on the wide hearth in the beamed living room after returning home from work, then settled into his green leather armchair to enjoy a gin fizz or a scotch and water. On the walls facing him were reproductions of portraits of his other scientific heroes, a head of Eddington by Augustus John, mezzotints of Newton by Kneller and of Halley by Murray. On the table and stand were his pipes, a half dozen or more, with plain briar bowls and straight stems, from Dunhill. The large tin of tobacco contained a special blend carried by the London Pipe Shop in Los Angeles. The second shelf of the stand held a small radio, which might be turned on for the weather report, the news, or, rarely, a political speech.

The dining room, like the living room, was also sunken, its French doors framing Mount Wilson in the background. The oak trestle table, eight matching chairs, and a massive credenza were the work of a fine craftsman named Gates, who fashioned the woodwork, doors, hardware, fireplace utensils, and other pieces of furniture.

In addition to a little-used bedroom next to Hubble's study off the living-room wing, Kuchera added maid's quarters along the tiled hallway leading to the small kitchen. Both Edwin and Grace had grown up in homes with servants, and they were thinking of hiring a full-time domestic in the future. At thirty-seven, Grace was approaching the age when bearing children was considered a significant risk, and it is possible that she was already pregnant with their first and only child. Edwin happened to be on the mountain when Dr. Robert L. I. Smith, her obstetrician, was hurriedly summoned. There was nothing he could do. The premature baby, thought to be a boy, was stillborn. Considering her husband's work too important to be interrupted, Grace forbade the doctor to contact Edwin until his run was over.[5]

A witty and much prized correspondent, Grace did most of her writing at a desk in the master bedroom on the second floor. Seated next to the window, she could look down on the patio or across the wild of the sloping hill to the Patton Ranch in the distance. Only steps away was the chaise longue on which she passed many hours reading her favorite novelists, mostly French and English. Off the other end of the bedroom was a raised bath adjoining the couple's separate dressing rooms.

Among their first guests was William Wright, Grace's former brother-in-law. The Captain, who had driven down from Mount Hamilton to spend a weekend discussing the "grinding of the celestial mills," came away deeply impressed. "Giving your place the once over in retrospect, I can think of no words which so fully express my awe and admiration as . . . its fastidious beauty."[6]

A proud Hubble, to whom the house was perfect, would listen to no theories from anyone suggesting imaginary changes. Even so, Grace feared he might lack the instincts of a true householder. She was pleasantly surprised when he displayed a lively interest in acquiring additional furnishings, including pots and pans, during their many visits to England. "With a princely generosity, his wallet or checkbook was always ready."[7]

As Grace soon discovered, however, there were limits beyond which he would not go. Detesting anything that smacked of manual labor, he spent long hours helping to repair the intricate wiring of the 100-inch while an electrician was called in to fix the lamp cord next to his leather chair. Away from the telescope, he never took a photograph or touched a camera. To Grace "it seemed incongruous to see him handling the equipment of the darkroom."[8] It was she who supervised the planting of flowers, shrubs, and trees and who worried over the two-hundred-year-old Engelmann oak in front of the house, Edwin who drew the attention of guests to the massive scar on its trunk, a legacy of the drought-stricken 1830s, when Spanish soldiers hacked off the lower limbs and fed them to their starving horses.

11

On the mornings Hubble was to head up the mountain on a run, it seemed to Grace that he was already gone. He ate his breakfast absentmindedly, a distant look in his eyes, and showed none of his usual interest in the newspaper or conversation. He packed quickly, stuffing his clothing, some books, a tin of tobacco, and a flashlight into a battered suitcase on which was stenciled "Major Edwin P. Hubble, 343rd Inf." For winter nights he carried a tan camel's-hair coat, a matching scarf, and a black beret whose soft contours did not interfere with observing. When he was ready, Grace drove him to Santa Barbara Street, where the observatory truck with Mike Brown at the wheel was already waiting in the driveway. "Take care of yourself," were his parting words as he got out of the car before hurrying away with a look of "absorbed enthusiasm."[9]

After lunch in the Monastery at 12:15, he strolled over to the 100-inch to arrange the plates for the night's observing, then to the laboratory to work at his desk. Before dinner he grew restless and took time for a walk out to the Deer Park or around the Rim Trail. Next to fly-fishing, hiking the High Sierras with Grace and their friends was his favorite outdoor activity, and he insisted on a schedule worthy of a platoon commander on bivouac. Shouldering a fifty-pound pack containing everything from books, a poncho, and puttees to mosquito netting, whiskey, and a fire permit, he invoked the army regulation of three miles an hour with a ten-minute rest, little appreciating Grace's observation that "the army, presumably, was not marching over the Sierras." About half past five, in the dim cold grayness that was not yet dawn, he delighted in rousing the camp with what Grace termed the "revolting" expression—"What ho!" In the evenings, as the night wind prowled about the campfire, he sat transfixed while the mule packer told Homeric stories of fatal accidents and unexplained deaths in the mountains.[10]

After dinner, when the shadows were beginning to lengthen, the astronomer returned to the great white cylinder and went inside through the shedlike entrance made of the same corrugated tin as the dome's exterior. He first climbed the long iron stairway to the concrete floor on which the telescope rested, then mounted the shorter steps to the platform surrounding the instrument like a balcony. From here he ascended the iron ladder leading to the observing platform.

The dome was open now, the tops of the tall pines visible against the darkening sky. The night assistant was ready at the wooden console near the sidereal clock below. Like a captain on the bridge of a great ship, the master mariner barked out his orders—so many hours or so many degrees. Then came the metallic whining of the traverse, a series of loud clicks, a final heavy clang of the Victorian machinery as the 100-inch was clamped. From his pocket he withdrew a small magnifying glass with which he examined the field at the eyepiece. If satisfied that all was well, he would sit back in the lone bentwood chair and deliberately fill and light his pipe. The remaining lights would be turned out, leaving the interior of the dome in blackness except for the soft glow of the many stars. Leaning over, he would slide the cover from the photographic plate and call out the exposure time to the assistant. "You can go out if you like." Doing the guiding himself, he had a casual and terrifying way of tipping forward on the chair, its back legs rising from the platform as he leaned into the void. When Grace commented on the danger, he attempted to reassure her by claiming that he could easily catch hold of the telescope, and that it wasn't going to slip anyway.[11]

To pass the hours he would sometimes reconstruct a poem, fragment by fragment, until he had it all. Old songs and sentimental ballads also came to mind. He spent the better part of a night recalling every verse of "Believe Me If All Those Endearing Young Charms" and, in a display of emotion that some would have found out of character, grew teary-eyed over such cowboy favorites as "Red River Valley" and "Strawberry Roan."[12]

Slipping, night after night, silent and alone, past the distant shoals of the nebulae would to many constitute the equivalent of a religious experience. Hubble purchased many books on the subject, including works of primitive and ancient religions, church histories, commentaries of the great theologians, the writings of mystics and saints. Yet he never attended church after Oxford and rarely discussed the subject. One morning, while driving north with Grace after the failed eclipse expedition of 1923, he broached Whitehead's idea of a God who might have chosen from a great many possibilities to make a different universe, but He made this one. By contemplating the universe, one might approximate some idea of its Creator. As time passed, however, he seemed even less certain: "We do not know why we are born into the world, but we can try to find out what sort of a world it is—at least in its physical aspects."[13] His life was dedicated to science and the objective world of phenomena. The world of pure values is one which science cannot enter, and science is unconcerned with the transcendent, however compelling a private revelation or individual moment of ecstasy. He pulled no punches when a deeply depressed friend asked him about his belief: "The whole thing is so much bigger than I am, and I can't understand it, so I just trust myself to it; and forget about it."[14]

Midnight lunch was no longer served in the chill bunker beneath the 60-inch. Adams, who was well acquainted with the rigors of the mountain, had authorized the construction of a small heated building located between the domes. Hale's ban on coffee had been mercifully lifted, and the usual fare of hardtack had given way to boiled eggs, toast, and fresh fruit. Eating in silence, the astronomers listened for the footfalls of a regular visitor. A lame doe, accompanied by her fawns in the spring, would climb the steps of the lunchroom to share their meal, while the foxes did their begging at the Monastery door.

Like his colleagues, Hubble found that the most difficult hours were those before dawn, when the numbing cold and drowsiness attacked his concentration. At length, the stars began to pale, signaling the end of the night's run. A button was pushed and the electric motor engaged a chain, causing the dome to close with a long rumbling sound like gathering thunder. The astronomer rose stiffly from his chair and collected his exposed plates in a box before retracing his steps down the three flights of stairs. He confided to Grace that the walk back to the Monastery in the brightening dawn was one of his greatest pleasures, for it conjured up intimations of what a mystic experience might be, a sense of something waiting to be revealed.[15]

III

Upon returning to the Monastery one morning Hubble encountered his avatar. The mountain lion stood quietly facing him, his eyes locked on those of the motionless astronomer. After several moments the tawny beast slowly turned and slipped back into the forest.[16]

A few years earlier, another member of its species had suffered an unenviable fate on the mountain. The cougar's demise was immortalized in doggerel by night assistant Wendell Hoge in the logbook of the 60-inch:

> A mountain lion got Dowd's goat
>   So Milton went a-hunting
> And now he has a lion's skin
>   To wrap up Baby Bunting.
>
> The lion was a monster beast
>   More than a hundred weight
> From tip of nose to end of tail
>   She measured five feet eight.
>
> Astronomers are very brave
>   To work up here all night
> With lions roaming all around
>   Most folks would faint with fright!
>
> They're brave allright but then we're sure
>   They'll all be glad to note
> No more will roam around the dome
>   The beast that got Dowd's goat.[17]

Milton Humason, the notorious "lion killer," was photographed with the slain beast draped over his shoulder. He described to a reporter from the Los Angeles Times how the lion had been feasting on goats belonging to his father-in-law, Merritt C. Dowd, chief electrical engineer on the mountain. Humason tracked the cat to its latest kill and set a large steel trap. Armed with a .22 rifle, he hurried down to the trap at dawn the next morning. No use. Suddenly, raising his head, he found himself staring into the angry eyes of the crouching lion. "Involuntarily, up went his hair and also the rifle. Bang! The .22 bullet landed precisely between the glaring eyes and the lion dropped dead at his feet."[18]

A natural storyteller, the likable Humason, whom everyone called "Milt," may have succumbed to temptation and gilded the lily. In light of his subsequent acclaim, it seems strange that so little is known of the background of the self-made astronomer, most of which is buried in the files of the Carnegie Institution of Washington, D.C.

Milton Lasalle Humason was born in Dodge Center, Minnesota, on August 19, 1891, to William G. Humason, a native of the state, and his wife, Laura, who had come to the upper Midwest from San Francisco. Milton was still a boy when his parents decided to resettle on the West Coast, perhaps at the insistence of his mother. The stocky, round-faced youth was fourteen when he was taken to a summer camp on Mount Wilson in 1905. After this experience, anything was better than returning to school. He talked his parents into letting him drop out for a year, during which he began his working life as a bellboy at the new Mount Wilson Hotel, settling guests, washing dishes, shingling cottages, and, most important, tending to the pack animals in the corral. Having gone only as far as grammar school, he made up his mind to remain on the mountain—and he did.

Sometime between 1908 and 1910, Humason, now in his late teens, became a mule driver for the pack trains plying the trail from Sierra Madre to the summit. Together with his fellow drivers, he brought up much of the lumber and other building materials for the observatory's support facilities, including the cottages and the Monastery. He also came into contact with many of the observatory staff, including the always gentle and smiling George Ellery Hale. But it was for Helen Dowd, the daughter of the mountain's engineer, that the young Humason reserved his deepest attention. Helen watched for the mules and listened to a cursing Milt as the animals wound their way along the summit of Mount Harvard, across a deep canyon from the observatory. The couple was engaged in 1910 and married a year later, shortly after celebrating their twentieth birthdays. Baby Bunting, who was christened William after his grandfather, arrived in October 1913.

A few months later, Humason took the position of head gardener on the estate of Pasadena landowner Samuel Storrow. After three years, he purchased his own "citrus ranch" near Monrovia, but abandoned the business after harvesting only a single crop. The pull of the mountain was impossible to resist. In 1917 he learned from his father-in-law that one of Mount Wilson's three janitors was about to leave. The 100-inch was soon to go into regular operation, and a relief night assistant for both it and the 60-inch would be needed. The salary was only $80 a month, but the benefits included a rent-free cottage, utilities, and most meals. As Dowd put it: "The trail is wide and the walking good." Humason assumed his janitorial duties on November 25, 1917.[19]

By the following year he was doing more than keeping up the buildings and grounds. A student named Hugo Benioff, who attended Pomona College and served as a volunteer summer assistant on the mountain, taught the janitor how to take plates of variable stars with the 10-inch astrographic camera. So adept was Humason that Benioff recommended he be allowed to carry on alone in the fall. When, in 1919, the astronomer W. H. Pickering published his calculations on the location of the mysterious Planet X, Humason began a photographic search of the region. In 1930, the year of Pluto's discovery, the astronomers Seth Nicholson and Nicholas Mayall went back over his early photos and found the planet, which was so blurred and off-center it had been overlooked.

Humason's skill had not gone unnoticed by Harlow Shapley, who was reminded of his Missouri roots when the janitor passed around a bottle of his notorious "Panther Juice" and broke out a deck of cards on cloud-filled nights. Humason began taking plates for him and was soon giving impromptu lectures to visiting dignitaries on the virtues of Shapley's ants. Shapley remembered that Milt believed in water divining with a forked limb and that "we had to shake that out of him," a charge Humason later denied, claiming that it was the astronomer Charles St. John who practiced water witching. "Dr. St. John," Humason chuckled, "was twirling his mustache all the time he was twirling the twig. People get funny at high altitudes." Shapley also claimed that he and Seth Nicholson taught Humason arithmetic and calculus, but there is no evidence to corroborate this statement either, and Shapley later allowed that it was mostly Nicholson who taught Humason what little mathematics he knew. Beyond doubt is Shapley's assertion that Humason "was one of the best observers we ever had." And it was partly at Shapley's urging that a skeptical Hale promoted the night assistant with an eighth-grade education to the staff of the observatory's nebular and stellar photography division in 1920, and thence to assistant astronomer in 1922.[20]

While Humason had not played a major role in Hubble's photographic work on the spiral nebulae, the time was nearing when his boatswain's skills would be enlisted for an even more distant voyage. A stickler for protocol, Hubble was well aware of Humason's humble background, his pride of workmanship, his enduring patience and easy way with others, especially his superiors. If anything, Humason was even more respectful of the expensive instrumentation, and was known for his meticulous cleanliness in the darkroom. From his experiences as a janitor and a night assistant, he knew all the weaknesses of those observers who were slipshod in their use of the telescopes and the developing equipment. He once complained that he could not understand how hyposulfite, the photographic fixing agent, "got on the [darkroom] *ceiling!*"[21] By the late twenties he was able to take a single plate, often representing a week's observing, and transform it into a beautifully crafted image unequaled by anyone else's work on the mountain, including Hubble's.

IV

The rebuff Hubble had suffered at the hands of the International Astronomical Union's Commission on Nebulae and Star Clusters had not been forgotten. Thus it was with a sense of triumph that he and Grace prepared to sail for Europe early in 1928 after learning that he had been selected to chair the commission's July meeting at Leiden. His career had great momentum now, as if propelled by the spiral arm of fate. Newspaper articles referred to him as "Major Hubble, the titan astronomer," and he posed frequently for photographs—studied compositions of tweed, pipe smoke, and distant eyes.[22] He suddenly found himself in demand as a speaker and began honing his skills at the lectern. Grace accompanied him to the University of California at Berkeley, where he was interviewed by the San Francisco press before addressing a record crowd, part of which stood behind ropes for lack of seating.[23]

At thirty-eight, the astronomer had recently been elected the youngest member of the National Academy of Sciences. On reaching England, he was informed by H. H. Turner that the Royal Astronomical Society had paid him the high honor of making him Foreign Associate.[24] In July at Leiden the Oxford don, who had visited the Hubbles during their honeymoon four years earlier, spoke of the American as "Nebulini" and told Grace, "It will be years before Edwin realizes the magnitude of what he has done. Such a thing can come only once to most men and they are fortunate." Grace relished the compliment but couldn't help thinking to herself that Edwin, although he had never said so, knew very well what he had accomplished. "[I]t is others, like Eddington, who will take the years to accept it."[25]

Doors opened for them as if they were royalty. The highlight of their five-month trip came during a leisurely drive across France. Their route led through Mont St. Michel, Bayeux, Caen, and on to Chartres, the couple betting one another on who would spot the spires of the great cathedral first. "Of course, Edwin did," Grace wrote, "he always sees everything first."[26] At Les Glycines they were greeted by an unkempt Abbé Breuil, the French paleontologist who was one of the first to record and interpret Europe's prehistoric cave art. The cleric was accompanied by his devoted assistant, Miss Boyle, a tall, auburn-haired Scot who believed in "the Little People" and in the influence of the waxing and waning moon at planting and harvest.

The next morning the couple followed the Abbé up a hill and through some woods to a hole in the ground, then down a ladder inside a passage so narrow that Grace brushed its sides with both shoulders. Dropping to the muddy floor, they watched the Abbé make tracings of mammoth, bison, reindeer, and masked human figures by the light of a miner's carbide lamp. In another cave they got down on all fours to reach the vivid form of a red rhinoceros, whose pigment would fade badly before their return twenty-five years later. That night the Abbé came hurriedly into dinner and was gently chided by Miss Boyle for not washing his hands. Looking down at the yellow mud stains on his fingers, he quipped, "The French bathe, in winter never, in summer perhaps."[27]

Mixing business with pleasure, Hubble delivered a number of scientific lectures and afterward talked a good deal of shop. As Humason later remembered, he returned home "rather excited" about the fact that two or three astronomers had suggested that the fainter the nebulae, the greater their distance from Earth—and the larger the redshifts. Though undoubtedly an oversimplification, someone or something in Europe had struck a tonic chord, causing Hubble to think back on the summer of 1914, when Vesto Melvin Slipher had brought his fellow astronomers to their feet by claiming that most nebulae are hurtling away from the sun. Humason was invited to Hubble's office, where he "asked me if I would try and check that out."[28]

V

Slipher's gallant fifteen-year assault on the radial velocities of the spiral nebulae had finally come to an end around 1926, not as a matter of choice but out of necessity. The astronomer had taken his plates with a finely crafted short-focus spectrograph attached to the Lowell 24-inch refractor. The equipment had performed beautifully when trained on the larger and brighter nebulae, but by the mid-1920s the old rules no longer applied. Probing farther into space than ever before, Slipher found that his telescope could not measure up to the task of photographing the increasingly smaller and dimmer objects he encountered. The problem of instrumentation was compounded by another, equally intractable obstacle. Vital data were missing from Slipher's results— the distances to the spiral nebulae. Without this information, the conundrum of the seemingly expanding universe could never be resolved.

Hubble's research on Cepheids had provided him with a wealth of distance measurements, but he was far from content. He thought, spoke, dreamed, and wrote of his work to date as mere "reconnaissance." The 100-odd Messier objects were as familiar to him as the alphabet, and he knew his own nebula (the Milky Way), with its complicated structure of bright and dark nebulosities, star clusters, and planetaries, as thoroughly as any pilot feeling his way through a treacherous system of channels, rip currents, and shoals.[29] But the Indies, supposedly somewhere far off in the distance, were yet to be sighted. In charting deep space, the period-luminosity relation for Cepheids and other stellar bodies would have to be exploited, as never before. Thus he continued his work on distances while Humason readied the telescope for the task of photographing the spectra of the nebulae. If, as Hubble suspected, a nebula's velocity of recession was truly an index of its distance, then the distances of nebulae far across the universe could be inferred by simply measuring their redshifts. Within weeks the unlikely pair hoisted sail and set a course for far-flung waters, Humason in the crow's nest and Hubble at the helm.

Poor but memorable was how Humason characterized his first plate. He intentionally picked a nebula whose redshift Slipher had not been able to obtain because of its distance from Earth. Photographing through a yellow prism which blocked the ultraviolet light, the assistant astronomer passed two frigid nights awaiting the result. He then developed the plate, and, with the aid of a magnifier, quickly located the so-called H and K lines produced by calcium atoms in the nebula. Though the spectrum was faint, the telltale vertical marks were shifted to the right or red end, as had been expected.

Humason immediately telephoned an elated Hubble, who was waiting for him at Santa Barbara Street when he finished his run. Hubble confirmed the redshift, calculating it at some 3,000 kilometers per second, about 1,800 kilometers per second greater than Slipher's highest value. When he was queried about his feelings at the moment of triumph years afterward, an educationally impoverished Humason claimed Byron as his model, though he could offer no reason as to why.[30]

Yet Humason's "adventures among the clusters," as Hubble characterized them, were nearly thwarted by his own mutiny. When pressed to do more, he balked. The "tremendously long exposures" hardly seemed worth the pain and suffering exacted by the mountain.[31]

At this point, Humason received a call from Hale, requesting him to stop by the solar observatory he had constructed in Pasadena after giving up the directorship of Mount Wilson. Having been briefed by Hubble—or perhaps by Adams on Hubble's behalf—Hale asked his visitor to continue with the project, promising him all the technical support he required, including a faster spectrograph and an improved camera. Humason, who was already beholden to the great man for promoting him to assistant astronomer despite deep misgivings, was both flattered and touched. After discussing a few more technical details, he agreed to push ahead.[32]

The promised spectrograph and camera were designed by J. A. Anderson of the physical laboratory and constructed in the observatory shop. Humason described the new equipment as "very fast—at least we thought it was fast compared to what I had worked with before." An exposure that had taken two or three nights to obtain was now ready in a few hours.[33]

Assuming the role played by a lonely Slipher over the years, Humason began by gathering spectra on many of the same forty-five nebulae photographed by his predecessor atop a pine-capped peak near Flagstaff—great spirals like M31, M33, M51, M101—the heart of Messier's catalogue, compiled a century and a half earlier. Not surprisingly, the redshifts were confirmed: in all directions, the nebulae appeared to be moving away from Earth, or Earth from them. On the basis of his calibration work with Cepheids, Hubble quickly established the first linear relation between the degree of spectral displacement and the estimated distance to the observed object—the greater the redshift, the more remote the source of light.

Caution prevailed. He wrote Shapley in May 1929 that his recently published paper, "A Relation Between Distance and Radial Velocity Among Extra-Galactic Nebulae," had been held "for over a year." He had wanted to wait even longer pending the accumulation of further data on fainter nebulae, "but we knew from past experience that others would rush into print the moment the new large velocities were known."[34] His distasteful confrontation with Lundmark over the classification of nebulae was still weighing heavily on his mind.

The Swedish astronomer had recently written Adams seeking permission to return to Mount Wilson for the purpose of "determining radial velocities and the spectrographic rotation of spiral nebulae." If possible, he would like to draw on the technical expertise of Milton Humason. "As you might have heard, Dr. Hubble and I have agreed this summer [in Leiden] to get into better mutual understanding."[35] Wasting no time, Adams cabled Lundmark, who had recently become director of the Observatory of Lund: "Cannot give you observing time with telescopes. Nebular program already arranged." In a follow-up letter Adams left no doubt that the Hubble-Lundmark Fault was still active, if only from the American side: "Dr. Hubble and some of the rest of us have been laying out for some time past a program of work on the radial velocities of spirals. . . . For this reason I do not feel that it would be desirable for you to undertake similar work at this observatory." Should Lundmark still wish to make the journey he was welcome to use the photographic archives.[36]

Though only six pages in length, Hubble's first paper on the velocity-distance relation represented a giant step in modern cosmology. Writing of Hubble for the *Dictionary of Scientific Biography*, the noted cosmologist G. J. Whitrow asserted that he had wrought as great a change in humankind's conception of the universe as the Copernican revolution four hundred years before. In place of a static picture of the cosmos, it seemed to many that the universe must be regarded as expanding, the rate of the mutual recession of its parts increasing with their relative distance.[37] Yet nowhere in his modestly drafted paper did the astronomer mention the expansion of the universe or, for that matter, the universe itself. "New data to be expected in the near future," Hubble wrote in conclusion, "may modify the significance of the present investigation or, if confirmatory, will lead to a solution having many times the weight."[38]

Truth to tell, Hubble was already in possession of large amounts of new data and pushing hard for more. Perched, like a monkey, on the small Cassegrain platform five stories above the observatory floor, his face grotesquely illuminated by red dark-vision lamps, a freezing Humason prodded and coaxed the reluctant beast through moonless nights, punctuated by staccato winds and the incessant ticking of the weight-driven clock. Swallowed up by the dark, the astronomer's presence was periodically signaled by chimes, indicating that he had pressed the button changing the drive rate to keep his guide star in place. If the mechanism balked, which happened all too frequently, he held the image in place by forcing his shoulder against the great cannon, and occasionally climbed onto its iron frame, bending his body at painfully awkward angles for the sake of the embryonic plate steeping in light from a nebula time out of mind. "You had to stretch out into nothing," he recalled, but sometimes even this was not enough. A whitish foam accumulated on top of the mercury in the float tanks, multiplying and roiling like a witch's brew until it set up an uncontrollable vibration. When this happened, the telescope had to be shut down while the troubled "waters" were stilled by skimming the fulminate from the vats.[39]

One week after Hubble informed Shapley of the need for more data, an exuberant Adams wrote Carnegie Institution president John C. Merriam, "We are getting some amazing results on the spectra of very distant spiral nebulae." Photographing in the great cluster centered in the constellation Virgo, Humason and Francis Pease, who is perhaps best remembered for his Mount Wilson collaborations with Albert A. Michelson, the Nobel laureate physicist, had obtained displacements corresponding to velocities ranging from 3,500 to 8,000 kilometers per second.[40] Having more than quadrupled Slipher's largest displacement, Humason was adding new velocities at the rate of ten a month, forging beyond Virgo to Pegasus, Pisces, Cancer, Perseus, Coma, and Leo, whose speed of recession was calculated at a staggering 19,700 kilometers per second.[41]

"Nick" Mayall, a graduate student at Berkeley who was working on the mountain, happened to be with Humason when he phoned Hubble from the dome informing him of the nearly 20,000 kilometers per second redshift, which Hubble himself had predicted. "I was so close to the phone I heard him tell Milt, 'Now you are beginning to use the 100-inch the way it should be used.'" Humason uncorked a bottle of his Panther Juice and they toasted the previous night's success. After gaining eminence as an astronomer in his own right, Mayall looked back on this period as "the most stimulating of my life."[42]

Leaving as little as possible to chance, Hubble kept strict tabs on Humason's schedule.

When I got back from the mountain, he would come striding down the hall to ask what luck I had had. . . . He was very fast in mathematical calculations. When I brought him new exposures of spectra he would pick up a pencil and a pad, and jot down figures as fast as the pencil could move, and have the distances in a few minutes.[43]

If Humason was successful, which was most of the time, he knew what was coming next. The procedure must be repeated in order to obtain an independent check on the first result. In using the term "weight" in the conclusion of his initial paper on redshifts, Hubble meant exactly what he said, the mind of the skeptical scientist echoing the teachings of the ancient prophet Isaiah, which he read in front of a crackling fire: "Precept upon precept, precept upon precept; line upon line, line upon line; here a little, and there a little."

The Cepheids, novae, and blue stars that had first guided Hubble deep into the universe were of little use in determining distances now. In the realms he plied with Humason, whole nebulae (galaxies) contained in giant clusters such as Ursa Major and Boötes were invisible to the eye, even through the 100-inch. Believing it reasonable to assume that clusters of nebulae are similar to one another, as individual nebulae are, he compared the brightest stars in the largest nebulae of the Virgo cluster with the brightest stars in the Milky Way. The idea of comparing the apparent brightness of two objects thought to have the same true brightness was then applied to the nebulae themselves. The most prominent members of the Virgo cluster turned out to have about the same true brightness as the Andromeda nebula. Once this was established, Hubble could compare a prominent member of any distant cluster with a prominent member of the Virgo cluster, just as he had compared Cepheids in the past. If the apparent brightness of the distant nebulae was one hundred times fainter than its Virgo counterpart, Hubble calculated it to be ten times farther away, having already demonstrated that apparent brightness decreases with the square of the distance. Soon, as Mayall had witnessed, Hubble was able to estimate the redshift Humason was after before the spectrum of the cluster nebulae could be captured on a plate.

"The Velocity-Distance Relation Among Extra-Galactic Nebulae," co-authored with Milton L. Humason, appeared in the year's first issue of *The Astrophysical Journal*, in 1931. What Hubble had termed the "rather sketchy" data contained in his 1929 paper was fortified by Humason's redshift measurements of another fifty nebulae, thirty-one of which occurred in clusters. The startling figures on the constellation Leo commanded the most attention: one of its clusters was receding from Earth at nearly 20,000 kilometers per second, placing the faint object 105 million light-years away, each light-year representing some 6 trillion miles. Amazing though it seemed, Hubble had every reason to believe that this sudden expansion of the cosmos was more the beginning of the journey than the end. He predicted that redshifts corresponding to distances at least three times as great were within the realm of possibility, and included a table with which to support his hypothesis.[44] This numbing realization added to the already considerable fear that Mount Wilson would run out of telescope before God ran out of universe. In July 1927, an article in the *Los Angeles Illustrated Daily News* had made known Hale's recurring dream of constructing a 300-inch behemoth, "the housing for which would stand taller than the Statue of Liberty." No less an enthusiast, Hubble was also quoted: "The $12,000,000 instrument has been declared possible and the subscription of funds for its erection is anticipated."[45]

Only tentatively developed in the late twenties, the now famous principle destined to bear Hubble's name appeared full-blown in his most recent article. According to *Hubble's law* or the *law of redshifts*, the distances and recessional speeds of nebulae are in direct proportion to each other. Double the distance to a nebula and the speed doubles; triple the distance and the speed triples. A nebula at a distance of 100 million light-years is moving away from Earth twice as fast as a nebula that is only 50 million light-years away. Thus, as Humason was destined to do many hundreds of times before his retirement, an astronomer takes a spectrum of a galaxy whose individual stars and globular clusters are too faint to be observed, determines its speed by means of the redshift, then plots its distance on a graph encompassing Hubble's elegant construct—the speed in thousands of kilometers per second on the vertical axis, the distance in millions of light-years on the horizontal axis.[46]

Yet to perform what would soon become a routine function, Hubble had to know the velocity or what he cautiously termed the "apparent velocity" of the expansion rate. His calibration of the distance to Cepheid variables and certain of the brightest gaseous nebulae led to the formulation of the K term. In 1929, he had calculated that for every million parsecs (1 parsec = 3.258 light-years), a nebula is receding at an additional 500 kilometers per second.[47] By 1931, with the accumulation of more data, he revised the K term upward, calculating it at 558 kilometers per million parsecs.[48] In time, astronomers would replace the K with the letter H, naming the value for the expansion rate *Hubble's constant*. Hence the formula used to calculate the velocity-distance relation, *Hubble's law*, is simply expressed: $V = Hd$.

As he had done in each of his previous papers, Hubble handled the theoretical implications of his findings with what one chronicler has called "long tongs."[49] "The present contribution concerns a correlation of empirical data of observation," the reserved astronomer wrote at the end of some forty revolutionary pages. "The writers are constrained to describe the 'apparent velocity-displacements' without venturing on the interpretation and its cosmologic significance." He had taught the man assisting him equally well. Long after Hubble was gone and their joint labors had shaken modern cosmology to its foundations, Humason told an interviewer:

> I have always been rather happy that my end of—my part in the work—was, you might say, fundamental, it can never be changed—no matter what the decision is as to what it means. Those lines are always where I measured them and the velocities, if you want to call them that or red shifts or whatever they are going to be called eventually, will always remain the same.[50]

VI

Beginning in the 1930s, the Hubbles could often be found in the main dining room of the Caltech faculty club, the Athenaeum, on Thursday evenings. Seated at a long table beneath a ceiling decorated in a style called *grottesca*, its fanciful paintings of human and animal forms copied from Nero's palace by Italian artists brought to Pasadena, they presided over impromptu dinner parties whose guests included English novelists and playwrights and what Grace described as "a few clever, pretty actresses." The building, an outstanding example of Italian Renaissance architecture and furnished with matching antiques, had first opened its doors in 1930, much to Edwin's delight. Like the proprietor of an Old World hotel, he took its inexperienced manager, Miss Bess Little, in tow, providing guidance on everything from the noted chef, who stayed twelve years, to menus, breakages, furnishings, and the budget. Miss Little later credited Edwin and Grace for helping her keep the establishment out of the red during its early years, their regular dinner parties adding to the profits.[51]

The creation of Caltech trustee George Ellery Hale, who raised the $500,000 for its construction single-handedly on the eve of the Great Depression, the Athenaeum was patterned after the famous Athenaeum Club founded in London in 1824, which describes itself as "an association of individuals known for their scientific or literary attainments, artists of eminence in any class of fine Arts, and noblemen and gentlemen distinguished as liberal patrons of Science, Literature, or the Arts."[52] Thus Hale envisioned a central club serving as both a social and an intellectual rendezvous where those most deeply involved in science, art, literature, history, and government would meet to discuss recent developments in their fields. Besides Hale, faculty member Arthur A. Noyes, the noted chemist and onetime president of MIT, was a familiar presence, as was Robert A. Millikan, for whom Hubble had served as a laboratory assistant during his undergraduate days. The distinguished, granite-jawed physicist had been lured from Chicago by his old friends Hale and Noyes in 1921. But before assuming his duties as president, Millikan had insisted that the lackluster name of the little-known Throop College of Technology be changed to California Institute of Technology. Two years later, at age fifty-five, and still a decade from reaching his full stride, Millikan established the tone of the institution by being awarded the Nobel Prize for his measurement of the charge on the electron and for his work on the photoelectric effect. In years to come a painting of the three founders of Caltech—Hale, Millikan, and Noyes—would grace the paneled south wall of the Athenaeum's main dining room, its subjects irreverently, but affectionately, dubbed "Thinker, Tinker, and Stinker."

Other academic lights were on the near horizon, and Hubble, though aloof, knew most of them. Like his colleagues at Mount Wilson, he found it difficult to keep up with the current advances is such fields as relativity and quantum theory. Looking for all the world like an Oxford don, the pipe-smoking astronomer frequently joined his friend Richard Tolman, a Caltech physicist and outdoorsman, for drinks at the Athenaeum. Tolman's deep interest in relativity theory was further stimulated by Hubble's recent discovery that redshifts are proportional to distance, and soon led to a series of studies on the applications of the general theory to the overall structure and evolution of the universe. Jeans, Eddington, and the Nobel laureate Albert Michelson, who repeated his epochal experiment for determining the speed of light through the auspices of Mount Wilson, all stayed at the Athenaeum and lectured both at Caltech and at Santa Barbara Street. Many others spoke as well, including the astronomers Kapteyn, Russell, Turner, Lundmark, Abetti, Stebbins, Plaskett, and Jan Oort. But it was the theoretical physicists that Nick Mayall remembered the most. "Hubble was really a tremendous attraction," he reflected. "[They] wanted his observations."[53]

Every Monday morning at eleven o'clock the Mount Wilson staff assembled in the "old government building" across the street from the main office. This gathering, known as the Journal Club, had been initiated by Hale as a means of keeping everyone abreast of the latest developments on the mountain. Hubble, like his colleagues, spoke once or twice a year, usually to an overflow audience which included distinguished visitors and fellow scientists from Caltech, several of whom were also invited to lecture. Extremely sensitive about his lack of a formal education, Milt Humason sought the advice of Harold D. Babcock, Mount Wilson's great solar observer, as the day of his first presentation neared. Babcock told the jittery astronomer, "Just be yourself and things will go all right," and they did.[54] More suspect than his verbal skills were Humason's qualifications as an author of advanced scientific papers. Playing sergeant to Hubble's major, there is no question as to who was doing the writing when it came to their co-authored works. When Humason published alone, it was almost certainly the editorial hand of Frederick Seares that saw him through.

Not long afterward the grammar school dropout joined a select company of astronomers, mathematicians, and physicists invited to the Hubble home on Woodstock Road. They met every month or so during the early and middle thirties, alternating between evening sessions and late breakfasts on Sunday mornings. A blackboard was put up on the living-room wall. On the dining-room table were either sandwiches, beer, and whiskey or the makings of an English breakfast, consisting of scrambled eggs, bacon and sausages, toast, marmalade, fruit, and coffee. The bowls and coffeepot were kept full by Alexander Ota, the Hubbles' Japanese-American housecleaner, formerly a valet to one of the Vanderbilts. Diminutive but erect, he was impeccably dressed in morning clothes, which he carried to work in a monogrammed briefcase after a large dog with amorous propensities pinned him to the ground, soiling his pants and shirt. After eating, the participants settled down before the fire and lit up their pipes and cigars. A discreet Grace retreated to her desk upstairs and knew little of what went on, except that equations were often left on the blackboard after Edwin's guests departed.

Those who attended were almost equally divided between Caltech —Howard P. Robertson, Richard Tolman, Fritz Zwicky—and Mount Wilson—Humason, Walter Baade, and, somewhat later, Rudolph Minkowski. Nick Mayall, a "mere" graduate student in whose eyes Hubble was an icon, also sat in."[55] Conspicuous by his absence was Adriaan van Maanen, with whom Hubble was still engaged in an intellectual blood feud over their conflicting data on the proper motions of the nebulae. By the time Mayall was preparing to leave Pasadena to continue his studies at Berkeley in the autumn of 1930, the two astronomers were passing each other in the office corridors like ships in the night. To Mayall had fallen the unenviable chore of making out the observing schedules of the telescopes. In doing so he had to shuttle between the combatants on cat's feet, expecting all the while to step on a career-ending mine.

Every time Hubble saw van Maanen's face he was reminded of Shapley, who had taken exception to his distance scale for redshifts, just as the Harvard astronomer had opposed virtually everything else he had accomplished. During Shapley's most recent visit to Pasadena, Mayall was incredulous when the man who had revolutionized the structure of the Milky Way by expanding its celestial boundaries bluntly remarked of the redshifts, "I don't believe these results."[56] In celebration of his friend's coming, van Maanen had planned one of his intimate dinner parties. Shapley was asked to invite five guests of his own choosing "with one exception only!"[57] Van Maanen supplied no name, but Shapley knew perfectly well who the exception was. Ironically, amid this friction of egos, all Grace noted was her husband's feeling of camaraderie: "Differences that might arise in the office, in the stress of the day's work, disappeared in an atmosphere of good fellowship and leisure. Something emerged of the teamwork and cooperation that [Edwin] always wanted."[58]

One clue involving the earliest of these informal seminars surfaced in the last paragraph of Hubble's first paper on redshifts, in which he refers to the Dutch astronomer Willem de Sitter, the theorist who, more than any other, left a permanent imprint on his observational approach to cosmology. His interest in de Sitter's work extended well back into the 1920s, and his visit to Leiden in the summer of 1928 had provided him the opportunity to beard the lion in his den. Lank, bald, with deep-set eyes and a meticulously groomed goatee, de Sitter was both professor of astronomy at the University of Leiden and director of the Leiden Observatory. According to Grace's notes recorded during the trip, it was de Sitter who had encouraged Edwin to extend Slipher's scale of velocity measurements.[59]

Yet the story begins much earlier, dating back to the mid-teens and the Kaiser Wilhelm Institute in Berlin, where the recently acclaimed theoretical physicist Albert Einstein was plying the cosmos via calculations jotted on the back of an old envelope. He was surprised to discover that according to his brainchild—the theory of general relativity—the universe cannot be static, but must be either expanding or contracting. When he turned to astronomers for confirmation he was even more surprised to learn that virtually all of them embraced the notion of no change on the grand scale. Forced to choose, Einstein, to his later regret, went against his intuition and modified his equations with the introduction of a "fudge factor" termed the "cosmological constant," thus postulating a uniform distribution of matter in static equilibrium, beauty and simplicity having been sacrificed in the name of empiricism.

In response to Einstein's "aesthetic blunder" the Russian mathematician and meteorologist Alexander Friedmann published two dissenting papers, the first in 1922, the second in 1924. Rejecting the cosmological constant, which Einstein had designated by the Greek letter lambda, Friedmann presented nonstatic solutions to the field equations of general relativity. What he failed to do, however, was to connect his equations with astronomical observations, thus opening the way for another.

In contrast to the wan emaciated Russian whose obscure work he did not know, the Belgian abbé Georges Lemaître seemed the very embodiment of bourgeois satiety. Stout and pink in his tight black cleric's suit, waistcoat of heavy wool, and high stiff collar, the young priest had distinguished himself on the battlefields of World War I before taking holy orders. More recently, he had been granted a professorship at the Catholic University of Louvain, from which he had received his doctorate in mathematics and physics in 1920.

Going over much of the same laborious ground that Friedmann had covered five years earlier, Lemaître, too, developed an expanding model of the universe. But the abbé did not believe that cosmology could be made a deductive science and favored a simple and direct approach to the study of the nebulae. Between 1925 and 1927, he came to believe that the light from distant objects would be redshifted due to the expansion of space. His model further predicted that the degree of the redshift would be directly proportional to the distance, the proof of the pudding dependent upon an observational test for expansion.[60]

Before Lemaître settled permanently in Louvain he had spent a fellowship year at Cambridge, where he came into contact with Arthur Eddington. Aware of the English physicist's abiding interest in relativity theory and the forces at work in the universe, he sent Eddington a reprint of his article, which, owing to its publication in an obscure journal, had caused not the slightest ripple. Unfortunately for the abbé, his bad luck held, for Eddington later admitted that he had either forgotten the article or misplaced it.

Meanwhile, Hubble, who knew nothing of the work of Friedmann and Lemaître, had read de Sitter, whose sometimes comic absent-mindedness masked a supple and inventive mind. Stimulated by one of Einstein's papers on general relativity penned in 1916, de Sitter entered into a fruitful correspondence with its author and soon produced three lengthy papers of his own on the subject. In the third article, published in 1917, he simplified his calculations by assuming that the universe is devoid of matter, a mathematical fiction he defended on the basis that the real universe is composed of mostly space anyway. He then conjectured that if two stationary objects were introduced into this void, light passing between them with respect to one another would be redshifted. Curiously, the redshift is not due to either the expansion of intergalactic space or the Doppler effect. Instead, it is the effect of the mysterious slowing down of time at great distances. As the frequency of light decreases through curved space its wavelength is elongated and a redshift results, causing a faraway clock to appear to run more slowly, the "de Sitter effect." Moreover, the amount of redshift in de Sitter's model was directly proportional to the distance between the emitting and receiving objects, a relationship that had only to be tested by someone with a telescope powerful enough to plumb deep space.

An incredulous Einstein demurred, arguing that only in a universe without stars could de Sitter's solution possibly apply. De Sitter responded by terming his hypothesis "Solution B," to distinguish it from Einstein's, which, as befit de Sitter's gentlemanly manner, he labeled "Solution A." The main difference between the two is that in Solution A the universe can contain matter and remain static; in Solution B it cannot.

For several years most observational astronomers, Hubble included, took little notice of either Einstein or de Sitter, whose mathematical abstruseness confounded almost everyone. But by the mid-twenties things had begun to change, thanks largely to the writings of Eddington, whose widely acclaimed *The Mathematical Theory of Relativity* appeared in 1923. The problem, as the revered physicist saw it, was that neither Solution A nor Solution B was intrinsically appealing, and he was overheard at scientific meetings complaining about the fact that no one had come up with a more convincing model. All but holding his nose, he gave his nod to Einstein's solution, which offered no explanation of the redshifts, a phenomenon that little interested Eddington at the time. Meanwhile, Lemaître's breakthrough paper lay unread.

Hubble was of a different mind. Humason's spectral lines were there in white on black, and not to be denied. "The outstanding feature," he wrote at the end of his 1929 article on redshifts, "is the possibility that the velocity-distance relation may represent the de Sitter effect, and hence that numerical data may be introduced into discussions of the general curvature of space."[61] For now, he was putting his money on Solution B, and would remain chary of all theories of cosmic expansion long after most astronomers and physicists had been won over. When queried about the matter as late as 1937, he sounded like an incredulous schoolboy: "Well, perhaps the nebulae are all receding in this peculiar manner. But the notion is rather startling."[62]

During this period of scientific crisis in the late twenties, the future astronomer and cosmologist G. C. McVitte served as Eddington's student research assistant. Twice a term the young Scot bicycled out to Cambridge Observatory on Madingley Road, where the maid showed him into Eddington's study, which mysteriously smelled of unseen apples. The great man would look up from his desk, and McVitte had the feeling that he was thinking, "Now, who is this young man and why does he come to see me?"[63]

By coincidence, Eddington put McVitte on a problem in astronomy that had already been solved by Abbé Lemaître. About this time Eddington attended a meeting in London that also drew the abbé, who was puzzled when the Cambridge professor mentioned the very problem that McVitte was struggling with. The cleric said nothing but, after returning to Louvain, he wrote to Eddington and enclosed another reprint of his unsung article, which had languished for nearly three years.

It was for Eddington, and many to follow, a parting of the mists. Solutions A and B did not hold a candle to Lemaître's dynamic theory of the expanding universe. De Sitter himself celebrated it as a "brilliant" piece of work, while Einstein, who had shown little interest in matters of cosmology after 1917, took out pencil and paper and began scratching away.

While Humason squinted at pinpoints of light against "black like black velvet," Hubble read about Lamaître's solution in a semipopular article authored by Eddington in 1930. Hanging back, he watched the chalk dust fly as his guests from Caltech took turns at the portable blackboard in his living room. As early as 1928 the cosmologist Howard Robertson had shown that, by a simple mathematical artifice, the de Sitter universe could be transformed into an expanding universe, albeit one devoid of matter. Tolman, with whom Hubble would collaborate in the future, had also rejected Einstein's solution before Lemaître's paper surfaced, and had expressed his reservations to de Sitter concerning Solution B when the Dutch astronomer addressed the Royal Astronomical Society in January 1930. After learning of Lemaître's work, Tolman spent much of the early thirties seeking to wed the concept of the expanding universe to Einstein's general theory of relativity, with Hubble's redshifts as the cornerstone of this union.

The irascible, some said crazed, Fritz Zwicky was present to cast flies in the ointment, which, to many, seemed the young astrophysicist's highest calling. Short and gnomish, with eyebrows that rose to a devilish peak, he hoarded slights, whether real or imagined, like so much alchemist's gold, never forgetting and never forgiving. Bulgarian-born, Zwicky believed that only he, not Einstein, with whom Zwicky had studied in Zurich, truly understood relativity theory, and Solution A was not for him. Neither was Solution B, nor, for that matter, Lemaître's explanation of the velocity shifts. Instead, Zwicky, his arms flailing as if at imaginary windmills, argued that photons of light on their journey to Earth from distant nebulae are subject to a gravitational drag. The consequent loss of energy would result in a redshift, since it would have a greater effect on light from more distant objects. With a combination of extreme self-confidence and a degree of intellectual legerdemain, he fashioned a "law" not unlike Hubble's. But to Zwicky's surprise no one embraced it but Zwicky himself.

Hubble drew on his pipe, listened, mused, but once the chalk dust settled he hadn't budged. Just as he never used the word "galaxy," the term "radial velocity" was soon purged from his lexicon. "It was the 'redshift,'" Mayall reminisced of his days with the mariner on the mountain. "That's what you measure."[64]

CHAPTER TEN

"YOUR HUSBAND'S WORK IS BEAUTIFUL"

I

For one of the few times in its history, Pasadena's Tournament of Roses was about to be upstaged. The New Year's Day extravaganza had suddenly taken a backseat when it was announced that the steamship *Belgenland*, after crossing the Atlantic and passing though the Panama Canal, was nudging up the California coast toward its appointed docking in San Diego on December 31, 1930. Its star passenger, together with his wife, would then be driven north to begin their two-month visit in Pasadena. The question most on the public's mind was this: Would the couple choose to reside in a guest apartment on the second floor of Caltech's Athenaeum or opt instead for a more commodious white bungalow? "No one yet knows," wrote Ransome Sutton of the *Los Angeles Times*, "for who can tell what a woman may do?" However, the reporter was betting on the bungalow, with its well-tuned piano. "For, besides being a mathematician, the professor is a 'tone-monologist' and without a piano he would get homesick for that little upstairs Berlin home about which we have heard so much."[1] A few days later a plump, wide-faced, and watchful Elsa Einstein decided upon what her husband called the "shingled gingerbread house" in South Pasadena.

The invitation to visit the city had been extended to Einstein in person in Berlin by Arthur Fleming, chairman of the board of trustees of Caltech. This was at the suggestion of Richard Tolman, whose work on relativity theory had reached a critical stage. Einstein knew of the Caltech professor's papers in mathematical physics and of the observational work being conducted at Mount Wilson concerning the nature and size of the universe. Anxious to discuss these matters firsthand with the men concerned, he had quickly agreed to visit Caltech as a research associate early in 1931.

The "lion hunt," as one perceptive observer termed it, began the moment news of the sojourn was announced. Fifty cables a day arrived from across the Atlantic, and mail postmarked in the United States soon outnumbered letters from Germany. Elsa, who assumed responsibility for handling the deluge of invitations, repeatedly informed the press and others that the professor would be traveling on holiday. Once the *Belgenland* docked in New York after the first leg of its voyage, her husband would remain on board.

The steamer sailed from Antwerp on December 2, 1930, the Einsteins occupying three flower-filled staterooms—"excessive and pretentious," he called them—their privacy guarded by a crew member stationed at the door. The closer they got to New York, the clearer it became that Einstein would have to disembark and make himself available to the press and others. The only alternative, it was argued, would be to lock himself in the purser's safe, and even then someone would want to take photographs of the safe.

According to biographer Ronald W. Clark, fifty reporters and fifty photographers swooped down on their bewildered victim, whose shaky command of English required the services of a translator, the German consul, Paul Schwarz. Elsa took the part of stage manager, valiantly steering her husband away from trick questions, explaining, mothering, elaborating when she thought it necessary. "Within one brief quarter of an hour," *The New York Times* reported the next morning, Einstein was asked "to define the fourth dimension in one word, state his theory of relativity in one sentence, give his views on prohibition, comment on politics and religion, and discuss the virtues of his violin."[2] At fifty-one years of age, with brooding brown eyes and tousled white hair brushed back from an imposing forehead, the Nobel laureate was perfectly capable of playing the fox at the wood's edge. "The reporters asked particularly inane questions, to which I replied with cheap jokes that were received with enthusiasm," he wrote of the occasion in his diary. Still, he was far from the infallible prophet he seemed to many. When queried about his thoughts on the growing prominence of political bully Adolf Hitler, he replied: "Hitler is living on the empty stomach of Germany. As soon as economic conditions in Germany improve he will cease to be important."[3]

Preceded by waves of publicity, Einstein reached a normally reserved Pasadena beside itself with anticipation. He was immediately besieged from all quarters by autograph seekers, reporters, publicity agents, photographers, would-be hosts and hostesses, swooning matrons, and crackpots of every description. His appearance on the street was the signal for the gathering of crowds of small boys, while invitations of all sorts were forced upon him. Caterers inquired as to his special tastes while shop owners ordered exotic merchandise worthy of the "planet's smartest man." Looking for all the world like one present at the creation, he weathered this unremitting assault with a quiet dignity and a quizzical smile, which endeared him to everyone. In the long history of lion hunting, seldom has the quarry been more equable.

Scheduled to divide his time between Caltech and Mount Wilson, Einstein was to take his first trip up the mountain in mid-February. The more he decried luxury and excess, the more it was thrust upon him, the observatory proving itself no exception. The usual transport over the zigzagging dirt toll road was by truck, a rude vehicle deemed unworthy of the gentle genius. After clearing the matter with Carnegie headquarters in Washington, Walter Adams, who was notorious for his penny-pinching ways, went out and purchased what was described as a "big Pierce Arrow touring car," the backseat of which enveloped the diminutive guest of honor in a leather cloud.[4] A much larger and more prominent Hubble, who donned plus fours for the occasion, sat on Einstein's right, Adams on his left, while fellow astronomer Charles St. John, the very picture of an English headmaster, rode with the driver in the front.

Upon its arrival, the party went to the Monastery to make plans for the day. This session was followed by a visit to the 150-foot tower telescope, used exclusively for the study of the sun. Much of the recent work confirming relativity theory had been undertaken with the instrument, and Einstein scrutinized the 17-inch image composed of dark spots set against a writhing cauldron of Dantean chaos. He then stepped calmly into an open steel box, somewhat resembling a miniature elevator, and was carried to the top by cables located on the side of the tower. Here the onetime patent office employee examined the mechanism of the telescope in operation, then paused to be photographed for the thousandth time while admiring the view of nearly the whole of Southern California, a strong breeze giving special prominence to the internationally famous Einstein hair.

After a luncheon in which Jupiter, the observatory cat, figured prominently, the party, which also included astronomers William Wallace Campell and Alfred H. Joy, the astrophysicist Arthur S. King, observatory photographer Ferdinand Ellerman, and Dr. Walther Meyer, Einstein's scientific assistant and interpreter, strolled over to the 60-inch and 100-inch telescope domes. Unlike his hosts, Einstein had never before been in the presence of instruments such as these, for the simple reason that they were not to be found in Europe or, for that matter, anywhere else in world. When Elsa Einstein, who seemed always to be on the defensive, was told that the giant Hooker telescope was essential for determining the universe's structure, she is said to have replied, "Well, well, my husband does that on the back of an old envelope."[5]

As he had that morning, Einstein displayed an extraordinary interest in the mechanical details and operation of the instruments. Much to everyone's alarm, he insisted on climbing over the steel framework of the 100-inch while rattling off an extensive knowledge of its every appliance, something Adams thought "remarkable even in a trained physicist."[6] He was photographed with Hubble and others on the great dome's catwalk, then escorted down to face the press, which could be held at bay no longer. More photographs were taken, many with Hubble at Einstein's shoulder, causing one indignant onlooker, who had witnessed similar performances in the past, to remark, "He sort of wormed his way or muscled his way in. That's where he wanted to be photographed, with the great man."[7]

Following an early dinner the members of the party returned to the 100-inch dome, their spirits only slightly dampened by the intermittent clouds pushing in from the west. Despite occasional interruptions, the evening's observing was deemed a success. Einstein was able to see Jupiter, Mars, Eros, various nebulae, and the faint companion of the bright star Sirius in the constellation Canis Major. This last object was of special interest to him because photos of its spectrum had confirmed one of his predictions based on relativity theory, leading to remarkable inferences regarding the density of matter among the stars. The enthralled night visitor remained in the dome until after one o'clock, finally retiring under protest to the Kapteyn Cottage, where he added his name to the illustrious register of its temporary occupants. As a weary Adams was departing for the Monastery, he called out, "Wake me in time to see the sun rise."[8]

On another dark winter evening Einstein was driven to a level field about forty miles south of Pasadena, where he was greeted by a bent and grizzled Albert A. Michelson. The Nobel laureates stood in the glare of a powerful arc lamp, the silence pierced by the shriek of a high-speed mirror, its thirty-two angled facets—polished to an accuracy of about one part in a million—spinning at hundreds of revolutions per second. Stretching out into the darkness was a steel pipe three feet in diameter and a mile in length. The air had been pumped out of this elongated chamber and its joints sealed in preparation for the demonstration to come. Michelson gave the signal to an assistant, who opened the throttle on a chugging compressor until the air striking the small vanes on the mirror produced a speed of precisely 730 revolutions per second. Light from the arc lamp reflecting off the mirror shot into the virtually airless pipe to be reflected back on the face of the mirror next to the one from which it was reflected originally. A proud Michelson, who had first measured the speed of light decades earlier, was able to claim that this was his most exacting measurement of all, accurate to one part in 200,000, or roughly one mile a second. Crucial to Einstein's work, the velocity of light is one of the most fundamental constants of nature and according to the theory of relativity is the limiting velocity in the physical world.

Adams, who looked on from a distance, was deeply moved by the silhouettes of the master experimenter and the master theoretician. Both had come a long way to meet in this mysterious and lonely place, and neither seemed anxious to leave despite the evening chill. Nor did the Mount Wilson director make an attempt to shorten what proved to be their final scientific encounter. Einstein, too, had his heroes and Michelson was one of them. When word of the American's passing reached him in Berlin two months later, he was stricken by the loss.[9]

Although the doors of the Athenaeum had been open for several months, the first formal dinner was deferred until Einstein's arrival. Fewer than three hundred persons received invitations to the black-tie affair, but millions of others in "radioland," including an ailing George Ellery Hale, tuned in for the national broadcast of the festivities. In his translated speech delivered by Caltech professor William B. Munro, Einstein singled out Hubble's work on redshifts and Tolman's use of the distance measurements "to formulate a dynamic conception of the spatial structure of the universe." The work of other physicists and mathematicians present was nothing short of "epoch-making," and "I account myself exceedingly fortunate to be able to break bread with you here in joyous mood."[10]

Grace Hubble later regretted the fact that she kept no journals during the period of Einstein's visit and therefore remembered only "little things." When visiting Santa Barbara Street he was given an office across the hall from Hubble's in an atmosphere bordering on strict quarantine. Only by locking the building and issuing the staff keys could the press and autograph hounds be kept at bay. As it was, Einstein worked at a desk suffering the presence of Gertrude Boyd Kanno, a local sculptress commissioned to do his bust.

The first time he and Frau Einstein came to Woodstock Road for dinner the Hubbles asked the couple what guests they would like to have. To adorn the table, Grace invited the screen actress Doris Kenyon, "blonde and lovely as a Fragonard." The physicist was in a gay mood, and at once took the lead in conversation. He quoted La Rochefoucauld and asked, "Have you *Les Maximes?*" Edwin produced the book, and he took it into the dining room and read to the guests, tossing back his mane as he laughed. Then he spoke of books. His favorite work, indeed the greatest of all novels, was *The Brothers Karamazov.* Without soliciting the opinions of others, he changed the subject to astronomy and cosmology, a discussion limited to the men.

Doris Kenyon turned to Frau Einstein and bestowed a compliment that would pay great dividends. "You are a singer; it shows in the quality of your voice."

"You are the first in this country to discover it," exclaimed the delighted Frau Einstein.

After they had all gone, Edwin said, "Did you notice how cleverly Doris went for Frau Einstein when she saw it was no use with him?" In the coming days the Einsteins would visit Doris when they refused others. Einstein presented her with manuscripts, signed autographs, and even wrote a short poem in a book he gave her.[11]

It was Frau Einstein who decided where they would go and whom they would see. She accepted almost all invitations as they came in and then, as the stakes increased, discarded most, as if she were playing a hand of gin rummy. Many a dismayed hostess was forced to tell her already assembled guests that the Einsteins were unable to come because Frau Einstein had a sudden headache. Meanwhile, the couple had slipped off to Charlie Chaplin's estate in Hollywood, where Einstein played his violin.

The mania to possess anything from the pen of the savant took some ironic twists. "I imagine you will be rather amused to have me appear in the role of an autograph-collector," the letter from John Barrymore to Hubble began. The actor wanted Professor Einstein's signature for a book he and Mrs. Barrymore were preparing for the baby. "Doris Kenyon intimated that you would be willing to do this for me."[12]

A procrastinator when it came to things distasteful or embarrassing, Hubble let the matter slide. The day before Einstein's departure the Hubbles were entertaining guests when the doorbell rang. Grace was surprised when confronted by Barrymore's private secretary and booking agent, both of whom had received wireless telegraphs from the Barrymore yacht at sea. That evening Hubble took Grace with him to the gingerbread cottage, where a bit meekly he explained his errand. An amused Einstein laughed and signed his name.

The Hubbles were also privy to Einstein's legendary ineptitude and preoccupation. The evening after they had moved into the cottage Einstein phoned the real estate agent in charge, asking him if he would please come at once. Deeply apprehensive, the man raced across town. Einstein himself answered the door. "I do not know how to use the can opener," he confessed sheepishly.

On the day of what was termed the "Long Beach earthquake," Hubble had just concluded a lecture in the large hall of Norman Bridge at Caltech. Einstein and scientist Beno Gutenberg were walking slowly away from the building, deep in conversation, when the trees began swaying. Having remarked, a few days before, that he would like to experience one of California's notorious tremors, Einstein regretted the fact that he had neither seen nor felt anything at all.[13]

11

Word of Einstein's change of positions echoed across the wires like the crack of a whip, catapulting Hubble into the eye of international fame. From the beginning, the German physicist had felt that his introduction of the cosmological constant had compromised the simplicity and elegance fundamental to all physical equations. Moreover, his basic insight in creating relativity had been to treat universal motion as a mathematical certainty. He had revolutionized dynamics by discarding Newton's ideas of "absolute space" and "absolute time" and replacing them with the concept of relative motion between two systems or frames of reference. Then, like an apostate, he had fudged in order to make relativity deliver the static universe that the astronomers had assured him fit reality. It was, he ruefully admitted, the worst blunder of his career.

As described by Associated Press reporter Walter B. Clausen on February 4, 1931, a "gasp of astonishment swept through the library of the Mount Wilson Observatory here today" when the Berlin professor Albert Einstein, "with a few simple words made this revelation." His original concept of the universe, as well as that of Professor Willem de Sitter, Dutch astronomer and friend, is no longer valid. The new concept of the universe—the expanding universe—hinges on the work of two great California scientists, the astronomer Dr. Edwin P. Hubble of Mount Wilson Observatory and physicist Dr. Richard Tolman of Caltech. New observations by Hubble and his associate Milton Humason concerning the redshift of light in distant nebulae make the "presumption near" that the general structure of the universe is not static. Theoretical investigations undertaken by Tolman confirm Lemaître's findings which fit "well into the general theory of relativity."[14]

Then, with a quizzical smile on his face, Einstein looked down at his watch. Seeing that he had exceeded his time, he grabbed his translator and rushed off to another engagement, leaving a startled Adams to face the bewildered press.

Clausen's piece, or one like it, ran in all of the country's major papers the next day. Back in Springfield, Missouri, where Hubble's spinster Aunt Janie and Uncle Edwin still lived, the headline in *The Springfield Daily News* read: "Youth Who Left Ozark Mountains to Study Stars Causes Einstein to Change His Mind." In a related article partly inspired by the deepening Depression but written tongue in cheek, the "red shift" was compared to the "red menace" in Russia and China and found even more threatening: "The universe, to use a nonscientific expression, [is going] hell-bent for chaos, ignoring the law of gravitation, flying ever outward, faster and faster. It looks as if the whole is breaking up and rushing into a limitless outer void. No good can come of this."[15]

More witty still was the poem that appeared in the May 27 issue of *Punch*. After establishing the proposition that Professor Einstein thinks Hubble "the brainiest of blokes," the author, who signed himself C.L.G., concluded:

> *Unmoved by spatial swerving,*
> *Or arbitrary views*
> *Which others find unnerving,*
> *He turns to spectral clues,*

And from his magic casement,
"Constant for red displacement,"
Predicts the near effacement
   Of Bolshevistic hues.

When Jeans grows too didactic
   Or Friedmann makes too free
Among extra-galactic
   Clusters of nebulae—
When life is full of trouble
And mostly froth and bubble,
I turn to Dr. Hubble,
   He is the man for me.[16]

Besieged from every side by reporters, Hubble provided them with all the details they needed to fill column after column based on his scientific discoveries. Rarely was Einstein's picture run without his own appearing beside it, while Tolman, the abstract theoretician, was more or less lost in the shuffle. Light-years measuring in the millions and nebulae racing away from Earth at tens of thousands of miles per second are what kindled the public's imagination. And if Einstein now believed that the universe was expanding, well, that was good enough for most.

Within months of their departure the Einsteins were contemplating a second visit to Pasadena. This time Robert Millikan himself traveled to Berlin to undertake the negotiations. His offer of $7,000 in addition to travel expenses and a handsomely furnished apartment in the Athenaeum was not accepted until November 1931. The couple arrived weeks later and Einstein lectured on space curvature to overflow audiences. Together with a visiting de Sitter, who had likewise rejected Solution A and Solution B, he issued a statement that further strengthened the case for an expanding universe.

Millikan asked Grace Hubble to serve as a kind of unofficial hostess, and she drove Einstein to seminars and conferences whenever the need arose. Given the opportunity to study her enigmatic passenger up close, she confided to a friend, "Einstein is like one of those trolls you find in a German beer garden; there is something inhuman about him."[17] He was silent sometimes, and sometimes he would talk in French or English, for Grace knew no German. One afternoon he broke his silence to say, "Your husband's work is beautiful—and—he has a beautiful spirit."[18]

Before leaving Germany, Einstein had written in his diary, "I decided today that I shall essentially give up my Berlin position and shall be a bird of passage for the rest of my life."[19] There would be one more year in his homeland before what Winston Churchill called the gathering storm suddenly broke over German Jewry. At the end of 1932, as the Einsteins were packing for their third and final visit to Pasadena, a surprised Elsa was told by her husband to take a good look at their country house on the lake. When she asked why, he replied, "You will never see it again."[20] The following February, Adolf Hitler, whose political demise Einstein had mistakenly predicted two years earlier, assumed dictatorial powers after a more than suspicious fire consumed the Reichstag. In mid-ocean came word that the pacifist's villa in Caputh had been sacked by an armed crowd on the pretext that it contained a secret arms cache, but the only thing confiscated was a common bread knife. "This is a very small star," a philosophical Einstein remarked to the British classical scholar Gilbert Murray. "All the universe's eggs are not in this basket now infested by the Nazis; and for a cosmogoner, that [is] convincingly consoling."[21]

III

The publicity generated by Hubble's discoveries together with much heralded visits by icons such as Einstein had combined to transform Mount Wilson from a quiet enclave of astronomers into a bustling tourist attraction. Smog was just beginning to creep up the 5,714-foot summit, but it had not yet impaired visibility. Beyond the stretch of valley and mountain ranges one could still take in the coastline of the Pacific, Catalina Island, like a mythic Avalon, thrusting out of a blue sea fifty miles away.

Many stayed the night in the Mount Wilson Hotel, a rambling wooden structure whose open-air porch provided diners with an inspiring view of the surrounding mountains. Those seeking greater privacy and a more rustic setting could rent one of forty rough-hewn cabins, called "bungalows," for $1.50 per person, only twenty-five cents more than a chicken or beef dinner with all the trimmings. Numerous trails wound their way around the mountain, bearing hikers to such quaintly named sites as Alpine Tavern, Barley Flats, and Big Tejunga. The less ambitious snoozed beside the newly constructed swimming pool.

But it was the telescopes and the increasingly public men who operated them that everyone came to see. A visit usually began with a tour of the small Observatory Museum, whose collection consisted mostly of pictures, followed by a lecture on astronomy by a member of the staff. Admission to the 100-inch reflector was limited to between 1:30 and 2:15 every afternoon and required tickets, which had to be obtained in advance by writing to the observatory offices, a rule unknown to many, who left the mountain disappointed, if not angry, at being turned away. On Friday nights the 60-inch, to which admission was also by ticket, was opened for public observation from dusk until shortly before 10 p.m. The astronomer whose run it was sometimes mingled with the tourists and answered questions; at other times he left it to the night assistant to handle the gawking crowd. The remainder of the facilities, especially the Monastery, were strictly off limits.

In 1935 the completion of the Angeles Crest Highway, a paved two-lane freeway, transformed the nerve-racking drive to the top into an inspiring afternoon's outing. Bright Sundays and holidays saw the number of visitors swell from a few hundred to between two thousand and four thousand.

The 100-inch dome was also opened to visitors on Friday nights in hopes of accommodating the increased numbers, and a more liberal policy on tickets was instituted. Still, the crowds could be overwhelming, and a new battle raged between the hotel management and Adams over who should receive priority when it came to access to the telescopes. Visitors with clout remained above the fray by going directly to the astronomers themselves, and the Hubbles were soon playing host to celebrities who wished to gaze upon stars of another order.

One of the most colorful was the English actor George Arliss, whose impersonations on-screen of Disraeli, Wellington, Rothschild, and Voltaire brought him worldwide fame. Grace was appointed to take "the civilized cynic," as *The Times* of London called him, up the mountain, where they would be met by Edwin, who was on a run. Arliss and his chauffeur, Baines, in livery, called for her at home. Soon they were on their way up the old toll road in the actor's Rolls-Royce limousine. As they ascended the grade, Arliss confessed to a fear of heights that led to faintness. He had learned to cope with the condition by getting out of the car at 2,000 feet and lying flat on the ground until it passed. "I [have] brought nothing for such an emergency," Grace was thinking to herself, "not even water."[22]

Arliss continued to talk up to the moment when he said calmly to Baines, "I must ask you to stop the car." The grade was too steep to suit him, so, with Grace at his heels, he walked on up the curving road until he found what he considered an appropriate spot. The actor was wearing a Savile Row suit from Poole complemented by dove-colored spats; his eyeglass swung from a chain on his waistcoat. Placing his "extraordinary hat" on the ground, he lay flat on his back in the dust and closed his eyes. Grace took a position between the prostrate Englishman and the edge of the cliff. She was bending over him solicitously when a car coming down the mountain rounded a nearby curve and braked to a grinding halt. Its startled passengers regarded the curious twosome with alarm. "Why was Mr. George Arliss lying, apparently dead, on a mountain road, who was I, how had we got there?"[23] Before anyone could speak, Arliss rose, bowed to the incredulous party, and took Grace's arm. They disappeared around the corner, where Baines was waiting with a clothes brush.

Once atop the mountain Arliss donned a shaggy black bearskin coat that belonged to the night assistant and cantered solemnly around the concrete floor of the 100-inch to keep warm. In the daytime of this first of several visits he was surrounded by tourists asking for his autograph while he kept a wary eye on a paper bag in Baines's possession. As the Hubbles soon learned, it held bottles of the best liquor and a cocktail shaker which, to see him use it, Edwin said, was as good as a play. Upon returning to London, he drafted an amusing letter of thanks in which he referred to himself with the editorial "we": "Of course, we do not believe anything we saw, or anything you told us, but we are able to talk about it to other people with great intelligence, and the knowledge we impart gives us a degree of importance, which we had never dared to hope we should attain."[24]

The cachet of adding a well-known astronomer to one's guest list finally brought the invitation, via telegram, Grace had long been hoping for. "Mr. Hammond says that you would be interested in seeing my ranch at San Simeon. I certainly would be very delighted to have you and Mrs. Hubble visit me here. Will you please set your own date at your own convenience." It was signed "William Randolph Hearst." Edwin took the message over the phone and, after repeating it to Grace, said with a smile, "Well, if you want, as you say, to ride fence in the Santa Lucias, we'll go for a weekend."[25]

From the sea's edge the Hubbles looked up at San Simeon, set like an Italian town on some hilltop south of Verona. The gatekeeper let them in, and they began the drive up the slopes of the coastal range. Scattered about in the distance were gnu, hartebeest, zebra, impala, and other transplanted species. An ostrich standing in the road temporarily barred their way. On the higher slopes the fog was closing in and the lions were roaring in instinctive anticipation of the nocturnal hunt that would never happen.

The flamboyant, highly controversial publisher, who was nearing sixty, had been alerted by the gatekeeper and stood out of doors waiting to welcome them as they drove up in front of the Casa Grande. They were taken to their room—"the room of the Moor's Heads"—in the Casa del Sol. Having some time to themselves before dinner, they took an exploratory walk in the fog, visiting the lions, bears, and other animals before returning to their room to change.

The paths leading to the great lighted hall of the mansion were lined with thickly blossoming flowers. The Hubbles joined some fifty other guests for cocktails and ate caviar spooned from blocks of ice and pâté de foie gras heaped on large silver trays. When summoned to dinner, they made their way into the dining hall and took their places before cards lining the long refectory table. Edwin sat at the center and to the right of the impish and irreverent Marion Davies, for whom Hearst had branched into film production, making her a major picture star. Occupying a chair directly behind them was Miss Davies's brown dachshund, Gandhi. Across from them sat Hearst, with Grace to his right. He told her the history of the ranch, of his boyhood there, and of the ways of the animals, wild and tame. It seemed to Grace that Hearst's normally vivacious paramour was badly out of her depth. As Edwin spoke to her, she "remained a down-glancing nymph, apparently stricken dumb."[26] Grace also recalled ices nestled on the backs of large swans of spun sugar, and a sufficiency of footmen. Only one wine, a dry champagne, was served, but both it and the food proved memorable.

Although it was the custom to change the seating for every meal, the Hubbles' place cards were not moved during the three days of their stay, enabling the central figures to become a "friendly foursome" after Hearst confessed the reason for Marion's reticence at dinner. She had sat next to Einstein on a similar occasion and he had completely ignored her, just as he had first ignored Doris Kenyon, while people watched, making her fearful of savants. Marion also feared horses, and declined Grace's invitation to accompany her on a morning ride. The accomplished horsewoman was given Canario, a strong young buckskin, good in coming down the steep grass slopes and in fording the swift streams of the Santa Lucias.

The Hubbles spent some of their free time in the library going through the volumes of catalogues that recorded the marbles, tapestries, paintings, furniture, carvings, and rare books which were already beginning to overflow into warehouses. There was never a withered flower or a fallen petal in the gardens. They speculated that the gardeners replaced the bushes in the night, like the gardeners in *Alice in Wonderland* who painted the roses. From their balcony they watched the sun disappear beyond the ocean's rim, while to the north the long coastline, with its redwood-bordered canyons, darkened. When it was time to go, Hearst, with Marion at his side, asked them to stay on, an invitation they reluctantly declined because of Edwin's next run on the mountain. "Come back, then," Hearst said, "and ride all the little horses," but they never did.[27]

IV

Twice a year representative astronomers from Mount Wilson traveled to Washington, D.C., to lecture at the observatory's parent Carnegie Institution before audiences admitted by invitation only. Hubble's sudden rise to scientific stardom made him a popular figure, and his 1928 lecture, titled "The Exploration of Space," which centered on the cosmological implications of his exploration with the 100-inch, was sold to *Harper's Magazine* for a tidy sum, winning him an even wider following.[28] When he returned to speak in 1930, Grace noted with pride that the list of acceptances posted on the board included the name of General Pershing, her husband's much admired commander in a war-ravaged France.

By now Hubble had developed his technique to the point of theater, and he had no more ardent admirer than Grace, who recorded the most minute details of his every performance. After being introduced, he rose slowly to his full height of six feet two inches and walked leisurely to the rostrum, where he arranged his notes, which were seldom referred to. He then unbuckled his wristwatch and laid it on the podium next to the loose pages. This act was followed by a pause during which he scanned the audience with a slight smile before launching into his subject without further preliminaries. Gestures were few but calculated. His hands traced the shape of the Milky Way and metaphorically spread apart to simulate vast distances; to make the points of a summary he might tap the large palm of one hand with the long fingers of the other. The voice did not match the imposing frame, but he mastered the speaker's skill of modulation, substituting color and warmth for power. A self-imposed rule limited the delivery to no longer than forty or forty-five minutes, which Grace characterized as tensionless, "with perfect timing and serene control." Edwin's secret, she later mused, may have been discovered by a friend who saw him walking home from his office one day and pulled over to offer him a lift. He was talking to himself as he walked and striking the pavement for emphasis with his ash stick. The offer was declined for the reason that he was lecturing before the Royal Institution.[29]

Hubble's second Carnegie lecture in two years, centering on the phenomenon of redshifts, was pronounced by the trustees to be the best ever given there. The next day the National Broadcasting Company carried speeches by institution president John C. Merriam and Hubble, as well as two others. That night at the formal reception attended by General Pershing, Elihu Root, who served as Secretary of War under Presidents McKinley and Roosevelt, and many from the diplomatic corps, Merriam announced that Hubble's lecture was the outstanding event in American science of the past decade.[30] Although the Depression was deepening, he was scheduled to receive a $500 raise, leapfrogging him to third on Mount Wilson's salary list behind stalwarts Adams and Seares.[31]

His stature was now such that invitations to distinguished lectureships began arriving. The first was extended on behalf of Princeton University by Henry Norris Russell, a frequent visitor to Mount Wilson and Hubble's champion in gaining the A.A.A.S. prize in 1925. He had been chosen to deliver the Vanuxem Lectures, a series of from four to six presentations on a subject chosen by the lecturer within his field of expertise. The material must then be submitted in book form for publication by Princeton University Press. The honorarium was $500 in addition to a royalty of 15 percent on the list price of all sales of the book. It was Russell's impression that his colleague's nebular work was "pretty well ready for presentation . . . and a book of yours on the subject would immediately become a standard."[32]

Hubble drafted a reply two weeks later in which he cited two potential problems, both relating to publication. First, illustrations would be important but potentially expensive. Second, he would pitch his lectures to a general university audience but the book might expand them into a more comprehensive and technically demanding treatise. "These questions, however, and others can be discussed later. Just now the important thing is to let you know that I accept the invitation."[33]

In late June 1931, J. Duncan Spaeth, chairman of Princeton's Committee on Public Lectures and a visiting summer professor at U.C.L.A., attended a presentation by Hubble in the Greek Theater under the stars. He wrote Russell of being "delighted" both with the astronomer's presentation and with his personality. Spaeth had introduced himself at the close of the talk, and Hubble invited the professor and Mrs. Spaeth to dine with him at the Athenaeum. Afterward, they were taken to Woodstock Road to discuss the details of the lectureship.

Hubble explained that he had been approached by several first-class publishers, including Macmillan, to write a popular book along the lines of Jeans and Eddington and wanted to know whether Princeton preferred such a work, or something more scholarly. Spaeth, who realized that Hubble's honorarium would little more than cover his expenses, opted for a specialized monograph, which would open the author's way to additional royalties by signing with a big-name firm. Hubble, too, preferred this arrangement, which Spaeth promised to bring before the lecture committee on his return home.[34]

Concerned about the fact that a popular Alfred North Whitehead had preceded him, Hubble, with Grace at his side, arrived in Princeton in November 1931. They were immediately relieved to learn that the demand for seats was so great that the Princeton Theater had been substituted for the customary lecture hall. Grace wrote that Edwin used no notes and that his voice was pitched in conversational tones easily heard throughout the packed house. Before the last lecture President and Mrs. John Grier Hibben had a tea in the Hubbles' honor, and Oswald Veblen, the noted mathematician, gave a dinner afterward.

Russell must have been in attendance, but evidently he was not privy to a conversation that seems to have taken place during the evening meal. Hubble stated that his research was progressing so rapidly that within a very short time his lectures would be too obsolete for publication.[35] On October 21, 1932, nearly a year after his visit to Princeton, an embarrassed Russell wrote, "Two or three people have asked me if I know anything about the manuscript of your lectures on nebulae which you were to send to the Princeton University Press, and at the risk of being a bore I will pass the inquiry on to you. Some of them seem rather anxious to see it."[36] Whether Hubble took the time to reply is open to question, for there is nothing further from him on this subject in Russell's files, much less a copy of a book bearing the imprint of Princeton University Press, or any other publisher.

In January 1932, Hubble was elected second vice president of the Astronomical Society of the Pacific and became its president one year later. It was a different person who lectured before his fellow scientists, during which other habits prevailed. He would lean back against the table in front of the blackboard or, sitting on the table's edge while he talked, absently toss and catch the piece of chalk with which he had been writing, not once looking at it, an eye-catching but safer version of his old match trick. Nor did he copy figures on the board in advance, or from a paper while discoursing, but wrote them down from memory rapidly while he spoke. When a question was asked, he looked directly at the questioner and ended his explanation with a query of his own: "Is that clear?"[37]

With scientific honors and offers to lecture came overtures of employment from various colleges and universities, giving Adams gray hairs. As early as 1926 Hubble had been offered the astronomy chair at Swarthmore College at an annual salary of $6,000, fully $2,000 more than he was earning at Mount Wilson. The director, in a memorandum to Merriam, lamented the fact that he could never match such an offer. Regrettably, the $500 increase he planned for Hubble and some of the other younger members of the staff, who were also being courted, might have to be taken from the projected raises of senior men unless Merriam could find a way to increase the budget.[38]

When, in September 1934, Thomas S. Baker, president of the Carnegie Institute of Technology in Pittsburgh, was stricken by a serious heart condition, board of trustees president Samuel Harden Church wrote Hubble a letter headed "Private and Confidential." Although the situation was most delicate, Church, who had met Hubble five years earlier in Pasadena and had thought carefully about how to approach him, wanted to know his feelings regarding the prospect of turning to other things: "It would perhaps be a congenial change for you to accept the presidency of a great technical institution where your studies in the field of law and literature and your profound acquaintance with astronomy would give you an unusual equipment for this new position."[39]

Hubble replied that "it is only after the most careful consideration that I have decided against an enthusiastic agreement. Ten years from now I may feel differently but today I am still possessed with a driving passion for research." He was writing a chapter in the exploration of space which, though not the whole story, "is a stirring [one] and I want to see it through." He further regretted having to turn down Church's telegraphed invitation to deliver the Carnegie Institute's Founder's Day address. It happened to fall on the same date as a reception and dinner to be given in the astronomer's honor by the associates of Caltech for what Hubble modestly characterized as "some recent foreign distinctions."[40]

V

On the morning of April 25, 1934, the Hubbles joined their fellow passengers bustling up the gangplank and onto the deck of the U.S.S. *Manhattan*. In less than two weeks they were being driven across Magdalen Bridge into Oxford's High Street, the ground still a rosy russet from the previous autumn's leaves.

Late the following afternoon, after tea with old friends and a fortifying whiskey and soda, Hubble put on his cap and gown and led his admiring retinue to a dimly lit lecture hall lined with dark wood. Just as he was about to enter the hall and deliver the Halley Lecture titled. "Red-shifts in the Spectra of Nebulae," he was told that Oxford was awarding him the honorary degree of Doctor of Science.

Grace, who had been tipped off in advance, described him as "quite overwhelmed." When he walked in to address the large audience minutes later he hadn't fully regained his composure. The adrenaline had taken over; he began not in his accustomed easy manner but with an undercurrent of emotion and intensity, reminding Grace of a Patrick Henry or a Sydney Carton: "I thought, what the hell?" But it was going better than she could have hoped. There was even occasional applause at the part about the valley lights seen from Mount Wilson and the startling slides of the spectra. And there were many kind words when it ended.[41]

The afternoon following the lecture, Edwin and Grace, who was wearing a gray tailored suit and silver fox, slipped away for a nostalgic reunion with the sites of his student days. They walked through the clipped gardens past the old city walls, then over to Queen's, which seemed just as he remembered it. He showed her his old digs and the bedroom turned into a library where the recurring apparition of the drowned don had once terrified the college. Together they climbed the haunted staircase and spoke of the irony that Edwin, a law student, should have returned to deliver a lecture named for an astronomer and one of Queen's most illustrious sons. In two weeks he would be the first Rhodes Scholar to receive an honorary degree from Oxford. What would John Hubble think of him now? he must have wondered. Grace wrote that he never spoke openly of his feelings for Oxford, but his love for the place was obviously genuine and deep: "Like a dream," she thought, "that one could watch and then re-enter."[42]

After arranging to do some fly-fishing on the river Test, Hubble returned to London with Grace to speak before the Royal Astronomical Society on the theory of the expanding universe, which had taken cosmologists by storm. They spent much of the time in the company of their well-connected friends Robert and Margaret Gore-Browne, who took them to the finest restaurants and secured choice tickets to the theater. Gore-Browne made Hubble an appointment with his tailor, Welch & Jeffries, who measured him for the new tweeds he always purchased while in England. On the drive back to Oxford for the degree ceremonies they stopped to examine the fire-gutted remains of a once great manor. The charred personal odds and ends that were still scattered about after decades put Grace in a melanchony mood. How exasperating, she thought, that futile things seem to persist for eternity when the things you want are gone in an instant.[43]

Escorted by the Gore-Brownes, Grace walked over to Oxford's Convocation Hall, a structure reminiscent of a large chapel with dark wooden benches parallel to the center aisle, and a dais and pulpit with a carved canopy at the end. Among the distinguished assembly were the Milnes, Plasketts, Daisy Turner, one of Edwin's former tutors, the provost of Queen's, and others, many wearing robes of various colors. Then came the procession, as it had for Einstein three years earlier. All stood while the vice-chancellor, preceded by four beadles bearing silver maces, made his way down the aisle to the pulpit. He spoke of the honoree in Latin, doffing his cap many times. Then the beadles momentarily departed and returned escorting Hubble, dressed in robes of scarlet and gray, his face slightly bronzed from fishing in the sun. He stood alone in the light slanting down from a high window while the public orator addressed the audience:

> We salute a former Rhodes scholar, who in his youth consecrated himself to astronomy, then led a legion in Gaul, and now, located on that California mountain which bears the name of a most illustrious President, has achieved world-wide fame. . . .
> Here is a man who has discovered worlds far removed from ours, and he defined the laws of their motion—for the more distant they are the faster they seem to be running away from us. To us lesser beings this most sagacious fellow demonstrates that the physical universe is far extended (if indeed that which is infinite can extend). This master, endowed with such authority, let us bind him to ourselves with stronger chains before he departs across the ocean—lest when he is far from us he move still further with increasing swiftness after the manner of his nebulae.[44]

In what Grace described as his "happy humility," he seemed for a moment like a knight in some old tale as he ascended the dais, where the vice-chancellor conferred the degree to enthusiastic applause. Then it was over, and her champion was again lost in that dream in which the past and present mingle to become one.

Although the Hubbles had been away for two months, their long absence from California had not yet reached the halfway point. Days later they were in northern France, driving along the edge of the forest of Compiègne. They stopped to look at the Pullman car where the Armistice was signed, quieting the guns along the Western Front. After Compiègne they followed the line of the advancing German army to Soissons and Château-Thierry, then into the Somme near Amiens, where, after twenty years, shell craters and foxholes still scarred the verdant fields on which adolescent boys who had suddenly become men had been swept away by the tens of thousands. That night Edwin pointed out Spica high overhead, the star the Romans used to test their soldiers' eyesight.

With the reminders of the war behind them, Edwin spoke of the pleasures of motoring in France compared with the drugstore civilization of America. "The French have the picture and we have the frame," he remarked.[45]

Followed by the Gore-Brownes in a second car, the party headed into a resurgent and swaggering Germany. In a Munich beer hall Grace marveled at the buxom waitresses, like Valkyries, and the raucous factory workers, tough as ever, who alternately quarreled violently and laughed loudly among themselves, intimidating the tourists.[46]

Gore-Browne, forgetting his vow to put his days as a German prisoner of war behind him, spoke of how the English had divided their captors into two categories—the swine and the eagles. The swine were simply gross and obscene, whereas the eagles were cruel and ruthless, but "we respected them."[47]

The two couples had left Munich and were on their way to Belgium when a local hotel clerk exclaimed, "You come from Munich! You were in the Revolution!" Only then did they learn of the rumors of a suspected plot against the self-styled Führer, who had arrived in Munich at four in the morning. The ringleaders were seized and summarily executed while, in Berlin, an old general and his wife had been shot in their house. Except for the announcement of certain arrests, the details had been withheld from the German papers, keeping foreigners, like everyone else, in the dark. Only with the arrival of a letter from Gore-Browne's worried mother did they begin to comprehend the magnitude of the so-called Blood Purge in which Röhm, Strasser, and many other disaffected Nazis fell.

Besides the Halley Lecture, the ostensible reason for Hubble's long stay in Europe was his selection as a delegate to the five-day meeting of the International Council of Scientific Unions in Brussels in early July. Although it had gone against the grain of Adams's Yankee up-bringing, the Mount Wilson director had written to Merriam in support of the extended trip. Merriam, who was under the impression that the astronomer would return in three months rather than the four Hubble eventually took, assented to the sojourn with pay, provided his vacation time was included in the bargain.[48]

Nothing further might have been said had Hubble not written a scathing letter to the Carnegie Institution president, complaining bit-terly about the fact that his latest Washington lecture had been pub-lished without his permission in *Scientific Monthly*, "a journal to which I do not contribute." Not only was the article full of errors but Hubble had planned to sell a revised version "for a good round sum" to help underwrite his expenses abroad, which were being paid out of his own pocket. "Such episodes stir up resentment, as I testify, and cannot but affect the harmony and confidence which should pervade the Institu-tion." He hoped to discuss the matter with Merriam during the pres-ident's next trip to Pasadena. "I will have cooled off by then."[49]

Merriam, who was not used to intemperate criticism, launched a quiet investigation into Hubble's claim. When it was concluded, he sent a seven-page memorandum to Adams, which Hubble never saw. The president regretted the fact that the article had not been proofread by its author, but Merriam had correspondence showing that Dr. Frank Bunker, head of the Institution's Division of Publication and a person whom Hubble had also attacked, had forwarded a copy of the proof pages to Pasadena. Moreover, Hubble had even agreed to furnish illus-trations for the article. "In other words, it would appear that Dr. Hub-ble was preparing to sell the article at the same time he was treating with Dr. Bunker regarding its publication through the Institution." Nor had the question of Hubble's expenses, which he had written to remind Adams about from England, been forgotten. As Merriam had promised, the matter would be taken up following his return. The stunned pres-ident could only conclude that

the comments of Dr. Hubble . . . are, both in the sense of adminis-tration and of human conduct, quite unusual and represent a very unfortunate attitude of mind. If enough members of the Institution staff were to insist upon an attitude toward the difficulties which are met in everyday life comparable to that represented by Dr. Hubble the Institution could not exist.[50]

As the Hubbles crossed the Thames in London, Edwin remarked that they had driven some 3,000 miles without a puncture or anything going wrong. Grace vowed never to forget the song of the black car as it went from first to second to third to fourth. Before departing the city they visited the zoo, which always drew them back. In the darkness of the aquarium the fish arranged and rearranged themselves in a series of exquisite friezes, drifting slowly against a background of green translucence and tiny silver bubbles. But it was the look in the eyes of the wolves that gained their attention. It was not a nice thing to remember, and both hoped for a time when all zoos would be abolished.[51]

VI

The Hubbles had just returned to England from the Continent when Virginia Lee James Hubble, who had been ailing for two months, died peacefully of heart failure on July 26. The seventy-year-old widow had been living in Alexandria, Louisiana, with her third son, Bill, the bachelor dairy farmer who had devoted himself to taking care of his mother and sisters after John Hubble's passing. According to Edwin's sister Betsy, Edwin had contributed little, if anything, toward Jennie's financial support, and had never visited her after she moved south in the twenties. The letters that passed between them have not survived because Grace either rejected or destroyed many of the family mementos offered her after Edwin's death.[52]

The body was shipped north to Springfield for burial in the Hubble family plot in Hazelwood Cemetery. In his record of the funeral, the Springfield mortician listed the deceased's seven children, noting that receipt for the charges, which were promptly paid, should be sent to William Hubble in Alexandria.

The lengthy journal entries of the period contain nothing about Jennie's passing, word of which had been cabled to her son in England. In all the many hundreds of pages written by Grace and edited by her husband for accuracy, the only references to family matters are those involving Grace's relatives, the Burkes. She never met her mother-in-law or any of Edwin's brothers and sisters. Whenever one of them came west on vacation, as Lucy had after her late marriage to Joseph Wasson, a civil engineer who would die in an auto accident, Edwin arranged a meeting at his office or elsewhere in Pasadena, never offering to take his visitor home. The girls thought that this intense desire for privacy was Grace's wish as much as their brother's. The only one who attempted to keep up a correspondence was Helen, who had married dairy farmer John Lane and had settled comfortably near El Paso, Texas.

Yet in reshaping his past Hubble lived in the perpetual fear that such stories as his alleged practice of the law and heroism under fire would be refuted by those who knew better. His childhood friend Albert Colvin claimed that even his wife did not know the facts you could write on two pages. Hubble had gone out of his way to keep quiet to the point of secrecy, an indication that he was embarrassed and considered his closest relatives a bit common.

Colvin gleaned this information while visiting friends in California during the mid-thirties. After an unsuccessful attempt to locate "Ed," who happened to be in England, he was referred to his father-in-law, John Burke. When the banker learned of Colvin's background in Wheaton he became very excited and asked him to sit down at his desk and record everything he remembered about Ed and his family. At first, Colvin protested: "I said if Ed has not told you, I will get in wrong with him." Burke replied that "Ed is a devoted husband. He has been wonderful to my daughter, but none of us know anything about his family. Don't worry about getting in wrong with Ed. I will see to it that you won't."[53]

Colvin sat down and wrote out almost everything he could think of, including details about Hubble's parents, his father's business interests, the background of his brothers and sisters, and his family life. Burke was so pleased he insisted that Colvin spend the weekend at his Los Angeles home, which would also please Mrs. Burke. Colvin, who was already feeling uncomfortable and did not wish to spend another two days reliving the past with wealthy strangers, excused himself by stating that he was due back in Norfolk, Nebraska, where he worked for the Canadian & Northwestern Railroad. He also turned down the banker's offer of a letter of credit for one hundred dollars.

As Colvin was about to leave, his host told him how grateful his daughter would be for the information he had imparted. Colvin wasn't so sure, and his suspicions were later borne out after his friend's death, when Grace, who treated her marriage like a spell, attempted to conceal all knowledge of her husband's past except for what few things he had told her. Musing philosophically on her brother's strange and sometimes hurtful behavior years later, Betsy, who knew better than Grace that life is not a work of art, remarked, "I always wondered if Edwin didn't feel guilty about not having done more. But great men have to go their own way. There is bound to be some trampling. We never minded."[54]

CHAPTER ELEVEN

"ALMOST A MIRACLE"

I

As the years passed, Hubble spent more of his time working at home when he was not on the mountain, encircled by relics and tools from distant centuries. He had ceased reading novels altogether after denouncing Tolstoy's shortcomings as a chronicler of military affairs, and Balzac for his gross exaggerations. Nor did the most advanced explorer of the cosmos care for contemporary poetry. The only writing of T. S. Eliot's that appealed to him was the alternately whimsical and poignant *Old Possum's Book of Practical Cats*. He continued to read Chaucer, whom he had discovered at Oxford, as well as Byron and Shelley, and could still repeat Walter de la Mare's "The Listeners" from memory. His newly won fame and emerging sense of his own place in history sharpened his interest in the biographies of great men, yet he was wary of those telling the tales, once remarking, "They drag the people they write about down to their own level."[1] When a question of fact arose during reading or conversation he attempted to settle it on the spot by consulting the eleventh edition of the Encyclopaedia Britannica.

On his desk beneath the stained-glass window lay an ivory-colored and faceted coup de poing purchased from the proprietor of the Hotel Cro-Magnon at Les Eyzies. "Feel how nicely it fits in your hand," he would say.[2] Lying about helter-skelter were chipped bones and flints from other Paleolithic sites in Europe, the curved incisor of a saber-toothed tiger found in the nearby La Brea tar pits, potsherds from various places and periods, fossil prints of timeless raindrops and ripples, leaves, shells, wood, and fish.

When mulling over a particularly difficult problem, he sometimes picked up a pack of cards. He shuffled the deck rapidly in the manner of a skilled gambler, snapped the cards together, and tossed them down hard as he dealt himself a hand of solitaire. The game, he said, freed his mind, allowing him to make the proper connections between the myriad blurred dots on the photographic plates and the chasm from which their light had sprung before life put in an appearance on Earth.

Almost as comforting as the surrounding walls of his study was the enclave that had become Woodstock Road. Directly across the cul-de-sac was the house of Caltech professor Clinton Judy, with whom Hubble had lived during his early days in Pasadena. Judy, who often fished and hiked with the Hubbles, had followed them to the exclusive Oak Knoll section of San Marino, engaging the same architect, Joseph Kuchera, to design his bachelor home.

Next door to the Hubbles were the Baldwins—Franklin, an attorney, and his wife, Florence, or "Gatesie," who, like her childhood playmate Grace, was reared in San Jose and attended Stanford. Dr. Robert L. I. Smith, the physician who saw Grace through her miscarriage, also lived on Woodstock Road, as did two of the Hubbles' closest friends, Homer and Ida Crotty.

Homer Crotty, a graduate of Harvard Law School, had come to Los Angeles in 1923 and joined Gibson, Dunn & Crutcher, the city's most prestigious legal firm. Unlike most of the partnership's other attorneys, he had grown up the hard way. His parents, who died before their son finished school, had an affinity for the classics and would have christened him Euclid had not the clergyman performing the service objected, forcing them to choose the poet over the mathematician.

As a senior in high school the orphaned youth became the night custodian of Oakland's newly constructed observatory. He was required to be on the premises from 5:30 p.m. to 7:30 a.m. every evening of the year. After graduating, he enrolled in the University of California, and kept up his janitorial duties another five years. With nothing else to do, he began to collect books and read voraciously in all subjects.

It was Franklin Baldwin, one of the six Harvard men in the firm, who steered Homer and his young bride, Ida, nicknamed "Idye," to Woodstock Road, where they moved in next to Clinton Judy in 1934. Pretty, strong-willed, but "very naive," Ida came from a family that had made its fortune in oil. Like Grace Hubble, she had attended Marlborough School and was immediately drawn to the older woman of the world, who treated her with a mother's concern.

The couples, together with their bachelor friend, met often for the cocktail hour; the men wore suits or sport coats, the women long dresses. Homer, who had endured the unremitting silence of an observatory more nights than Hubble himself, usually took the lead in conversation. The stocky, ruddy-faced lawyer loved to argue. A teetotaler, he possessed an extraordinary knowledge of vintage wines, and though he never learned to drive, could expound in detail on the differences between fuel injection and carburetion. He loved riding to work with others, which allowed him to read while someone else drove. But when he rode with Hubble reading was forgotten. A lead foot and cavalier regard for stop signs and cross traffic kept him on the edge of his seat, as it did most others.

Ida thought Grace neither beautiful nor pretty but admitted "there was something there." One afternoon, while playing a recording of *The Love for Three Oranges*, it suddenly occurred to her that Grace was "a Prokofiev person," controlled, precise, yet strangely magnetic. When she told her neighbor about her insight, she was amazed when Grace replied that a noted symphonic conductor had said the same thing.[3]

In contrast to her husband, Grace was not one to rely on clothes to establish her identity. She usually dressed the part of an Englishwoman, favoring the "no-nonsense look" of tailored tweeds and simple but elegant blouses. For special occasions she sometimes borrowed a formal dress from Ida, who shared her diminutive stature. She was, in Ida's estimation, "a man's woman," attractive physically but even more appealing intellectually. "Many of the people in the journals were just as much interested in conversation with Grace as they were with Edwin. She would never fawn on anybody, never go soft or use feminine wiles. That would be ridiculous! Both of them measured you," and neither suffered fools. Grace was even more demonstrative than her husband when it came to stuffed shirts. Homer, her fellow bibliophile and "buddy," who lent her anything she wanted to read from his extensive personal library, loved it when Grace took the measure of some bore, assuming what he described as her "cold potato look. Watch out, because she could verbally annihilate you."[4]

Edwin was everything to Grace, and the Crottys were told the stock tale of how he had made a small fortune practicing law, giving it up for his greater love of astronomy. Ida knew of Grace's previous marriage only because Gatesie secretly told her about the ill-fated Earl Leib. She also learned that John Burke and Howard Hughes were good friends and that, between marriages, Grace rode horses out of Flintridge with the millionaire aviator and future movie producer. For a time Grace played Chopin and Mozart on a grand piano lent to her by Gatesie, but Ida never knew how skilled she was, for she would not allow anyone but Edwin to listen. Sharing her husband's agnosticism, Grace, who disliked domestic tasks, made no special preparations for the holidays, including Christmas and Easter, which were occasionally spent in the homes of friends. Being "a true intellectual" took time, and, aside from Edwin, the rest mattered very little.

From the time they met, Ida knew that Edwin would distinguish himself in any group, aided, in the first instance, by his physical stature and rugged good looks. Besides his height and commanding physique, she was most impressed by his "kind, kind eye" and warm smile. But he could also show a firm jaw when something or someone displeased him: "There was nothing about him that wasn't very, very masculine." He was soft-spoken in the company of others and retained the Briticisms acquired at Oxford. But what many considered an affectation Ida thought charming and quite natural in one educated abroad. "Grace and Edwin were great Anglophiles, and I didn't fault them for that."[5] Both drank scotch and soda and both smoked cigarettes; Grace carried hers around in a long holder, while Edwin preferred them in a social setting to his more cumbersome pipe. Normally relaxed, he could become tense and impatient when something threatened to interfere with his work. Once, when Ida was visiting Grace, he strode in at lunchtime, wanting things to happen "right then and there." As her embarrassed friend scrambled to prepare the forgotten meal, Ida hurried home, promising herself to pay close attention to the clock in the future.

II

The Greeks, whom Edwin Hubble admired and was frequently compared with, both physically and intellectually, would have said that he had been woven into the tapestry of the blessed. A contemporary historian would more than likely write of a dash of genius, a good deal of luck, and a full measure of self-confidence. Yet there were times, Grace later admitted in private conversation, when her husband would come home and curl up on his bed with stomach pains resulting from his ongoing quarrel with Adriaan van Maanen.[6]

More than anything else, he abhorred challenges to his scientific thinking and was ever fearful that some rival would attempt to steal his thunder. His sharp public exchange with Knut Lundmark had been followed by a bitter, if little known, flare-up with Willem de Sitter, whom Grace had credited with spurring him to pursue the velocity-distance relationship with the great telescope at his command. When, in August 1931, the Dutch astronomer sent him his recently published article from the *Bulletin of the Astronomical Institutes of the Netherlands*, Hubble became enraged at de Sitter's casual statement that several astronomers had commented on the relation. While it was true, he wrote in a stinging five-page reply, that the idea "has been in the air for years," and de Sitter himself had been the first to mention it, such considerations were now beside the point. Hubble's preliminary note of 1929 contained the first presentation of actual data, and long months of additional labor of the most arduous kind had steadily confirmed it. "For these reasons I consider the velocity-distance relation, its formulation, testing and confirmation, as a Mount Wilson contribution and I am deeply concerned in its recognition as such." Not satisfied that he had made his point, Hubble went on to upbraid the venerable astronomer as if he were a naughty schoolboy, echoing his criticism of Lundmark.

> We have always assumed that, where a preliminary result is published and a program is announced for testing the result in new regions, the first discussion of the new data is reserved as a matter of courtesy to those who do the actual work. Are we to infer that you do not subscribe to this ethic; that we must hoard our observations in secret? Surely there is a misunderstanding some where.[7]

Still, he was willing to give de Sitter the benefit of the doubt and not go public with his charges, at least for the time being. De Sitter's reply is lost, but when Hubble wrote to him several weeks later he was clearly satisfied with its contents. "Mr. Humason and I are both deeply sensible of your gracious appreciation of the papers on velocities and distances of nebulae," he wrote.[8] Two months later, de Sitter arrived in Pasadena, where he joined Einstein in proclaiming Hubble's laurels.

By the mid-1930s the conflict with van Maanen had been seething so long and had taken so many twists that it would eventually become the central theme of a book.[9] Still hoping to gain the support of Shapley, whose place in the history of science was assured by virtue of his pioneering work with Cepheids and the expansion of the Milky Way galaxy, Hubble penned the following in the margin of a letter to his hard-bitten nemesis at Harvard: "Van Maanen's work is not included. He has excluded us from his Cosmos."[10]

The work to which Hubble alluded was the ongoing nebular program at Mount Wilson; the "us" included not only himself and fellow "dark man" Milt Humason but also Walter Baade, a gifted German émigré who had joined the observatory staff in 1931. Feeling himself the victim of a conspiracy led by Hubble, van Maanen refused to endorse the recommendations of the program committee concerning the apportionment of observing time on the 100-inch. He not only withheld his signature from the annual reports but composed long handwritten dissents in which he complained bitterly about the number of nights assigned to him: "Van Maanen [had] 12 less than his share." Hubble gleefully wrote at the bottom of another document, "Van Maanen refuses to sign the memorandum and hence offers no program."[11]

Relationships had become so strained that this escalating game of tit for tat took on farcical overtones. A new sign, still in place and yellow with age, suddenly appeared on the blink machine in the basement near the vault: "Do not use this Stereocomparator without consulting." It was signed "A. van Maanen."

Hubble's chance for one-upmanship had come on the mountain itself and was witnessed secretly by stellar spectroscopist Olin Wilson. Van Maanen was the 100-inch observer and therefore entitled to sit at the head of the table at dinner. Hubble came in early and walked directly into the dining room, which was considered a breach of etiquette. Wilson, who came in shortly after him, was curious and peered around the corner to see what he was up to. Hubble snatched van Maanen's napkin ring from the place of honor and replaced it with his own. When the bell rang, Wilson, who couldn't wait to see what would happen, slipped in ahead of the others and watched van Maanen march up to the head of the table, where he looked down, dumbstruck. "It says 'Hubble' here. He looks over and it says 'van Maanen' over here. Hubble's a big guy. Van Maanen is a little guy. So he just has to go over there where his napkin is."[12]

Early in his investigations, van Maanen had stated that it would take him ten years to determine whether or not his discovery of proper motions in spiral nebulae was the result of some type of systematic error. Yet every time he went back over his original plates and obtained new ones his measurements remained internally consistent. Meanwhile,

Hubble fumed in the background, until, as Grace recalled, he was spurred to action by a speaker at the Royal Astronomical Society, who commented that if it were not for van Maanen's measurements, Hubble's results might be accepted.[13] It was then that he told a friend, "They asked me to give [van Maanen] time. Well, I gave him time, I gave him ten years."[14]

In a confidential memorandum drafted by Walter Adams for John C. Merriam in August 1935, the Mount Wilson director chronicled the resulting series of events, which he termed "one of the most difficult problems with which the observatory has had to deal." Hubble had visited Adams's office to make a request: since van Maanen's motions stood as the outstanding discrepancy in his distance measurements, he wanted access to the photographs van Maanen had used. Adams replied that the primary concern of the observatory was simply to learn the truth and that he also thought it desirable that the photographs be measured by other observers. However, Adams asked Hubble to consult with van Maanen on the subject in an attempt to work with him, but to no avail. "Due to the attitude and temperaments of both men there was no cooperation in the matter but much feeling developed."[15]

Hubble countered by having Humason take new plates of four major spirals, M33, M51, M81, and M101. These were then compared with earlier plates taken by George Ritchey, which van Maanen had also used. With the backing of Adams, Hubble asked Seth Nicholson and Walter Baade to conduct independent measurements. As Adams noted, their results were essentially negative in character and opposed to van Maanen's conclusions.

Thinking that he had proved his point, Hubble drafted a long statement for publication in the Mount Wilson *Contributions*, which was only one of many hitherto written assaults on van Maanen's work, none of which was published. When Adams and editor Frederick Seares read this latest offering, both blanched. "Its language," Adams wrote, "was intemperate in many places and the attitude of animosity was marked."[16] Nor would Hubble stand for any material changes in the wording.

Seares, who thought the matter serious enough to keep George Ellery Hale posted on a day-to-day basis, decried the fact that the two astronomers could not settle their differences like gentlemen. Fearing that the rift would become a public scandal that might detract from the hard-won reputation of the observatory, Seares had labored to keep the feud out of print. But the situation had now reached the point where it might no longer be possible to hold Hubble back without being accused of outright censorship. "The institution has, I think, the right to enforce this procedure; but in certain cases it may be wiser to waive its technical right and say to a dissatisfied individual, 'Print what you like, but print it elsewhere.' "[17]

An only slightly less pessimistic Adams next suggested that the antagonists discuss their differences with Seares, who would then write a statement and submit it to both parties before publishing it in the *Contributions.* After both agreed, the assistant director undertook what Adams described as "a long and careful analysis" to be published under the names of van Maanen, Hubble, Nicholson, and Baade. When the paper was circulated, all consented to its publication with minor changes except for Hubble, "who opposed it violently." Adams, who was also reaching his limits, noted, "I do not feel that Hubble's attitude in this matter was in any way justified."[18]

Adams was especially angry at Hubble because he had everything to gain; the balance of scientific evidence, which was strongly in his favor, was about to knock the last pillars from beneath van Maanen's universe, just as it had destroyed Shapley's vision of the cosmos a decade earlier. With both Lundmark and de Sitter in mind, the director further noted, "This is not the first case in which Hubble has seriously injured himself in the opinion of scientific men by the intemperate and intolerant way in which he has expressed himself."[19]

Still feeling it essential that the new measurements be published, Adams took a firmer hand. After several conferences with Hubble, he got him to agree to the publication of a brief statement of his results in the *Astrophysical Journal.* In less than two pages, Hubble not only cited his own remeasurements of the four spirals but also included the names of Nicholson and Baade in his scientific brief. On the source of van Maanen's error, which remains a mystery to this day, he said nothing.[20] But buried in his unpublished manuscripts was the charge that van Maanen had read his personal expectations into his data, which, had it been published, would have brought censure and ridicule down upon a sincere, albeit misinformed colleague.[21]

An even shorter statement by van Maanen directly followed Hubble's article. Van Maanen acknowledged the probable existence of systematic errors in light of the new photographs and their independent analysis by his Mount Wilson colleagues. He also spoke wistfully of new investigations to resolve the issue, when, in truth, there was nothing left to resolve.[22] The most van Maanen could claim was a somewhat tainted moral victory, and even that was concealed in Adams's secret report: "The attitude of van Maanen in the matter was much superior to that of Hubble."[23]

In her memoir titled *Edwin Hubble: The Astronomer,* Grace wrote that her husband was a man without rancor when it came to Adriaan van Maanen or, for that matter, everyone else. She quoted their neighbor Franklin Baldwin, who had also said that Edwin took an "almost astronomical attitude towards human affairs."[24] Yet not until 1969, after historian of science Michael Hoskin initiated a scholarly query into the scientific controversy, did she add this dubious bit of information to the record. Unaware of Adams's confidential memorandum and much else, she further sought to protect her husband's reputation when writing to Hoskin: "It was, the contradiction, not very important in the long run. [Edwin's] work had become so obvious and so extensive."[25]

Grace rarely, if ever, experienced the full fury of her husband's temper, nor did any of their neighbors. Yet others besides those he had already crossed swords with were reluctant witnesses to it. Martin Schwarzschild of Princeton University claimed that "Hubble was the worst. I have suffered under a couple of sermons from him, ranting in the most unreasonable way against Shapley that you can imagine," although "Shapley was no angel either."[26] And while Harvard astronomer Bart J. Bok later became friends with Hubble, he gritted his teeth every time he visited Mount Wilson in the "early days." "Ooff! Hubble would have nothing to do with anyone associated with Harvard or with Shapley."[27] A rare exception was Cecilia Payne, then a graduate student, who, over Shapley's objections, headed west to consult with Hubble. "Hubble spoke of my work with kindly, royal condescension," Payne recalled. She later heard that he had remarked of her, "She's the best man at Harvard."[28]

Such arrogance was best captured in the story told by the Oxford astronomer H. H. Plaskett, who likened Hubble to the noted physicist H. A. Rowland. During his cross-examination as an expert witness in a trial, the don was asked by the opposing attorney, "Well, Professor Rowland, who is the greatest physicist in the world?" Rowland replied, "I am, I suppose." When his friends later questioned this statement, the scientist protested, "But I was under oath."[29]

III

In the summer of 1932, having completed the course work for his Ph.D. in astronomy at Berkeley, a ruggedly handsome Nick Mayall packed his bags and headed for Lick Observatory on Mount Hamilton, where he was to spend a year on fellowship writing his dissertation under the supervision of Dr. J. H. Moore. In October, Hubble took the rather unusual step of writing to Moore to offer his unsolicited advice on a possible research topic. Considering Mayall something of a protégé, he also composed a three-page memorandum in which he outlined the method of approach. "The exploration of extra-galactic space is a new field of major importance," he began. "The fundamental problem is the determination of the general characteristics of the observable region considered as a sample of the universe at large."[30] The 60- and 100-inch reflectors at Mount Wilson had thus far indicated that the nebulae are scattered throughout space in random fashion, that the universe is homogeneous and isotropic—that is, invariant in all directions. If this promising hypothesis could be proved, accurate numerical estimates of the order of density of matter would become available for the first time, exerting a profound impact on cosmology and future nebular investigations.

The way was long and tedious, so the astronomers at Mount Wilson could not hope to shoulder the entire burden. "It is of some importance that the solution of a problem of such wide interest . . . represent the contributions of various independent and competent investigations." Lick's Crossley reflector, equipped with a 36-inch mirror, was well suited for the task. Indeed, Mayall could draw on one of the greatest collections of nebular photographs in existence while filling in the gaps with plates of his own. As a further incentive, Hubble concluded, "[it] is both expedient and appropriate that the authoritative analysis should proceed from the Lick Observatory."[31]

What Hubble did not say is that he had every intention of bringing Mayall to Mount Wilson once his degree was in hand. One strategy involved packing up the 60-inch for relocation in the Southern Hemisphere, "with me," Mayall wrote, "along to help trail-boss, ramrod, and round up observations, principally in the field of extragalactic research."[32] Neither was Hubble as forthcoming as he might have been concerning the level of his own interest and involvement in the problem at hand. As both he and Mayall hoped, Moore soon gave the nod to go ahead.

In January 1934, some fifteen months later, the longest scientific paper of Hubble's career appeared in the Astrophysical Journal. It was titled "The Distribution of Extra-Galactic Nebulae" and covered a full 69 pages.[33] Of the 80,000 extragalactic nebulae photographed on Mount Wilson with its two large reflectors, Hubble based his analysis on 44,000 of these gleaned from a staggering 1,283 plates, each of which had been examined at least three times under both high and low power. All images not definitely stars or obvious defects were marked as nebulae. Several duplicate plates were included as a check against visual error, and this comparison of pairs indicated that "the mistakes tended to balance the misses," rendering the counts valid for statistical purposes.

As always, no nebulae were to be found across the heart of the Milky Way, where they are hidden in the zone of avoidance by a great irregular band of obscuring matter, varying in width from 10° to 40°. Along the borders of this band nebulae remain "scarce," strong evidence that the obscuration is due largely to isolated clouds rather than to a uniform layer of diffuse material. It is in the higher latitudes north and south of the galactic plane that the nebulae flare, rich in stars, some hidden in their own obscuring veils. Toward the polar caps the picture becomes even clearer, as the nebulae themselves sometimes cluster in the hundreds to form huge aggregations of galactic matter. With the aid of charts and graphs, the author, who was still referring to his work as "reconnaissance," declared his hypothesis confirmed—at least provisionally. On the broadest scale available to humankind, the universe is indeed homogeneous, for even the clustering nebulae are randomly dispersed throughout. The universe that astronomers see is the universe writ large. Emboldened by this vision, Hubble calculated that the average density of material in space is represented by a single gram of matter distributed uniformly throughout a volume a thousand times that of Earth.[34]

His grand synthesis had no sooner been published than Mayall wrote to thank him for reading over a draft of his dissertation, which was about to go to the printer. "I have made a number of changes in writing up the results from the counts, principally suggested by the desire to make my results comparable with yours." "The members of the thesis committee," he added, "are of the opinion that the thesis should be published as a piece of work essentially completed before your paper came out, so I have not made any reference to it in the main body of the text."[35]

Safely in print and assured of the glory, Hubble replied that he was delighted with Mayall's progress and in full agreement with the committee that the dissertation should be published without reference to his own paper. In so doing he created the appearance of neutrality and the illusion of keeping faith with Moore concerning Lick's central role in the investigations. "Actually," he admitted, "I have a personal interest in this decision for I am eager to have as much independent investigation as possible so that the accumulating evidence will eventually convince the eastern group of the importance of careful work."[36]

Like the dreams of many, Mayall's hope of becoming Hubble's "man Friday" in South America sank in the wake of the Great Depression, as funding disappeared. To the east on the Great Plains, the prairie winds were sweeping up tons of multicolored soil—red, yellow, black, and brown—carrying it from Oklahoma to the Dakotas and beyond, blotting out the sun, creating darkness at noon. People, like their withered crops, became straws in the wind. Men, women, and children of all ages, races, and descriptions suddenly found themselves a part of a new social dimension someone once called "the great underground." By day they sunned themselves on park benches; by night they cooked their meager fare in blackened tin cans, then slept with one eye open, wrapped in old newspapers in nameless hobo camps and roadside ditches. Many begged their way across the country; others stole anything that was not nailed down. Some worked, but rarely for long; the jobs did not last, and the pay was always too low to kindle hope.

The dispossessed who succeeded in reaching California were rudely stripped of their last dream of nirvana by jackbooted state troopers and sheriff's deputies wielding clubs and brass knuckles. "Scum Not Wanted!" read the signs, "Okies Go Home!" But on Woodstock Road life remained unchanged. Thanks to Homer Crotty, Hubble became a member of the Sunset Club, a prestigious all-male order located in downtown Los Angeles. The Hubbles and their neighbors frequently dined at Perino's on Wilshire Boulevard, which Ida Crotty described as "the" restaurant at the time. They favored the French cuisine served in the private dining room upstairs.

It was only by chance that Ida glimpsed the social consequences of economic collapse. Wearing a fur coat and accompanied by her mother and two sisters, who were similarly dressed, she had gone downtown to obtain a passport. Across the street was a breadline, which, despite her attire, gave her the chills.[37] Homer sometimes spoke of the possibility of a socialist revolution, while Hubble, whose politics were more conservative than his science, fumed against Franklin Roosevelt and the New Deal. Grace read *The Grapes of Wrath* without emotion and wrote in her journal of dining at the Baldwins' with friends who had made a survey of the Okie camps in California and Arizona. All agreed that there were bound to be great changes for the worse affecting taxes and living conditions. "The adults [are] mostly hopeless, something could be done for the children if the parents could be eliminated. Squads of men have to clean up the lavatories (so on Steinbeck's mind) every half hour."[38]

Nick Mayall was only a week away from joining the ranks of the unemployed when a last-minute grant equal to the salary of an assistant janitor permitted him to hang on at Mount Hamilton. Hubble, together with Humason, who kept up a steady correspondence with the young astronomer, persuaded him to join in their nebular program on redshifts. The Crossley telescope to which Mayall had access was an excellent instrument for photographing faint objects of considerable surface brightness. Moreover, it was capable of observing nebulae north of declination +64°, the limit of the 100-inch's range. Thus, by mutual agreement, Humason and Mayall would photograph different nebulae, the latter concentrating on the dispersion velocity of a group of fifty spirals of apparent magnitude 12.5.[39] The Wisconsin astronomer Joel Stebbins also agreed to participate in Hubble's ever more ambitious undertaking.

The strain on Humason was beginning to tell. In a class of his own where nebulae were concerned, the gifted observer was forever competing against himself. Although the cameras and spectrographs mounted at the Cassegrain focus were undergoing steady refinement, as was the quality of the film provided by the Eastman Kodak Company, such gains were largely offset by the ever fainter observational targets assigned him by Hubble, who was pushing both the instrument and the man to their limits. In January 1936, Humason wrote Mayall that he was about to head up the mountain for a five-night run to seek out a nebula in the Hydra cluster. "It is the faintest thing I ever have seen and probably there isn't a chance in the world to get a direct plate but if my nerve holds up (and the seeing also), [I] may try it. Hate to think of 4 or 5 nights and then end up with nothing to show for it. Hubble," he added, "claims I always tell him it is impossible and then come down with a velocity but this time I think not."[40]

In an attempt to keep Mayall's flagging hopes for a Mount Wilson appointment alive, Humason penned a *"very confidential"* postscript. Hubble had recently turned down a second offer from the trustees of the Carnegie Institute of Technology to become its president at four times his current salary. He had done so with the understanding that he would become the director of the new 200-inch telescope under construction atop Mount Palomar. Should things go as planned, "you will be asked to join the staff. We have talked about that several times and both of us think [that] Baade, Mayall, Hubble and M.L.H. ought to make a pretty fair combination."[41]

IV

A depressive George Ellery Hale was bedeviled by starlight. The astronomical equivalent of pure gold, it falls on every square foot of Earth's surface, yet the best that could be done was to gather and concentrate the rays striking a reflecting mirror 100 inches in diameter. It was during the contemplation of these vast and squandered riches that the hypochondriacal genius had begun to daydream of a still larger telescope than his intellectual offspring at Yerkes and Mount Wilson. "I believe," he wrote a friend, "that a 200-inch or even a 300-inch telescope could now be built and used to the great advantage of astronomy."[42]

In Hale's case, thought was tantamount to deed. He turned to instrument designer Francis Pease, one of Mount Wilson's oldest, most reliable hands, and asked him to put some ideas on paper. Carried away by the possibilities, Pease soon produced sketches for a 300-inch monster capable of bringing yet unseen nebulae to within the touch of human fingertips. As rich in imagination as in starlight, all the two dreamers needed was a few million dollars and, like Archimedes, a mountaintop on which to stand.

Charles Yerkes, John D. Hooker, and Andrew Carnegie had passed on. Few other wealthy men in the country were sufficiently interested in astronomy to bankroll the construction of a telescope so large that Hale could not absolutely guarantee it would work. He decided to test the philanthropic waters by contacting Wickliffe Rose at the Rockefeller Foundation early in 1928. Would the International Education Board consider making a modest grant for the purpose of determining how large a mirror it would be feasible to cast?

Weeks later Hale traveled east for a private meeting with Rose, who, much to the astronomer's surprise, eagerly asked how much the telescope itself would cost and to whom it would be given. A nonplussed Hale had no ready answer to the first question, but he replied to the second that the instrument should be under the jurisdiction of the Carnegie Institution of Washington, the greatest observatory in the world.

Rose balked at this, arguing that the telescope should go to an educational institution, preferably the California Institute of Technology, of whom Hale was a founding father. Because Mount Wilson's expertise in astronomy was unquestioned, Rose further insisted that its staff enter into a cooperative agreement with Caltech for the purpose of planning the telescope and carrying on research at the new observatory.

With millions at stake, Hale was in no position to dicker. He departed New York after promising to provide Rose with an estimate of the costs.

As reality set in during the ensuing months, Hale and Pease reluctantly concluded that the casting and mounting of the 25-foot mirror of their dreams was fraught with too many technical risks. They settled on a scaled-down version measuring 200 inches in diameter and a construction figure of $6 million, with the option of requesting additional funds should this sum prove inadequate. At its meeting of May 25, 1928, the members of the Rockefeller board voted their unanimous approval.

Even to astronomers the projected capabilities of the yet to be cast mirror, dubbed the "Big Eye," were awesome to contemplate. Its optical range would be approximately one billion light-years, an eightfold increase in the observable volume of space. Its light-gathering power—the equivalent of a million human eyes—would enable an astronomer to see a candle at 10,000 miles, and to photograph the object at three times that distance. To the estimated hundred million nebulae within the range of the 100-inch Hooker would be added hundreds of millions more scattered singly, in groups, and in giant clusters. In the coming years, as construction progressed and anticipation mounted, Humason would sometimes hear Hubble speak of the possibility of reaching the "limit" or "observational horizon," where the light of Einstein's universe theoretically curves back upon itself and the stars cease to swarm in their fainter and fainter billions.[43]

Passing from a 100- to a 200-inch mirror required the casting of a disk unlike the ordinary plate-glass object made at St. Gobain in 1908. From the very beginning Hale had been thinking of Pyrex, a glass whose low coefficient of expansion made dishes composed of it a favorite among housewives, who transferred them from refrigerator to stove and back without shattering. In conjunction with scientists at New York's Corning Glass Works, a series of test disks ranging from 30 to 120 inches were produced, all of which were pronounced usable though far from perfect.

The problems of producing a satisfactory blank were compounded by the tremendous weight of the glass and the need to anchor it in such a manner that it would not slip and crack because of excessive stress. As a solution Pease came up with the ingenious idea of the ribbed back or "honeycomb" design. In contrast to a solid disk of the old type, deep grooves interspersed by thirty-six equally spaced ridges were to be forged on the back. These ridges, or "support points," would then be attached to the frame by thirty-six individual mechanisms, each containing 1,100 parts. When operational, an intricate series of levers and balances would adjust each support automatically, ensuring that the proper stresses and strains would be compensated for and the mirror would always hold its desired shape.

After a long period of experimentation, the 200-inch mold was finally completed and the casting of the disk scheduled for March 25, 1934. The tank, containing 65 tons of molten Pyrex, had taken fifteen days to fill, and another sixteen days to heat to a temperature of 1,575° C, when melting began to occur. The melt was transferred from the tank to an igloo mold by means of steel buckets, each containing 750 pounds, suspended from overhead by monorail trolleys guided with long handles by a crew of workmen. Many hours were required to fill the mold, as only about half of the liquid glass was poured from each bucket; the rest, which hardened on the sides of the massive containers, was broken away with heavy rods and returned to the melting tank. The annealing took four weeks, a rate ten times faster than was considered safe. Yet well before the 20-ton disk had cooled, Humason wrote Mayall that the casting was "a total failure."[44]

Several of the prismlike cores fixed to the bottom of the casting had come loose because the metal pins holding them in place had melted. The cores gradually floated to the surface and had to be fished out with crowbars. Even as the annealing continued, a second mold was under construction. G. V. McCauley, the technician in charge, was fixing the new cores with bolts of chrome-nickel steel to assure against history repeating itself.

The second pour took place on December 2, with less publicity and fewer visitors present. This time ten months were allowed for the annealing. Notwithstanding a minor earthquake and a flood on the Chemung River that forced the temporary shutdown of the temperature-control equipment, all went well.

The 40,000-pound disk was encased in steel and mounted vertically in a special low-slung railroad car for its journey to Pasadena. Clearing the tracks, tunnels, and bridges by no more than a few inches, it lumbered westward only by day at twenty-five miles per hour, its vibrations automatically recorded. A nostalgic Hale was on hand to meet the train, which reminded him of the arrival of the 40-inch mirror at Yerkes Observatory in 1897. Amazing as it seemed, the glass from his first great reflector would fit perfectly in the center hole of his new giant.

Twenty workers waited in the recently constructed Caltech optical shop, where a grinding and polishing machine weighing 160 tons would first grind the disk flat, then deepen the center nearly four inches by removing 5.25 tons of glass. Interrupted by war, the process, which began in April 1936, would not be completed until October 1947, at a cost of 180,000 man-hours and some 31 tons of abrasives. When finished the mirror would be ground and polished within two-millionths of an inch of a perfect concave surface.

In the 1920s, Lick astronomer W. J. Hussey, armed with a portable 9-inch Clark refractor, had begun testing the atmospheric conditions on mountains stretching from the Mexican border northward beyond Mount Wilson, which had been ruled out as a possible home for the 200-inch due to the increase of electric lighting in the San Gabriel Valley. After five years of study and much deliberation, land was acquired from a rancher on isolated Mount Palomar, a 6,100-foot tabletop thirty miles long and ten miles wide in the San Jacinto range. As it was located ninety-three miles southeast of Pasadena and fifty miles north of San Diego, there was no danger of light pollution, at least in the foreseeable future. Although the site was geologically active, the telescope would literally be anchored to the mountain, subject to minimal danger from earthquakes.

Hubble, who was a member of the Telescope Advisory Committee, enjoyed slipping away with Grace on weekends and driving down to the construction site, where, by 1936, crews were engaged in building two roads, one up the southern face of the mountain, the other about four miles long across its top to the observatory grounds. Cottages and camp quarters were going up simultaneously, accompanied by the laying of water pipes and electrical conduits. The footings of the dome, equivalent in scale to the Pantheon, were in place by the next year, and bids were being taken for a dozen other projects ranging from the gargantuan telescope mountings at $600,000 to a parking lot at $2,000.[45] Ironically, the project was aided by what seemed an endless Depression: companies desperate to keep their top technical people and workmen together willingly accepted many jobs at cost or even less, stretching precious funds well beyond expectations. As one thankful participant reflected, "I know we came out very well on a lot of things."[46] Meanwhile, an anxious Hubble told Grace that it would take a decade's observing with the 200-inch to resolve the major cosmological problems. "We don't say it's perfect because we don't use the word perfect, but it's almost a miracle."[47]

Hale always knew that he would not last long enough to relive the thrilling experience of that November night in 1917, when Vega first danced in the eyepiece of the 100-inch. All he could hope for was to push his most ambitious project beyond the point of no return. He had done so by 1936 when his health rapidly began to fail, calling into doubt his role as chairman of the Observatory Council. By July, matters had gotten to the point where a loyal and reluctant Adams felt it necessary to draft a confidential report addressed to Merriam in Washington.

The 200-inch was in need of an astronomical director, and Hubble would almost certainly be the first choice if Caltech could come up with the money. "Personally I think there is little doubt that [he] would accept the appointment from the Institute if the salary were satisfactory." Adams, who had lately been at crossed swords with Merriam, was contemplating stepping down as director of Mount Wilson. Perhaps Hubble should be considered for that post as well; it was not for Adams to say.[48]

Merriam had other ideas, at least for the time being. He dispatched Carnegie Institution physicist Max Mason to Pasadena in September, where he was appointed research associate at Caltech and vice-chairman of the Observatory Council. By February of the following year Humason wrote Mayall that "Dr. Hale . . . is probably out of it for all time. Mason is now the active head of the whole project." Exactly what had happened to Hubble's prospects is not entirely clear, but his previous flare-up with Merriam over publication and travel as well as a recent prolonged absence in England doubtless weighed in the decision, as did the van Maanen controversy. Mason, too, developed a distaste for Hubble's patrician manner, which was conveyed to Merriam. Promising to fill in a downcast Mayall at their next meeting, a no less disappointed Humason wrote only, "[B]elieve it or not, there are many people who dislike Hubble and in every way possible they are trying to prevent him from being named director. If Baade and I come up there in the spring . . . I'll tell you the whole thing then."[49]

Their hopes were briefly rekindled in June. An upbeat Humason wrote that "Hubble is going about things in a different sort of way and the future looks pretty bright for him as far as the 200-inch is concerned."[50] Once again he refused to provide details, which he termed "too complicated" for a letter. It was the last Humason wrote of the matter, and nothing further is contained in the Adams and Merriam files.

V

An inveterate fly-fisherman, Hubble was haunted by waters, especially those he wasn't fishing at any given time. Among his favorite books was a charming volume in the collection of the Huntington Library entitled *A Quaint Treatise on Flees and the Art of Artyfichalle Flee Making . . . By an Old Man.* When the rare books librarian saw him coming he automatically unlocked the cabinet and placed the work on the oak reading table. Hubble would reverently turn the pages to which real century-old flies were attached above their pictures.[51]

Too impatient to master the art of tying his own lures, he purchased most of his collection from a nearby sporting-goods store. They were frequently lined up on the table next to his green leather chair on the pretext that he was sorting them out, but a half-smiling Grace knew better. On one side were the "generals," with which a fisherman can imitate many insects in various stages from larval to winged. On the other were arrayed the "specials," flies that imitate specific hatches such as honeybees, mayflies, spruce beetles, and winged ants.

A wader of swift waters and an excellent caster, he favored northwestern Colorado's White River, to which he and Grace were first taken as guests of the Crottys in the summer of 1935. The private and secluded Rio Blanco Ranch, the property of thirty-seven shareholders, had been carved out of old homesteads by wealthy professionals who had to approve of a prospective buyer before a membership could change hands. The Hubbles rode the train to Rifle, where they were met by ranch manager Harry Jordan, then driven north to Meeker. There Edwin purchased a fishing license and supplies before beginning the forty-five-mile trip eastward into the mountains.

Located at an elevation of more than 8,000 feet on the White River Plateau, an area known locally as the "Flattops," the ranch, seven miles long and half a mile wide, spanned both sides of the river. Five miles upstream lay Trapper's Lake, accessible by a makeshift road and numerous game trails frequented by deer, elk, horseback riders, and an occasional bear. The log and chink cabins—Spruce, Balsam, Timberline, Sage, Lodge Pole, Aspen, Willow, and more—overlooked Big Fish Creek. Each had a large fireplace and a radiator for the less hardy as well as a respectable population of mice whose numbers were held more or less in check by half-wild cats, the last of which had to be collected at the end of the season, or perish.

Harry Jordan and his wife, Cleo, were in charge, but most of the domestic duties were performed by high school students from Meeker happy for the chance to earn some money during a period of few economic opportunities. The bell rang for breakfast, lunch, and dinner, all of which were served in the lodge. Cleo Jordan raised chickens that supplied fresh eggs before being sacrificed in the frying pan while Harry slaughtered and dressed beef from a herd trucked up for the summer.

John Powell Hubble
(COURTESY OF LENA JAMES JUMP)

Virginia James Hubble
(COURTESY OF LENA JAMES JUMP)

Paternal grandparents Martin Jones Hubble and Mary Jane Powell Hubble (COURTESY OF ELIZABETH HUBBLE AND JOHN F. LANE)

Maternal grandparents William H. James and Lucy Ann James with their grandchildren. Edwin is third from the right, his arms too long for his jacket. His brother Henry, a head taller, stands behind Edwin, while their sister Lucy is next to the window behind Grandmother James (COURTESY OF LENA JAMES JUMP)

Sixteen-year-old Edwin's high school graduation picture (COURTESY OF ELIZABETH HUBBLE AND JOHN F. LANE)

University of Chicago star athletes on the town (THE HENRY HUNTINGTON LIBRARY, SAN MARINO, CALIFORNIA)

Towering in the confidence of being twenty-one, the Rhodes Scholar is England-bound, September 1910 (COURTESY OF ELIZABETH HUBBLE AND JOHN F. LANE)

The Oxford athlete perfectly posed (COURTESY OF ELIZABETH HUBBLE AND JOHN F. LANE)

The family in Louisville after John Hubble's death. Elizabeth "Betsy" stands between the widowed Jennie and Lucy. Helen is to the right of her mother, Emily in front. Henry, who stayed home, stands next to Lucy (COURTESY OF ELIZABETH HUBBLE AND JOHN F. LANE)

The first known photo of Edwin Hubble with a telescope. It was taken in 1914 on the carriage sweep of the Hale house in New Albany, Indiana, following Hubble's return from Oxford (PHOTO BY JOHN R. ROBERTS, COURTESY OF JOHN R. HALE)

World War I: Major Edwin Hubble and his sister Lucy, a Red Cross nurse (THE HENRY HUNTINGTON LIBRARY, SAN MARINO, CALIFORNIA)

The Yerkes Observatory group at Williams Bay, Wisconsin, September 1916. Above is the giant 40-inch refractor; below, the great movable oak floor. Hubble, wearing a suit and tie, is fourth from the right in the last row. To the right of him, wearing a light gray suit, is Yerkes director Edwin B. Frost [COURTESY OF YERKES OBSERVATORY]

The 100-inch Hooker telescope with its tube 40 degrees from horizontal. The bentwood chair Hubble occupied is visible through the railing [THE HENRY HUNTINGTON LIBRARY, SAN MARINO, CALIFORNIA]

The millionaire and the dreamer: Andrew Carnegie (left) and George Ellery Hale outside the 60-inch dome atop Mount Wilson, March 1910 [THE HENRY HUNTINGTON LIBRARY, SAN MARINO, CALIFORNIA]

Hubble in plus-fours on Mount Wilson in 1923. The photo was taken by the visiting astronomer Margaret Harwood near the time Hubble discovered three Cepheids in Andromeda [AIP EMILIO SEGRE ARCHIVES]

" 'The world's smartest man' visits the Monastery library in January 1931." (Left to right, front) Charles St. John, Albert Einstein, Walter S. Adams, Walther Mayer; (back) Edwin Hubble, Arthur King, and Alfred Joy [THE HENRY HUNTINGTON LIBRARY, SAN MARINO, CALIFORNIA]

Grace Burke Hubble in 1931 [THE HENRY HUNTINGTON LIBRARY, SAN MARINO, CALIFORNIA]

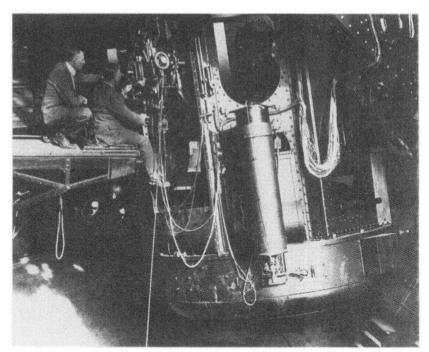

Edwin Hubble and Sir James Hopwood Jeans at the 100-inch Hooker
telescope [THE HENRY HUNTINGTON LIBRARY, SAN MARINO, CALIFORNIA]

Group portrait: the Mount Wilson Observatory research staff in front of the Santa Barbara Street
offices, Pasadena, March 1939. (Back rows, left to right) Ralph E. Wilson, Milton L. Humason,
Robert S. Richardson, Seth B. Nicholson, W. H. Christie, Arthur S. King, Theodore Dunham, Jr.,
Edwin Hubble, Robert B. King, Rudolph Minkowski, Walter Baade, Edison Pettit, Olin C. Wilson,
Edison R. Hoge, Roscoe F. Sanford. (Front row, left to right) Joseph O. Hickox, Adriaan van Maanen,
Gustav Strömberg, Harold D. Babcock, Frederick H. Seares, J. A. Anderson, Walter S. Adams, Paul
W. Merrill, Alfred H. Joy [THE HENRY HUNTINGTON LIBRARY, SAN MARINO, CALIFORNIA]

Deep thoughts: Hubble and Aldous Huxley [THE HENRY HUNTINGTON LIBRARY, SAN MARINO, CALIFORNIA]

The master mariner examines a photographic plate [THE HENRY HUNTINGTON LIBRARY, SAN MARINO, CALIFORNIA]

Ready to wet a line at Rio Blanco Ranch, Colorado [THE HENRY HUNTINGTON LIBRARY, SAN MARINO, CALIFORNIA]

Edwin and Grace
[THE HENRY HUNTINGTON LIBRARY, SAN MARINO, CALIFORNIA]

Hubble at the helm of the 200-inch Hale telescope, Mount Palomar (THE HENRY HUNTINGTON LIBRARY, SAN MARINO, CALIFORNIA)

Large as a boxcar, delicate as a moth: the Hubble Space Telescope photographed from the Space Shuttle *Discovery* during the instrument's deployment on April 25, 1990 (NATIONAL AERONAUTICS AND SPACE ADMINISTRATION)

Anne Crotty, who, like her older brother Dan, called the Hubbles "Aunt Grace" and "Uncle Edwin," remembered fresh cream so thick it poured in clots and newly churned butter on rich homemade bread.

Whiskey was banned from the lodge, but nothing was said about the cabins. The outside world could be reached by a single telephone when the line wasn't down, a frequent occurrence, and you could bring your own radio if you wanted to. Newspapers had to be specially ordered. Mostly, Anne recalled, the world just rolled away as nature burgeoned and blossomed beneath purplish-blue skies.[52]

Captivated by these surroundings and the company of such influential shareholders as Eugene Holman, president of Standard Oil of New Jersey, the Hubbles soon became shareholders themselves, favoring Colorado for vacations and England for culture. Grace spent her time riding the trails on a rugged mount named Meeker, covering as much as thirty miles a day, or she wrote of her encounters with nature in her journals. Edwin, meanwhile, could be seen studying a "hole" in the White, the fast rapids at its head, the big turn called the pool, and the quiet water or tail below. He would wade in up to his waist, his arms spread wide for balance, pick his spot, then raise the feather-light rod to the twelve o'clock position, unfurling the line in a widening arc before bringing the fly gently to rest on the surface of the water. Within moments he was lost in the poetry of geology as he was so often lost in the vast oceans of sky formed in that time before time. Reading through Grace's journal one night, he declared in a disappointed voice, "Next time at camp there must be more about fish."[53]

Within weeks Hubble was contemplating other waters from atop a round stone tower overlooking picturesque Gloucester Harbor in far northeastern Massachusetts. The narrow stairway spiraled upward to a room with a large fireplace, ancient dark furniture, and armor on the walls. At dusk the lighthouse on Cape Ann gleamed red, the open Atlantic lying just beyond. As he did every morning and afternoon for a month, the astronomer returned to the sanctuary at night, his own light sometimes burning into the small hours.

Lookout Hill was part of the estate owned by the legendary American mining engineer John Hays Hammond, a onetime employee of Cecil Rhodes in the goldfields of South Africa. One of the four leaders of the reform movement in the Transvaal, Hammond had been arrested following the famous Jameson Raid (with which he did not agree) into the Boer colony and had escaped execution only after paying a $125,000 fine. Banished from South Africa, he returned to the United States via England and became associated with some of the country's most important financial groups. His connections gained him lectureships at Columbia, Harvard, and Yale, his alma mater, and he was elected a fellow of the American Academy of Arts and Sciences. The Hubbles had met their host during a recent visit to Washington, D.C., where Grace marveled at Hammond's Tudor mansion, supposed to have been the first of its kind in the capital.

Hubble spent most of September 1935 refining eight Silliman Lectures to be delivered at Yale the following month for a stipend of $2,500. Named in honor of Hepsa Ely Silliman by her husband, Augustus, who established the Silliman Memorial Fund, the annual lectures were supposed to confirm "the presence and wisdom of God as manifested in the Natural and Moral World" without recourse to "Polemical or Dogmatical Theology."[54] The science had remained, but the theology had largely vanished in the thirty years since the English physicist and Nobel laureate Sir Joseph John Thomson delivered the first series of lectures in 1903. Among other distinguished names gracing the list were Ernest Rutherford, Thomson's successor as director of the Cavendish Laboratory at Cambridge, geneticist William Bateson, John Scott Haldane, an expert on gases and deep-sea diving, and Niels Bohr, who took the podium in 1923, the year after receiving the Nobel Prize for his seminal work in quantum theory.

In October, at the invitation of Yale president James Rowland Angell, the Hubbles moved into the guest suite of Saybrook College on the New Haven campus. They attended the Yale-Navy game along with some 61,000 others, and went to a Frank Lloyd Wright lecture on wooden architecure. Grace wrote her sister Max, who had married Clarendon Eyer, a courtly businessman and weekend sailor, that there was practically a riot at the end with everybody insulting each other. "Clarendon would have loved it."[55]

Contrary to her usual practice, Grace went to the lectures alone and sat in the back row. The doorman, not knowing who she was, expressed surprise at the size of the audience. One of the previous lecturers had addressed only three people his final night, which would explain why no one had been asked to speak three out of the previous seven years. In contrast, Hubble's audience, which was large to begin with, increased with each offering, climaxing with a packed auditorium and prolonged ovation.[56]

In the space of a single month, laboring day and night, Hubble had written much of the text of *The Realm of the Nebulae*, the book he had first promised to Princeton University Press in 1931 but had never delivered. In a letter to Nick Mayall, he described it as "a cross between

popular lectures and a text book," making it "a little difficult to estimate," something to worry about once it reaches the hands of reviewers
after its publication by Yale University Press.[57]

Aimed more at the college-educated than at scientists, who had re-
course to the technical literature, *The Realm of the Nebulae* spans the
period from 1922 to 1936, the most creative years in any astronomer's
life since Galileo, who had prepared the way three centuries earlier by
addressing *The Starry Messenger* to literate men of the Renaissance.
And like Galileo, Hubble first discusses the historical backdrop against
which he had labored before rendering what are often elegantly drafted
accounts of his solutions to four critical problems in cosmology, any
one of which would have assured him a position of the first rank in
the annals of science.

He wrote of how, beginning in 1922, no satisfactory classification
system for nebulae existed. Following the exhaustive examination of
thousands of photographic plates, the astronomer was able to fit virtually all nebulae into a few major types barely four years later. "Order
has emerged from apparent confusion," he observed, "and the planning
of further research is greatly simplified." The nebulae are closely related
members of a single family, constructed on a fundamental pattern
which varies systematically through a limited range. Falling naturally
into an ordered sequence of structural forms, they are readily reduced
to a standard position in the sequence, the luminosity of each directly
related to its type.[58]

But it was the discovery of a Cepheid in Andromeda in October 1923
that the audience had come to hear about, and the astronomer did not
disappoint. The hunt was on as the mainspring of eternity began to
unwind some one million light-years into the void. With a swiftness
that in hindsight appears ineluctable, Cepheid led to Cepheid, cluster
to cluster, nebulae to nebulae: "The most conspicuous nebulae were
. . . independent stellar systems in extragalactic space. Further investigations demonstrated that the other, fainter nebulae were similar
systems at greater distances, and the theory of island universes was
confirmed."[59] No longer was creation merely "a congeries" of Galileo's
stars, but a congeries of stars amid a congeries of galaxies untold.

In a rare instance of personal revelation, Hubble next wrote of the
year 1929: "Expectations ran high. There was a feeling that almost
anything might happen."[60] With nebulae now recognized as independent stellar systems scattered through space, his deft colleague Milton
Humason began photographing the great spirals in an attempt to expand the work pioneered by Lowell's Vesto Melvin Slipher. A galaxy at

a distance of 900,000 light-years was discovered to be moving away from our planet at 125 miles per second; at 135 million light-years another nebula in the Gemini cluster was speeding outward at 14,300 miles per second. "The velocity-distance relation emerged as the mists receded." By the time of the Silliman Lectures, the faintest cluster for which a velocity had been measured was estimated to be more than 240 million light-years distant, hurtling through space at 40,000 kilometers per second.[61] Having pushed the 100-inch to its limit, Hubble and Humason could only await the completion of the Palomar giant before taking up the chase again.

Finally, Hubble had succeeded in demolishing the last underpinnings of the hierarchical universe that appealed to ancient, medieval, and early modern minds alike. Successive studies confirmed what preliminary reconnaissance had indicated: "uniformity throughout the whole of the observable region."[62] The realm of the nebulae is the realm of scientific democracy—proportional, homogeneous, nonsectarian—a republic over which consistency rules and number holds sway above the flux.

Yet missing from Hubble's soon to be published little classic is the final ingredient that made Galileo such a compelling figure. Unlike his Renaissance hero, the astronomer revealed almost nothing of his inner universe. Neither in his private conversations nor in his writings did he discuss the impact of his discoveries on such concepts as stellar evolution or a grand "creation event," first termed the "Big Bang theory" by the astronomer Fred Hoyle in 1950. In the minds of many astronomers, Lemaître's cosmology and Hubble's redshifts had already replaced a vast, static universe with a vast, expanding one, shedding new light on Einstein's relativistic speculation.

Like the secretive trout of his daydreams, the poetic fly-fisherman chose to remain elusive:

> [T]he explorations of space end on a note of uncertainty. And necessarily so. We are, by definition, in the very center of the observable region. We know our immediate neighborhood rather intimately. With increasing distance, our knowledge fades and fades rapidly. Eventually, we reach the dim boundary—the utmost limits of our telescopes. There, we measure shadows, and we search among ghostly errors of measurement for landmarks that are scarcely more substantial.
>
> The search will continue. Not until the empirical resources are exhausted, need we pass on to the dreamy realms of speculation.[63]

CHAPTER TWELVE

"NOW WHOM DO WE WANT TO MEET?"

I

Fred Hoyle, the colorful, unconventional English astronomer, was a frequent dinner guest at the Hubbles' during his visits to Pasadena. A practitioner of the grand gesture, the Yorkshireman was drawn to the terrace at sunset, where the snowcapped peaks, glowing red in the distance, seemed to float above the earth. "They might be the Himalayas," he exclaimed.[1] After dinner and talk of astronomy, he would pick up a volume of Shakespeare's plays and read a particularly moving passage, such as Henry V's address to his greatly outnumbered troops on the eve of Agincourt. During a visit in the 1930s, he informed the Hubbles that it was well known in England that the Nobel Prize committee had discussed the legal possibility of amending the statutes, which contained no provision for astronomy, so that science's highest award could go to Edwin. Guglielmo Marconi, the inventor of wireless telegraphy, was told much the same thing by Caltech's Robert Millikan as the Nobel laureates strolled atop Mount Wilson with the Hubbles.[2]

The highest and most valued of the many awards Hubble received was the Barnard Medal, granted to him in June 1935 at Columbia University by the National Academy of Sciences. The medal, initiated in 1895, was awarded only once every five years. All eight of the previous recipients, including Roentgen, Rutherford, Einstein, Bohr, and Heisenberg, were Nobel Prize winners. Hubble had the double distinction of being the first American as well as the first astronomer to receive the medal. He was cited for his "important studies of nebulae, particularly of the extragalactic nebulae which provide the greatest contribution that has been made in recent years to our observational knowledge of the large-scale behavior of the Universe."[3] The following June, Grace accompanied her husband to commencement exercises at nearby Occidental College, which awarded Edwin his second honorary degree, a doctor of laws. The next week she drove him to Santa Fe Station after dark. They walked up and down the platform, awaiting the arrival of the late train to the East Coast. Feeling terribly low after he got aboard, Grace drove home crying, and found it difficult to see the road in the glare of the oncoming headlights. A few days later at Princeton, moments before he was hooded a third time, Hubble was proclaimed "a Ulysses embarked with his telescope . . . beyond the utmost bound of human thought."[4]

Grace had stayed at home in anticipation of their upcoming trip to England, in September 1936. Before leaving, the Hubbles entertained the balding and bespectacled novelist Sir Hugh Walpole, regarded by contemporary critics as on a level with Thomas Hardy and Henry James. "What American novelists do you read?" Walpole asked Grace over dinner at the Huntington Hotel.

"None," she replied snobbishly.

"Why?" he asked, a startled look on his pinched visage.

"Because their adolescent prattle bores me."

"Yes, I can see that. Everything you say is always right."[5]

Under the topic "Special Data" at the beginning of her journal for 1936, Grace wrote a single sentence: "Los Angeles is a place that is merely a long way from anywhere."[6] Writing to her mother from England later in the year, she expressed her deepest feelings, undoubtedly shared by Edwin: "Aside from all of you and the telescope and some friends, I should treacherously resolve . . . to live the rest of my life in England. Everything about it so completely fulfills my ideals of what a civilization should be, there is such a fine sense of rightness and dignity and beauty."[7]

The Hubbles sailed from Hoboken aboard the *Statendam* in early

September, embarking upon what both would later agree was "our apogee. We said we could never surpass it," a wistful Grace wrote two years later, "the best we could hope would be to have something as good."[8]

She had pronounced a dinosaur and its egg "the finest things in New York" during her honeymoon visit in 1924. This time she used the same words to describe Edwin's recently published book, which they had stopped at Brentano's to look at before boarding their ship. Although Edwin was feeling queasy, as he usually did until he gained his sea legs, the couple accepted an invitation to dine at the captain's table the first night out, after which Edwin beat a hasty retreat to their comfortable cabin on B Deck and the bottle of medication prescribed by his doctor.

On reaching Plymouth, they rented a car and motored through the southern counties on their way to London. Keeping a long-standing vow from his Oxford days, Edwin stopped at Dorchester to show Grace the decimated abbey containing the grave and epitaph of the young woman who had died "a martyr to excessive sensibility."

In London they stayed in a freshly decorated flat owned by the Gore-Brownes, luxuriating in the service provided by the butler and maid. "I believe that the first step in European civilization," Grace wrote in her journal, "was taken when Homo Sapiens discovered that it was easier to coerce a weaker, not so intelligent H.S. to wait on him, than to invent labor-saving devices for himself." It was the English aristocracy that had originated the traditions of how living was to be carried on, and they saw to it that they were observed. She suffered a slight pang of conscience on seeing a charwoman crawling on her hands and knees and dragging a bucket of soapy water along Regent Street, while the well-dressed crowds dodged her as they would an animal. "However, let us not turn Cobbett!" she hastened to add, in reference to England's radical political reformer.[9]

By October they were back in the familiar surroundings of Oxford, where, on the recommendation of astronomy professor H. H. Plaskett, Hubble had been chosen to deliver the Rhodes Memorial Lectures, the first American to be so honored. While her husband polished his offerings, Grace attended a series of talks by Dorothy Sayers, creator of the nobleman-detective Lord Peter Wimsey and one of the first women to receive an Oxford degree. The mystery writer's stark appearance surprised Grace, who likened her to "a Peter Arno drawing of a President of Confederate Women's Clubs." Her lecture, titled "Are Women Human Beings," prompted Grace to remark afterward, "The great body of women could be divided into two classes, parasites and satellites." Grace met with the novelist following her second presentation and was impressed by Sayers's criticism of psychoanalysis and its jargon: "Shakespeare knew as much about complexes etc. as is known today."[10]

Hubble spied the Oxford University Press edition of *The Realm of the Nebulae* in the window of a local bookshop and went inside to purchase a copy. The volume on display was the last, nor were any others to be had in the city. His welling nostalgia led to a reprise of the visit to his old haunts two years earlier, where he evoked the ghosts of days long gone in an atmosphere once again permeated by talk of war. German troops had occupied the Rhineland, while Hitler claimed 99 percent of the popular vote before inaugurating his Four-Year Plan. The Führer and his ally, Italian dictator Benito Mussolini, fresh from the conquest and annexation of Ethiopia, had recently proclaimed the Rome-Berlin Axis even as construction of the Siegfried Line began a new stage in German rearmament. All this amid mounting gossip concerning the recently crowned Edward VIII, who was said to be so deeply smitten by the American divorcée Mrs. Wallis Simpson that the King had lost his head.

Wearing a black gown with heavy black embroidery ordered from his Savile Row tailors, Welsh & Jefferies, Hubble delivered the first of three lectures in the late afternoon of October 29, a dark and misty day. Titled "The Observable Region as a Sample of the Universe," it contained much material from his Silliman Lectures, the Oxford University Press edition of which had sold out before he took the rostrum. "The Role of the Red-Shifts," also largely drawn from previous publications, followed two weeks later, while the final lecture, simply titled "Possible Worlds," was delivered on November 26 to a standing ovation both before and after the speech. Protective of her husband's aura of infallibility, Grace described him as being "perfectly at ease." Indeed, such was his "warmth and colour" (Grace had long since adopted British spelling in her journals) that "you felt that you would vote for him no matter what he was talking about," which was more than either of the Hubbles could say for Franklin Roosevelt, who, much to their displeasure, had just been elected to his second term by a landslide.[11] In scientific lectures, she concluded, "an English audience is far more en rapport than an American audience."[12] Thus it seemed right that the Rhodes Memorial Lectures, titled *The Observational Approach to Cosmology*, should be published by Oxford the following year.

11

While the Hubbles were fending off a deluge of invitations and par-taking of Europe's prize wines courtesy of Edwin's unabashed admirer Philip Henry Kerr, the eleventh Marquess of Lothian and former sec-retary to Prime Minister Lloyd George, an embittered Walter Adams was still smarting over his latest failed attempt to bring the astronomer into line. Having reconsidered his plan to resign the directorship of the Carnegie Observatories, Adams had expended considerable energy the previous year trying to persuade John Merriam that Hubble was not entitled to his full salary during his upcoming visit to England.

While preparing the Silliman Lectures at Lookout Hill, Hubble had intentionally snubbed Adams by writing directly to Merriam, asking for another extended paid leave so that he could deliver the Rhodes Lec-tures. Merriam in turn wrote Adams, seeking his advice. Taking his cue from Merriam's concern that Hubble had been absent from Pasadena some eleven months during the past three years, the angry director immediately dispatched a night letter to Washington recommending that Hubble be granted a leave of absence without pay, except for the normal vacation period of one month. He then drafted a more detailed document in which he discussed Hubble's interest in lectures offering liberal honoraria and requiring long absences: "I confess that I have not liked for some time past the casual way in which he assumes he is entitled to privileges of this character." Merriam concurred and wrote Hubble at Lookout Hill to inform him of their joint decision.[13]

With Grace at his side, Hubble composed a masterly reply. To begin with, the audience for the lectures would be exceptional in terms of public opinion and educational value. The honorarium aside, the prize was going to an employee of the Carnegie Institution, which, in and of itself, was reason enough to keep him on the payroll. Morever, the trip would provide an excellent opportunity to establish prolonged and intimate contacts with the leading theoreticians in the field of nebular cosmology. This benefit, too, would "seem to justify the continuation of salary quite apart from the question of honors." Finally, the hono-rarium of 500 pounds would just about cover the Hubbles' expenses. "With no salary we would be out of pocket to the extent of the upkeep of our home in Pasadena and the savings we customarily place in our reserve funds against the future. To those with our modest incomes, these items seem considerable and the question of salary is important." Surely Merriam would agree with certain of these views, and "I hope you agree with them all."[14]

Adams, who was provided with a copy of Hubble's letter, refused to be swayed by its arguments, all of which he had heard before and considered disingenuous. The long absences were only part of the problem. Hubble had spent three months writing the Silliman Lectures before leaving for New England, and Adams assumed the Rhodes Lectures would require at least as much, if not more, advance preparation: "It would be necessary to count on a period of at least six months during which Dr. Hubble would be carrying on no original investigations." And while it was pleasant for him to be accompanied on his trips by Mrs. Hubble, her presence was by no means essential. If he went alone his expenses would be greatly reduced and more than compensated for by the honorarium.[15]

Adams, it seemed, was about to prevail. In January 1936, Merriam drafted a long memorandum raising issues that had concerned him in the past. Hubble had never been a team player, considering himself a freelance. The 1934 Halley Lecture was a perfect example. "It . . . seems to have been almost solely a matter of Dr. Hubble personally and no reference appears to be made in publications to the fact that he was a member of the Observatory or of the Institution," which had helped pay his way.[16] Nor had Merriam seen any mention of the Carnegie Institution in publicity surrounding the Silliman Lectures. The president had also been informed that Hubble had not so much as acknowledged a seven-month-old invitation to lecture at the Carnegie Institution itself.

In his most candid letter to date, Adams attempted to put Hubble's behavior in perspective.

In Dr. Hubble's case I find very great difficulty in understanding his attitude toward the organization which has made his scientific work possible. I do not think it in any way deliberate but results from an extreme form of individualism and personal ambition, together with a type of obtuseness regarding his relations to the Institution and other scientific men. I have recognized this curious "blind spot" in almost every important dealing I have had with him. It has injured him seriously among astronomers since it results in a failure on his part to join in cooperative plans in which the Observatory should play an active role. At the Cambridge, Massachusetts, meeting of the International Astronomical Union in 1932 he lost his chairmanship of the Commission on Nebulae because he neither attended the meeting nor prepared a report, and at the Paris meeting last July he contributed nothing as a member of the Commission. I could see that this was resented by many astronomers.

Adams closed by reiterating his position that Hubble should receive nothing beyond his vacation pay:

> If he had not visited Europe recently the situation might be different, but in view of his stay in England at the time of the Halley Lecture and his long continental tour a year or two previously he could have little to bring back to us through his contacts with scientific men abroad. Furthermore his scientific interests are limited so closely to his own field that the observatory would not benefit in a broad way from this trip.[17]

Like a good soldier, Adams closed with the promise to do all within his power to convey Merriam's point of view to Hubble should the president decide not to accept Adams's recommendation.

Once again, Merriam sided with Adams. Yet as the moment approached to face Hubble down, he began to temporize, as he had in the past. Though Hubble had received a recent salary increase he was not being paid as much as Merriam would like or the astronomer deserved. "Would it not be possible to suggest that Dr. Hubble have at least one month's leave with salary and his vacation, as opening the way for doing all that Dr. Hubble desires?"[18]

Realizing that the game was up and that he had been checkmated again, Adams threw in the towel. During a long talk with Hubble in March 1936, he told him of his decision to recommend a leave of three months with pay. Still, the director did not hesitate to remind him that he had received many more privileges from the observatory than any other member of the staff and that he must set a limit "at some point" regarding invited lectures. "Also that it would be desirable to settle down to a period of steady investigation . . . since the Rhodes lectures must merely be a restatement of the material of the Silliman lectures before a different audience." While Adams could not be entirely certain, he came away from the meeting with the feeling that his comments had had "a salutary effect."[19]

The director's impression was correct, if only in the short run. On the eve of his departure for England, Hubble provided Merriam with a synopsis of the Rhodes Lectures, a more technical version of which had already been submitted for publication in the Mount Wilson *Contributions*. He also agreed to stop in Washington on the way home for the purpose of delivering a lecture at the Carnegie Institution.[20]

The prospective collaboration Hubble wrote Merriam about more closely resembled hero worship. Grace was told by Lady Willetts that a scientist friend had pronounced Edwin the greatest man in the world today, the greatest astronomer and the greatest philosopher; "he could put Jeans and Eddington in the fire."[21]

Two days later the Hubbles were driven to Cleveland Lodge in Dorking to dine with the Jeanses. They were both enchanted by the Viennese-born Lady Jeans, Sir James's second wife. She was tall and lithe, and her long oval face was framed by thick brown hair worn down to her shoulders, over which was draped the fur of a white fox. Susi's childlike nature provided an intriguing contrast to her profession as a concert organist. A heavy accent and amusing grammatical slips reminded Grace of a character out of *The Constant Nymph*. While the men retired to the library after dinner Susi played Bach for Grace on her new custom-built organ: "It was like music in a Cathedral, not the orchestral attempt of the modern organ."[22] An accomplished organist himself, Jeans often played for three or more hours a day, but never for other people.

Once back in their hotel room, Edwin told Grace that Jeans had informed him that he had written a review of *The Realm of the Nebulae* for *Nature*, the first he had consented to do in a decade. Then a slight smile had crossed the theoretician's normally inscrutable face: "If your last chapter and the material in your Oxford lectures are as sound as I have every reason to believe them, you have destroyed Eddington's work of the last five years."[23]

The following Sunday the Hubbles were taken by taxi to Cambridge Observatory for high tea with Eddington and his sister. White-haired Sir Arthur stared silently into the distance, his deep-set eyes seemingly fixed on the infinite. An annoyed Grace, who could not stand being ignored, "attacked him on three strategic points," but without success. She envisioned herself as a savage who bows in obedience before a stone image, lays a flower at its feet, and then runs shrieking into the forest after muttering a prayer.

She tried another tack after learning from the younger Miss Eddington that her brother was about to depart for London and a meeting of the Royal Society. "Perhaps you will read a mystery on the train?" Grace inquired.

"Bulls-eye!" she wrote triumphantly in her diary. Eddington's favorite was Father Ronald Knox, and he preferred Agatha Christie to Dorothy Sayers. It would have been very nice, he observed half jokingly, if Dr. Millikan had really been murdered by Sir James Jeans in *Murder in the Church* instead of merely being under suspicion.

Satisfied with this modest success, Grace suggested a taxi, whereupon Eddington rose, saying he wanted Edwin to come into his study for a discussion. While his sister explained her mastery of the Torquemada crosswords in the *Observer* to Grace, Sir Arthur told Hubble that he had assembled all of the astronomer's recent papers and had checked up on several things that he had opposed. Hubble, it turned out, was right in each instance. This feat was accomplished in just three days, whereas it would have taken most mathematicians weeks.[24]

That same November evening the Hubbles attended a dinner and late party in their honor in Cambridge, where they were introduced to Charles Darwin, the grandson of the great evolutionist. Also present was the brilliant young Indian astrophysicist Subrahmanyan Chandrasekhar and his bride, Lalitha, shivering in native costume. The wife of the future Nobel laureate murmured to Grace that she had read Edwin's book and had deeply enjoyed it.

His Nobel Prize securely in hand, the atom-splitting physicist Ernest Rutherford took Grace in tow. Hearty and full of vigor, the ruddy-faced icon sat her down on a sofa and began expounding on American politics. He considered Franklin Roosevelt a great man who, in the face of massive opposition from big business, was conscientiously attempting to improve the lot of the working class by providing benefits long since enacted in England. The only alternative was revolution.

The conversation then shifted to the Hubbles' host that afternoon. Eddington's last three books, Rutherford volunteered, "are queer; he is like a religious mystic, and is not all there. I don't pay any attention to him." The two physicists still walked together late at night along the Backs behind the colleges, planning how to murder each other while establishing an alibi. "I've sat by him for thirty years at lunch, I've never heard him say a thing worth remembering. It all came out in his books, for of course he was a brilliant man."[25]

Their last two weeks were spent in London, where all the talk was of the King and Mrs. Simpson. The papers sold out as fast as they appeared and thousands walked the streets reading, oblivious to everything else. Rumors, all of which Edwin dismissed, were rampant.

A light snow was falling on December 10 when word came of the abdication. Edward, his voice sounding flat and dispirited, addressed his former subjects the following day. "His words," a disgusted Grace wrote, "were those of a stubborn, infatuated man incapable of realizing the implications of what he was saying."[26] For relief the Hubbles visited the zoo and their friends the gorillas, who obliged by striking poses of "The Thinker" and the Emperor Napoleon. Then Hubble, already planning ahead, sat down and drafted a letter that must have astounded John Merriam and would have enraged Walter Adams, had he seen it.

The lectures at Oxford had gone very well, and had aroused an unusual amount of interest. "Moreover, the prolonged contact with English astronomers has been extraordinarily profitable. . . . I am so impressed . . . that I should like to talk with you concerning the possibility of making occasional tours a routine part of the Institution's program of research."[27]

III

On the night of March 4, 1937, the Hubbles attended the annual Motion Picture Academy Awards ceremonies at the Biltmore Bowl in Los Angeles as guests of the director Frank Capra. Capra had already won an Oscar for It Happened One Night and was on the verge of receiving another for Mr. Deeds Goes to Town. He was also president of the Academy and saw to it that the Hubbles were seated at his table, along with actresses Jane Wyatt and Luise Rainer, comedian George Jessel, and actor turned playwright Clifford Odets. Capra began the ceremonies at nine o'clock by introducing the world's greatest living astronomer and asking Hubble to rise. As he did so, three huge spotlights converged on him, while the auditorium resounded to applause. "Next to invoking the Deity," a star-struck Grace gushed, "this seemed for some mysterious reason the correct way to begin."[28]

Driving up the new Angeles Crest Highway to peer through the 100-inch with Edwin Hubble had become a popular diversion among the famous, who had learned of his work via the popular press and the Hollywood grapevine. Usually notified of such visits in advance, Hubble made certain Grace was on hand to serve as hostess. His recent discovery of Comet Hubble, the first object of its kind to be found with a Mount Wilson telescope, had prompted the British actor Leslie Howard, who was filming Gone With the Wind, to drive up with his son "Winkie," a Cambridge undergraduate. They were joined on the mountain by John Balderston, the handsome Beverly Hills journalist turned playwright, and his wife, Marian, both of whom had spent much of their lives in England and had become close friends of the Hubbles. The show included Jupiter, a globular cluster in Hercules, and a planetary nebula. Howard especially requested a spiral nebula and Hubble obliged, urging the actor to be generous to his imagination when looking at it.[29]

Also present was H. H. Plaskett's assistant, a young Englishman named Paige, who attempted to emulate Hubble by staying up all night and matching him pipe for pipe. When he admitted to failure on both counts the next morning, Grace, who had spent the night in Kapteyn Cottage, quipped, "You theoretical men can talk prettily, but you see you just can't take it." When someone else remarked that Western Europe was dying and that it was up to America to carry on the torch, she thought to herself, "Dubious outlook for torch."[30]

One day in early 1937, Anita Loos, the pixie-like screenwriter and playwright whose best-selling flapper novel *Gentlemen Prefer Blondes* became one of Broadway's biggest musical hits, wrote a letter to the observatory asking about visiting privileges. Hubble, who recognized the name, called the startled writer at M-G-M Studios and arranged for Loos and her husband, John Emerson, to visit Mount Wilson at the end of April. They arrived after dark and could barely make out the mechanical paraphernalia, heightening the dramatic tension. Wearing high, laced boots and a flannel shirt open at the collar, a "very tall and muscular" Hubble was working far overhead on what appeared to be a rickety little platform which swung about to follow the movement of the telescope. To Loos he was so much the picture of how a great man ought to look—and seldom does—that had he been auditioning for the part of Edwin Hubble he would have been turned down as un-realistic by the casting director: "You can't have him look like a bloom-ing Clark Gable!"[31]

Loos wrote a note of thanks and a request for Hubble's continued indulgence. Cole Porter, the composer of *Anything Goes* and *You're the Tops*, was coming to town with his wife, Linda Lee Thomas, and both were hopeful of visiting the mountain. Could she bring them up on Sunday? "They are enormously rich and their house in Paris is one of the great rendezvous. Cole, by the way, is greatly admired by [the conductor Leopold] Stokowski," who Loos understood was also a friend and admirer of the Hubbles.[32]

Soon to become one of Grace's few confidantes, Loos smoothed the way for others. In 1934, after appearing opposite a string of leading men that included Ronald Colman, Gary Cooper, Clark Gable, and Brian Aherne, the Oscar-winning Helen Hayes stunned Hollywood by renouncing film for the theater. Accompanied by Loos, the writer Al-dous Huxley, and her irreverently flamboyant husband, the playwright Charles "Charlie" MacArthur, who had dragged Hayes to Hollywood, the actress visited the observatory. "We were intrigued," she wrote, "by the fact that a tiny bit of glass, which could fit into the socket of one man's eye, could swell out to embrace the whole universe. It seemed to put us in hailing distance of all eternity."[33]

Huxley joked that, in not too long a time, lenses would become so strong that a man would be able to sit in front of a telescope, cut through the atmosphere and the galaxies, and see his own backside. Everyone laughed except the fun-loving Charlie, who was suddenly the serious one. "No, that's not at all what's going to happen," he said softly. "We're . . . going to see God's reproachful finger wagging at us—telling us not to be so fresh."[34]

Grace wrote of meeting Ethel Barrymore and Sir Anthony Eden, the future Prime Minister, at the home of Douglas Fairbanks; at the Balderstons' were Walter Lippmann, Igor Stravinsky, and the playwrights Robert Sherriff, John Van Druten, and the Pulitzer Prize-winning Sidney Howard. They also became friends with Salka Viertel, a Polish-born screenwriter at M-G-M, whose home was a haven for Hollywood's more intellectual foreign visitors—Thomas Mann, Bertolt Brecht, and Sergei Eisenstein.

In October 1937, the Hubbles accepted the first of many invitations to spend a Sunday at the Loos house located on Santa Monica's Ocean Front, part of the "Gold Coast" inhabited by the Hollywood elite. To the south lived Norma Shearer, Mervyn LeRoy, Darryl Zanuck, and Louis B. Mayer; to the north Hearst's paramour, Marion Davies, had recently moved into her grandiose "beach cottage." On the ground floor of Loos's white stucco home was a large living room with windows facing the Pacific, a library, and a dining room that opened onto a garden and a swimming pool. The main hall contained a curving staircase, which led to two guest rooms and two master suites, a fully equipped gymnasium and a steam room. Occasionally, when the fog closed in early, the Hubbles would spend the night. Edwin would gaze into the impenetrable gray mist from the balcony adjoining their quarters. The simple furnishings, consisting of wicker chairs, chintz curtains, and sisal rugs, were selected by the comedienne Fanny Brice, then moonlighting as an interior decorator. Loos dubbed the style "early Baby Snooks."[35]

A low fog hung over everything that first Sunday, while the ocean hissed quietly. Aldous and Maria Huxley were members of the small party, which included the actress Tilly Losch, whom Grace described as looking very much like Undine, the water sylph from the fairy romance by Fouqué. At six feet five inches, Huxley possessed what Anita, who was no larger than a minute, described as the height of a giant with the look of an archangel drawn by William Blake. With his severe myopia, his eyes made him appear to be focusing on things above and beyond what ordinary mortals saw. The Belgian-born Maria, a tiny brunette in her late thirties, still had the figure of the ballerina she had studied to be as a girl. But the dream was shattered when she was sent to London with a group of children to escape the Germans at the beginning of World War I. Maria later became the teenage bride of the dashingly handsome genius, with whom every prominent English bluestocking was in love, or thought she was.[36]

Aldous did most of the talking over Chinese food, likening Hubble to a saint showing them a glimpse of heaven—so different from most of the crackpot population of Southern California and its bizarre religious cults: the "Four Square Gospel" of Aimee Semple McPherson, something called "The Great I Am," and the mass writhing on the floor of the Holy Rollers. Grace was taken aback when Maria spoke of her belief in palmistry, and for once spoke without thinking: "I said, how surprising it was that the future was revealed by lines in the hand." A protective Anita shot a wicked glance at Edwin.[37]

The Huxleys left California in November to undertake a lecture tour with Aldous's fellow intellectual and alter ego Gerald Heard. But Heard had fractured a leg in Iowa, forcing Huxley to carry on alone. In late January 1938, they returned to Hollywood, where Aldous, who had published five books in the previous two years, planned to continue writing novels while doing scripts for M-G-M, the first based on the life of Madame Curie. At three on a Sunday afternoon in March, Maria called Grace and asked the Hubbles to come over in an hour.

They found the Huxleys in a sparsely furnished apartment on North Laurel Avenue. Maria made tea and sat on the floor with Grace while Aldous asked Edwin questions. He had taken off his thick glasses, and his eyes had the tragically blind look of a modern-day Tiresias. Otherwise, Grace thought his face, beneath a mass of thick dark hair, quite beautiful and youthful, very white, the mouth large and sensitive. A sunburnt Edwin, who had finally met his physical and intellectual equal, appeared serene, displaying what a theosophist might call an aura of kindness and ease—but still very much the recumbent lion, ready at an instant's warning to spring to the alert. Grace wrote that together the reformed Missourian and the scion of English high culture "would have given a painter a chance."[38]

The Hubbles were pleased to learn that the writer's views on the lower classes mirrored their own. Huxley described *The Grapes of Wrath* as "a dull, long book, written so badly, [it is] impossible to read it all." Its author "is so ignorant he doesn't even go as far as utilitarianism."[39]

Grace's journals also reflected the ingrained racism of the day. When last in England she had engaged a radical Labour M.P. in a political debate, arguing that the Mexicans—part Indian, part Spanish, part Negro—were in her eyes "a mongrel race." When her opponent countered that a Negro child has the same reflexes and physiological constituents as a white child, and is a victim not of nature but of its environment, Grace, who favored the term "darky" or occasionally "Uncle Tom," retorted, "Take a thoroughbred colt and a common colt, the same can be said of them. Try to make a race horse out of the second."[40] When she and Edwin were visiting the Supreme Court building in Washington, she could barely stand the sight of Associate Justice Hugo Black, the former Ku Klux Klansman turned civil libertarian: "Black, rocking in a rocking chair, looking with his low insignificant rat-face, as if he should be, not judging, but in the Dock."[41]

Through the Huxleys, the Hubbles were introduced to Aldous's good friend Jake Zeitlin, a Los Angeles bookseller and manuscript dealer with Jamesian restraint and a colorful past that included spells as a hobo and bakery supply salesman. When Grace promised Zeitlin that they would visit his shop, her husband protested, "My dear, we can't afford to go to his place."

"Oh yes, we can give him a back cheque."

"Do," Zeitlin parried, "and I'll sell your autograph."[42]

Zeitlin had been commissioned by the widow of D. H. Lawrence, Frieda von Richthofen Lawrence, to negotiate the sale of the writer's manuscripts. As frequent summer visitors to the Lawrence ranch in the mountains above Taos, New Mexico, the Huxleys repaid Frieda's hospitality by inviting her to Hollywood, where she was introduced to the Hubbles at dinner. Grace alternately pictured her as "an unbeaten Valkyrie" and as one of the women Tacitus wrote of who followed their warriors through the ancient forests of Europe—"a broad-shouldered, deep-chested, shapeless bulk in a burgundy-red chiffon dress. Her blue eyes danced . . . straight through one, and she laughed a great deal." Over dinner Frieda spoke of the poet Robinson Jeffers and his wife, Una, whom she cautioned the Huxleys against: "You won't like him, Aldous, and you, Maria, won't like Una. His poetry is all about incest —old stuff and it doesn't make you shudder at all."[43]

Feeling "rather crazy," everyone drove over to Harpo Marx's after dinner. On the walls were a Rembrandt rendition of the four brothers, a Gainsborough caricature of Harpo as the Blue Boy, and another of Harpo by Hals or Vandyke. Huxley talked of writing a script for the frenetic quartet; Groucho as Karl Marx, Chico as Bakunin, Harpo as Engels, with Zeppo yet uncast. Or perhaps an astronomy film, with the comics pressing the wrong buttons and looking into the 100-inch and seeing Harpo's grinning face. The latter had recently given a dinner which Huxley described as a perfect reflection of the host. First some-one upset a platter of canapés, then Huxley in his blinded way put his hand deep into a dish of fish and mayonnaise. The china was frightfully expensive, yet all through the meal one could hear things being broken and smashed in the kitchen.[44]

In Hollywood as in the heavens, there were stars of varying magni-tudes, and none outshone the mythical Charlie Chaplin. Though the same age as Hubble and destined to outlive him by decades, the forty-nine-year-old film actor, director, producer, writer, and composer was completely gray when the Hubbles first met him at the Huxleys' in November 1938. Chaplin's blue eyes were watchful and aware, his smooth face rippled with expression, like a pool to a breeze. A tense nervous carriage belied a shy self-consciousness, for he was always thinking of what effect he might be having on a guest while registering secretly to himself the effect the guest was having on him.

At Chaplin's side was his wife, the actress Paulette Goddard, consid-ered by some to be the most beautiful woman in Hollywood—and one of the most clever. Anita Loos remembered that before Paulette, Chaplin led a meager social life. She, on the other hand, enjoyed com-pany, and parties at the Chaplin home on Tower Drive became small, celebrity-studded gatherings with after-dinner entertainment provided by the host.

Grace wore her new black dinner dress, but she couldn't begin to compete with the dark-haired actress, who was sheathed in velvet and bedecked in jewels. After dinner Charlie did an imitation of the Sheik to demonstrate the "old style," and guest Charles MacArthur, an ex-reporter, told horrible stories about the hangings he had covered. Dur-ing lunch some months later, Grace watched as Chaplin's accent and mannerisms became more and more British until, by the end of the meal, he had reverted to the complete Englishman, charming all pres-ent, including a radiant Lillian Gish, who had brought along her small white terrier. Helen Hayes, wearing gray gabardine slacks, gray sport shoes, and a flowered shirt "looked better than last year," according to an unusually catty Maria. Hayes said the finest performance of her life was given one night in Columbus, Ohio. Afterward she had said to herself, dazed, "I was really *great* tonight, why did it have to be in Columbus, Ohio?"[45] After the first of what were to be many meetings with Chaplin, the Hubbles drove home in the moonlight saying to themselves, "We have at last met Charlie, now whom do we want to meet?"[46]

IV

On the morning of January 27, 1938, Hubble drove down to Mount Palomar alone. After returning late in the day, he confessed to Grace that he had left home expecting to be matter-of-fact, but the sight of it got to him anew. "The empty dome is like a church."[47]

George Ellery Hale, now sixty-nine and seriously ill, could only envision the new landmark in his mind's eye. Day and night he dreamt of a last visit to Palomar, but it was not to be. Arteriosclerosis had forced his wife, Evelina, with whom he had fallen in love at age thirteen, to hospitalize him in Pasadena's Las Encinas Sanitarium. With the exception of Adams and Seares, few had seen him since his retirement in 1924. Indeed, to many members of the staff he was almost a mythical being.

According to the old axiom, death comes in threes. Francis Pease, Hale's fellow dreamer and architect of the blueprints for a 300-inch telescope, succumbed in 1938, as did thirty-nine-year-old Sinclair Smith, who had helped design the drive system for the yet unnamed Palomar giant. At one-thirty in the afternoon of February 21, George Ellery Hale passed away after slipping into a coma. Two days later Hubble, who happened to be on a run, drove down Mount Wilson to serve as a pallbearer at the funeral, and then went back to work at six in the evening, something Hale himself would have appreciated.[48]

On the day of the funeral, *The New York Times* published an editorial suggesting that a fitting monument be erected to Hale's memory. "If the two-hundred-inch mirror could be called the 'Hale mirror,' or better still, if the whole plant which is to be erected on Palomar Mountain could be called the 'Hale Observatory' both astronomers and the public would be perpetually reminded of their debt to one of the most eminent men of science this country ever produced."[49] At its dedication ten years later, the astronomer's most ambitious brainchild was christened the Hale telescope.

The telegram Hubble had been told to expect after Hale's death arrived one month later, on April 30. He had been unanimously chosen to fill the astronomer's seat on the board of trustees of the Huntington Library and Art Gallery. In May he took his place at a massive, ornately carved table in the great neoclassical library, joining his fellow trustees: Herbert Hoover, the former President; Dr. William B. Munro, a professor of history and government at Caltech; the rarely seen Archer Huntington, son of the institution's founder; and Caltech's Robert A. Millikan, who had recently assumed the position of chairman. Also in attendance was Allan C. Balch, chairman of the Finance Committee, whose millions in stocks Hale had tapped to underwrite the construction of the Athenaeum on the eve of the lingering Depression.

Few knew the Huntington's graveled walks and countless botanical specimens, covering some 250 acres, better than the Hubbles. Accompanied by the Crottys and other friends such as Max Farrand, the Huntington's director and Edwin's weekly bridge partner, they spent Sunday mornings strolling the perfectly manicured grounds. The oldest roses, dating back to the Middle Ages and the Renaissance, were in constant bloom in the Shakespeare garden, not far from other species first introduced into Europe from China about 1800. A short walk away but a world apart lay the Japanese garden, distinguished by undulating pathways, a five-room imported house, gurgling streams, and small, silent lakes around which painstakingly trained plants had replaced the aboriginal tangle of wild grapevines and poison oak. Elsewhere Mexican and Japanese cycads, ancient seed-bearing relics that were once fodder for the dinosaurs, abounded, along with palms and cactus of every variety and description. Populating the terraced lawns and graceful loggias were imported sculptures of gesticulating gods and cavorting nymphs, creating the impression of a Versailles among the palms.

It was all the handiwork of Henry Edwards Huntington, whose electric-railway empire, famous for its bright red cars and punctual service, had spread tentacle-like out of Los Angeles in the mid-1890s. To this were added vast real estate holdings that bordered the expanding trolley lines, including a 600-acre private estate in San Marino, where nature was tamed by master gardeners and art and rare books worth millions were collected.

A man of simple habits, Huntington rose at six, washed and trimmed his elegant handlebar mustache, breakfasted at seven, and declared himself open to the business of the day no later than eight. Abstemious to a fault, he was haunted by the pennies wasted when an unused light was left burning, yet he would not so much as bat an eye at paying $10,000 or more for what Arabella Huntington, his beloved but irrepressible second wife, called "an old book."

Whereas most collectors sought out individual works, Huntington acquired entire libraries, often purchasing them on the eve of a major auction, no matter the cost. A passionate Anglophile, he soon amassed a collection of documents unequaled in the United States for the study of medieval England and a collection of rare books amounting to some 40 percent of the 26,000 items published in England, Scotland, and Ireland between 1475 and 1641. To the Ellesmere manuscript of

Chaucer's *Canterbury Tales* were added numerous Shakespeare folios, illuminated Books of Hours, William Langland's *Piers Plowman*, rare first editions of Spenser and Milton, a copy of the Gutenberg Bible, the original manuscript of Benjamin Franklin's *Autobiography*, and so many thousands more it would take cataloguers decades to catch up with the frenzied collector.

At the same time, the railroad magnate used new money to acquire old masters. Concentrating on the English school of the eighteenth century, he filled his newly constructed Beaux Arts mansion overlooking the San Gabriel Valley with twenty full-length portraits by Joshua Reynolds, Thomas Lawrence, George Romney, and Thomas Gainsborough —the latter's masterpiece, *The Blue Boy*, serving as the heart of the collection. The paintings were complemented by British sculpture, French provincial furniture, Boucher tapestries, and fireplaces of Italian marble in which only the wood from orange trees was burned, a tribute to Louis XIV, the Sun King, who literally bathed his rancid court in the scent of orange blossoms.

Homer Crotty was not long out of Harvard Law School when, one summer afternoon in 1925, the young attorney was taken to the Huntington estate by a senior partner of Gibson, Dunn & Crutcher. There he witnessed the signing of the historic documents that would soon open the Huntington Library and its invaluable collections to scholars for all time. Next to an ailing Henry Huntington himself, no one was more deeply pleased by this development than George Ellery Hale, who had been urging his millionaire friend to enter into such an arrangement for over twenty years. As head of the newly established board of trustees, Hale had laid out the program of research and was largely responsible for hiring Dr. Max Farrand, a Yale University professor and Benjamin Franklin scholar, as the first director of research.

Hubble's universe now consisted of Mount Wilson and the nebulae at one end, the Huntington and great literature at the other. The Hubbles became close friends with Godfrey Davies, a member of the permanent research staff responsible for the collection in English history. Davies sometimes joined Grace and Edwin during their Sunday walks, when the deserted grounds were closed to the public, creating the illusion that the pipe-smoking astronomer was surveying his own country estate in the company of his lady and chief overseer. Indeed, the only things missing were the trout.

Adding to this sense of proprietorship was Davies's willingness to show the Hubbles' friends through the library at a moment's notice. One such guest was the novelist Susan Ertz, Grace's childhood playmate back in San Jose, who now lived in England, where she occasionally entertained the Hubbles at her fifty-acre farm in Sussex, Pooks Hill. Ertz wrote that Edwin's affection for the country was "a hearthlove, almost a filial love." She remembered him stooping while hiking a path on the Sussex Downs and picking up a piece of flint, which he cradled in the palm of his hand as if it had been a bird. "What did Edwin not know? Birds and rocks and Downmen, fishes and prehistoric cave-paintings and wines and the best thoughts of the best thinkers and so much else besides. But there was never any display of knowledge; nothing that did not come out spontaneously in its own right and natural context."[50]

Grace took her friend to the Huntington when she came to Pasadena, and they wandered the normally closed stacks, thumbing through rare first editions of the Brontë novels and reading aloud from the Shakespeare folios. As if on cue, Max Farrand put in a surprise appearance and began reciting Act I, Scene 1 of *King John* from memory, the very play Grace had been reading the previous night. Time, as always at the Huntington, seemed to stand still.[51]

v

When announcing Hubble's appointment to the Huntington's board of trustees, a facetious reporter for the *Los Angeles Times* wrote that "Dr. Edwin P. Hubble . . . is facing a serious situation, but at last report was bearing up bravely." The astronomer, it appeared, "has next to no chance to win further honors. Already he has won them all with the possible exception of a minor medal or two."[52]

The most recent recognition had come two months earlier, in March 1938, when Hubble was awarded the Bruce Gold Medal by the Astronomical Society of the Pacific at its headquarters in San Francisco. He was the thirty-third recipient, his name added to a distinguished list that included astronomers from nine nations, among them Sir Arthur Eddington of England, Henri Poincaré of France, and Mount Wilson colleagues George Ellery Hale, Walter Adams, and Harold D. Babcock. Yet except for his formal address, during which he spoke to an enthralled audience about remote nebulae so bright they could be identified at a distance of one billion miles, things had not gone well.

The next morning, the *San Francisco Chronicle* announced that Edwin Hubble had been given the "Astrology Medal." And though reporters had taken numerous flashbulb pictures and copious notes, the story was all but crowded out by accounts of violent crimes, causing a miffed Grace to bristle over breakfast: "If you want to get your picture in a later edition, try slapping the waitress as we go out of here. That is news, cosmic things don't count."[53]

During the return drive, Edwin had insisted on stopping off to visit the place where Grace had spent her childhood. Though filled with misgivings, she reluctantly agreed. They arrived in San Jose to find what Grace described as a "rubbishy, derelict, and confused" city. The red house once owned by her parents survived, but those of their neighbors, the Hatches and Brasseys, had been torn down and replaced. "The whole thing," she wrote in her journal, "is like a very bad dream; I can't precisely think why. Just like Cousin Charles Moore appearing out of some dim Hades at the talk the night before last, muttering, 'Cousin Grace, you are just as lovely as ever.' "[54]

If Grace was haunted by the ghost of Earl Leib she never told Edwin, let alone mentioned it in her journals, which he read for accuracy with her approval. Yet her open revulsion on visiting a place of such painful associations was a way of reassuring her husband—as well as herself— that he was the only one who had ever mattered. Nor was the outwardly poised maven always so self-confident as she appeared. Grace some- times had nightmares of talking dogs and wrote of awakening every morning "humiliated by the mental and moral lowness of my subcon- scious mind."[55]

The disappointing trip north concluded with an overnight stay at Carmel, on the Monterey Peninsula, as guests of Robinson and Una Jeffers. The personification of the alienated twentieth-century man who knew too much, Jeffers had moved so far west that he came to the end of the land. There he single-handedly built Tor House, his granite castle overlooking the sea, in whose observation tower the poet worked in splendid, albeit tortured, isolation.

The Jefferses, as usual, were hardly speaking to one another. This caused the Hubbles to focus on their other guest, the Indian giant Tony Luhan, husband of the heiress Mabel Dodge, who had once competed with Frieda Lawrence for D.H.'s affections. A friend and patron of leading artists and writers, Dodge had moved to Taos, New Mexico, where she met Luhan and erected her famous compound modeled after the ancient Indian communal dwellings. Grace described Tony as "broad as a barn-door, and very thick as well." He was dressed in English-cut riding breeches, new riding boots, a coat, and a rather odd shirt and tie, which seemed to have come from a trading post. His hair, in two tight braids, was plaited with rose-colored ribbons. "Very still and dignified and simple." When the Hubbles reported this bizarre meeting to the Huxleys a few days later, Maria told them that Tony fancied himself a Don Juan and had a superstitious hold on Mabel: "Whenever he needs money he puts 3 owl feathers on her pillow."[56]

With war an ever more distinct possibility given Hitler's recent move into Austria and Chamberlain's obsequious performance at Munich, English artists and intellectuals took advantage of what might be the last opportunity to visit the United States in a long time. Sensing that their lives were also about to undergo a profound change, the Hubbles purchased a new La Salle, whose multicolored dashboard lights they found almost as entertaining as a Disney movie, and decided to go out even more often than in the past. Having located a transplanted London tailor in Los Angeles, Edwin ordered new suits, shirts, and ties. "There is a certain moral satisfaction in being properly dressed," he remarked to Grace as she was modeling dinner dresses at home.[57] Though Grace almost never wrote of her husband in the negative, she conceded that he had "a naive vanity" about his clothes and loved having them fitted.[58]

Among their guests at the Athenaeum was the Shakespeare scholar A. L. Rowse, who regaled them with stories of his early poverty and desperate struggle to succeed, which he later recounted in A Cornish Childhood. When Rowse arrived at Oxford as a scholarship student, he knew so little of etiquette he had to be shown what forks and spoons to use. He asked the Hubbles if they knew his old nemesis Gladwyn Jebb: "He is a snob, a cruel snob." The next evening at dinner at Douglas Fairbanks's, Grace was seated next to Jebb, who volunteered, "I hear you met Rowse; I'll bet he told you about those forks and spoons."[59]

During a dinner at the Huxleys', Grace was seated next to the novelist's brother Julian, the biologist and writer against whom Edwin had played soccer in his Rhodes Scholar days. Edwin was given the place of honor beside Bertrand Russell, who confessed to a weakness for Agatha Christie mysteries. During one of their visits to England, the Hubbles had concluded that the three elements of "the good life" as expounded by the philosopher were well worth embracing:

1. Creative work, or some effective contribution.

2. Freedom of movement, which would imply economic security and independence.

3. Living a full life, on terms of equality anywhere with anyone.[60]

Hoping to pass the evening on a high intellectual plane, the Hubbles were disappointed by Russell's preoccupation with his fate. The long-standing pacifist and his wife would soon have to go at great expense to Mexico, provided the authorities would admit them, and then, if they got there, the United States might not let them come back. Grace offered to contact a certain well-connected friend on the Russells' behalf, but the annoying lament continued. "[He] insisted on being persecuted." She followed Edwin's longing gaze across the room where the actor Ronald Colman, who had a head cold from sitting in a massive refrigeration unit during filming, was reverently listening to a discussion dominated by Aldous.[61]

Seventy-four-year-old H. G. Wells was sent over to lunch at Woodstock Road by Maria in the Huxleys' smart new Cadillac. Possessed of a small roundish head, a clipped mustache, and sandy hair, which he parted very low on the side to compensate for a bald pate, Wells stared apprehensively at Hubble through blinking blue eyes. The author of *The Time Machine* and *The War of the Worlds* seemed utterly terrified of the large man who had pushed back the boundaries of space with a less imaginative device than the ones with which his characters traveled across the centuries and battled one another for control of the universe. He spoke in a nasal, squeaky voice, repeating, "Yiss, yiss!"

Taking the offensive, Wells opened fire by attacking science, saying it shouldn't be called science; it should be called research. True science was political science. When Edwin refused to take the bait, a less self-contained Grace rose to the surface: "And Mary Baker Eddy and Christian Science should be called science too," a retort that secretly pleased Edwin.[62] His ego deflated, Wells turned "very cozy," and began telling stories of Henry James, George Moore, and their days at Rye House. When the time came for him to leave for a meeting with fellow novelist Upton Sinclair, he protested before reluctantly donning a beautiful overcoat lined with sealskin, a far cry from the ill-fitting clothes the old socialist had once worn while apprenticed to a draper in his youth.

Although Hubble liked the theater and the company of actors, he was no great fan of the movies and went only on occasion. Accompanied by Anita Loos, her husband, and others, the Hubbles attended the opening of Chaplin's *The Great Dictator* at the Cathay Theatre in November 1940. During the drive home, Edwin dissected the film. The first half, a vicious satire, was the best, but the second half, too long by far, turned into a tedious farce, with a bad speech at the end. The barber should have been given the last word.[63] Grace was surprised when he "shamefacedly" announced one afternoon that he was going alone to see *King Solomon's Mines*, based on the H. Rider Haggard adventure story of his youth. He returned home enthused over the quest for the Congo's mythical lost city of Zinj, and even admitted to having eaten popcorn, something she had never seen him do.[64]

A few days before Christmas, the Hubbles were met at the entrance of Disney Studios by a young woman who introduced herself as "the Publicity." She led them up a flight of stairs to Walt Disney's office, where the film pioneer greeted them warmly, his dark brown eyes and gentle manner reminding Grace of one of his unassuming animals. After talking for a while, he led them to a screening room, where, for the next hour and a half, parts of films in progress were run by an invisible projectionist—Dukas's *The Sorcerer's Apprentice* conducted by Stokowski; a ballet of flowers, toadstools, fish, and birds to Tchaikovsky's *The Nutcracker*; and a film on evolution from nebulae to the dinosaurs against the background of Stravinsky's *Rite of Spring*. Disney then guided them through the workrooms, where some 1,200 sketch artists and animators were laboring over slanted wooden desks. The walls burgeoned with pictures of volcanoes, dinosaurs, hippos, ostriches, elephants, and mermaids. A pensive Disney explained that he was trying to advance from simple cartoons and dialogue to a higher art form. All the while a photographer with a flash camera followed them around, patiently taking pictures for the Publicity Department, which had plans for Hubble's image when the time came to release the films.[65]

Hubble's sentimentality on hearing certain cowboy songs and poems carried over to the theater. Grace, who never saw him shed a tear over his own troubles, watched in fascination as he wept during a London performance of *Saint Joan* with Sybil Thorndike. His other favorites were the husband-and-wife team of Alfred Lunt and Lynn Fontanne, however bad the play. Leaving the theater one night in New York after seeing Maurice Evans in *Richard II*, he confessed to being practically intoxicated. Though he rarely told a joke and hated the off-color stories common on the cocktail circuit, a rousing comedy caused him to reach for his handkerchief with which to wipe away the streaming tears. With some notable exceptions, such as *The Magic Flute*, he avoided opera, whose theatrics annoyed him: "I wish they would stand still and sing, or just stand still," he complained. As the Hubbles drew closer to the Huxleys, Edwin and Aldous began listening to records together. Among their favorites was a Mass by Guillaume de Machault that had been sung at Rheims Cathedral at the coronation of Charles V. When Stravinsky, who was in town while his music was being recorded for Disney, came by, they played it for him.[66]

On many levels, Hubble and Huxley presented a study in contrasts, the one quiet, at ease, nonapologetic, and unruffled by criticism, at least outwardly; the other introspective, temperamental, argumentative, melancholy, and occasionally cynical. Though she was drawn to the cosmopolitan and physically attractive Englishman, Grace compared Aldous to D. H. Lawrence, who had no great love for his fellow man. "He has a lot of rancour to get out of his system, also like D.H.L. He has read too much." Nor could the Hubbles understand the almost childlike naiveté with which Aldous and his friend Gerald Heard approached Eastern civilization, especially Indian religion. Grace compared them to small boys attracted to a conjuror's box of parlor tricks. "They are looking for magic and power, for the secret word, the 'open sesame' that rolls back the door. It is not religion, it is magic."[67]

While they never said so publicly, the Hubbles must have talked between themselves about Heard's association with such fellow homosexuals as the English writer Christopher Isherwood. It doubtless helped that the witty, blue-eyed, red-bearded Heard harbored a profound respect for Edwin, who, like Aldous, considered him the ideal man of the Renaissance, the type described by Francis Bacon. When Heard had attempted to read *The Realm of the Nebulae*, he found it tough going, yet was convinced that its author was "the master of his own subject; he knows everything else and is fully developed in every way, the most civilized man I know."[68] This was saying a great deal considering Heard's umbilical-like attachment to Aldous. When Maria was asked by a friend whether she found the relationship trying, she replied that she loved and respected Gerald, who was one of the very few capable of stimulating her husband intellectually.[69]

Yet the deepest strain on the Hubble-Huxley friendship was only beginning to be felt. Aldous had become active in the pacifist movement in 1935; two years later, shortly before coming to the United States, he penned *An Encyclopedia of Pacifism*, decrying a world "lousy with armaments . . . utterly lost to all reason and decency." After reaching American shores, he promptly began to propagate his views before enthusiastic audiences gathered at such prestigious sites as the Los Angeles Philharmonic Auditorium. Hubble, on the other hand, was seething over Franklin Roosevelt's delay in readying the nation for a war he considered inevitable. When his friend Lord Lothian attempted to defend Chamberlain's actions as the only reasonable course for an England unprepared for conflict, he replied acidly that "all bluffs should have been called beginning with the first bluff."[70] On the mountain they were calling him "the Major" once again.

CHAPTER THIRTEEN

LANDLOCKED

I

Edwin and Grace sat glued to their radio for almost five hours on the night of June 27, 1940. The Republican National Convention in Philadelphia was about to enter its sixth round of balloting. With his support eroding, New York attorney Thomas Dewey, who had won a national reputation for "racket-busting," withdrew from the race, leaving Senator Robert Taft of Ohio and dark horse industrialist Wendell Willkie, a recent convert to the Republican standard, to fight it out for the party's presidential nomination.

Most of the world was now at war. Earlier in the year, Chamberlain had resigned in disgrace while his successor as Prime Minister, Winston Churchill, had attempted to rally the beleaguered English with his "blood, toil, tears, and sweat" speech. Belgium and Holland had been overrun, and the Hubbles mourned the destruction of Rotterdam and the *Statendam*, the graceful liner on which they had crossed the Atlantic during their idyllic sojourn of 1936. The mass evacuation of Dunkirk had followed in early June, and only two weeks prior to the convention the goose-stepping Germans had entered a Paris denied them in World War I.

Hubble wired the chairman of the California delegation, urging him to switch from Dewey to Willkie, but it was rumored that Taft, the insider, would almost certainly prevail, the key being Pennsylvania. But when the roll call reached Kansas, former governor Alf Landon, who had been mauled by Roosevelt in 1936, delivered the delegation to Willkie. Michigan, at the urging of favorite son Arthur Vandenberg, did the same, and, as Grace put it, "Pennsylvania cracked through for Willkie." Taft, faced with the inevitable, moved that the nomination be made unanimous, and the voice of CBS commentator Elmer Davis, Hubble's old friend from their days as Rhodes Scholars, was drowned out by chants from the gallery of "Willkie, Willkie, Willkie!" "For once," an excited Grace wrote, "the politicians were not getting their way; a fight was coming that would be worth fighting."[1]

As Hubble saw it, the fault was all Germany's. No matter how flawed the Treaty of Versailles may have been, each signatory was duty bound to abide by its terms, especially the vanquished. Whenever anyone voiced skepticism he reminded him that the Allies could have dismembered Germany but they left her to her word: "If a nation gives its word the nation has to stand by it or there is no confidence or hope of international integrity."[2]

While the raw-throated Republican candidate went on the stump, covering some 19,000 miles and thirty-four states in less than two months aboard the "Willkie Special," Hubble became politically active. Together with such distinguished figures as Robert Millikan, he became a member of the William Allen White Committee to Defend America. Named for the Emporia, Kansas, newspaper editor who had made his reputation as a spokesman for small-town life and liberal Republicanism, the committee favored massive aid for Britain in the form of "every ship, every plane, every tank, and every gun that she needs." When Clark Eichelberger, the committee's national director, addressed a gathering of a thousand at the Riviera Country Club, he shared the podium with Hubble, Millikan, and Judge John Perry Wood, chairman of the Southern California chapter. When Hubble's turn came, he spoke with uncharacteristic passion, waking everyone in the room, according to his friend John Balderston, and quoting Browning at the end:

> We have watched the explosion of total war in Europe. We have seen the little nations, striving to maintain the decencies of civiliza-

tion, suddenly blotted out by the stamp of an iron heel. We have seen the shambles of Rotterdam.

We may turn to Hitler's conversations, as reported by Rauschning, to learn what he plans for us. We have watched those incredible fantasies materialize and then congeal into the hard facts of history. Hitler has called his shots. His armies have swept over Europe all according to plan. He has failed just once; he has been stopped at the English Channel by the magnificent courage and effectiveness of the Royal Air Force, and by the vigilance of the British Navy.

So, in this emergency, we lack the one essential factor—time, which waits for no man. And we have one shield to shelter us while we prepare. That is the armed force of Britain. Does anyone suppose for a moment that, without this force to hold Hitler at bay, we shall be given the chance to become an obstacle to his plans? So, for the survival of our country, it is obvious that our best insurance is aid to Britain, holding our own front line of defence. . . .

It is this people who are defending us with their blood, toil, tears and sweat. Let us say to each other,

> Here and here did England help us,
> How can we help England.[3]

The Hubbles cast their votes for Willkie at a nearby park on Tuesday, November 6, then spent another anxious, and ultimately disappointing, evening next to the radio, listening to the returns. Yet considering how poorly Hoover and Landon had done against Roosevelt, the Hoosier's loss by fewer than 5 million votes in a record turnout was trumpeted by Republicans as a great moral victory. Indeed, the President himself was so impressed by his opponent's showing that he accepted Willkie's offer to embark on a diplomatic mission to help unify the nation in support of the war effort, and to rally the flagging spirits of the European allies.

The Hubbles were in Washington in December, where Edwin was scheduled to speak at the Carnegie Institution, when they met Willkie at a midnight buffet for members of the Gridiron Club. The occasion had all the elements of an opposition party. When Willkie arrived, he immediately became the focus of attention among the ladies, and it took some time before the Hubbles were introduced to him. In build, posture, and gestures he reminded Grace of the central character in Booth Tarkington's play The Man from Home, set in Willkie's native Indiana. "When he shakes hands with you he has a most kind, sympathetic look, direct and searching, of both feeling and intelligence, so that one responds to him at once as a human being."[4] The Hubbles had hoped to sit with the erstwhile candidate, but he was soon spirited away by a bejeweled matron, leaving them to dine with Robert Taft and Tom Dewey.

Also present was fluffy-haired, blue-eyed Alice Longworth, the iconoclastic daughter of Theodore Roosevelt's tragic marriage to the beautiful star-crossed Alice Lee. She made straight for the Hubbles, immediately endearing herself to Edwin by claiming to have read his book. She was delighted with the astronomer's "warmongering" and asked Grace all sorts of questions about him. The subject then switched to the First Lady and her talks on the radio. Alice volunteered that Cousin Eleanor's high seesaw voice was the "root voice" of the Roosevelt women, but Eleanor used nothing else. When Alice's Aunt Corrine used it, they called it her "elbow in the soup voice."[5]

Several of the same dinner guests were present at the Carnegie Institution the night of Hubble's lecture in the newly completed Elihu Root Hall, whose construction Grace rather too generously attributed to her husband's ability to fill the old auditorium to overflowing. Hubble's name was familiar to readers of the Eastern press, which had covered his recent address on the occasion of receiving the Franklin Medal from Philadelphia's Franklin Institute. Calling him the "celestial commuter," *The New York Times* gave him four columns in the news section, and then, three days later, editorialized in a manner reminiscent of Einstein: "A lecture by Dr. Edwin Hubble . . . leaves one confused and solemn. This is not Dr. Hubble's fault. It is the universe's fault. The layman . . . cannot help wondering where he is going, and why."[6] Hubble was accorded similar treatment after it was announced that he had been chosen to receive the Royal Astronomical Society's Gold Medal for 1940. As foreign secretary, Eddington wrote, "I think it is not indiscreet to mention that at the selection the voting was unanimous." The otherwise joyous occasion was dampened by the secretary's observation that while all seemed well for the moment, "one has the feeling that the real war conditions have scarcely begun yet."[7] Nor, unfortunately, would Hubble be able to collect his medal in person. Three months later, he received a large envelope addressed to "the Honorable Secretary of State," bearing the seal of the U.S. Embassy in London. Inside, in a brown box resting on a satin lining, was the gold medal containing the head of Newton in high relief.[8]

Grace pronounced her husband in good voice, magnetic and natural as ever. Thus, she was surprised during the drive back to the Hay-Adams Hotel when Edwin confessed that he "couldn't get into it." He explained that his subject, supernovae, was not truly his own. Though he had discovered one of the massive erupting stars in the Virgo cluster while using a 10-inch refractor in 1937, the mercurial Fritz Zwicky had since laid claim to the field by locating a dozen more. Grace attempted to console him by repeating Turner's remark that it would be years before Edwin realized the magnitude of what he had done. While her concern may have helped, what he really needed was more data on the seemingly expanding universe, and for that, he told Lawrence Davies of *The New York Times*, his hands were tied until the 200-inch was operative.[9]

11

Before the Hubbles left Washington for New York, Edwin was picked up at the hotel by a military car and taken to see three admirals and a lieutenant. He told Grace that they had discussed some secret affairs not to be written down. Only later did she learn that he had been summoned by Admiral Chester Nimitz, who asked him to look into codes and ciphers. There would be more such meetings in the future, as well as mysterious calls after which he would sometimes leave the house for an unannounced destination.[10]

In New York they were reunited with a fit and prosperous Elmer Davis, whose fears about having to settle for first in the provinces had never materialized. The CBS newsman had spent ten years as a reporter for *The New York Times* before entering broadcast journalism. His calm and steady voice, still bearing the twang of his native Indiana, became familiar to millions of eager radio listeners, many of whom also bought his novels and short stories. Davis took Edwin and Grace to his seventeenth-floor office in CBS's glass-and-chromium headquarters at Madison Avenue and Fifty-second Street, where they were introduced to an eloquent young reporter named Eric Sevareid, just back from London, where too many air raids had unsettled his nerves. They talked of war, international politics, and personalities. The liberal Davis professed his dislike for Willkie and the crowd around him.[11]

Charlie Chaplin, who was visiting the city with his Japanese valet Frank, insisted on taking them to dinner at "21" and then on to the Stork Club, where they were joined by Helen Hayes and Charles Mac-Arthur. Grace was surprised at the size of the famed nightclub, which loomed much larger in her imagination than the light blue room filled with tables crowded around a small dance floor. The orchestra and a young woman who wriggled and sang made a horrid noise to her ears, but the parting of the waters and accompanying buzz of recognition when they entered the room more than made up for it. They were seated at the best table next to the dancers, and their waitress immediately rushed up with a cart containing bottles of chilled champagne.

It was a different Chaplin than the one they had last seen in Hollywood. The pacifist was now a warmonger, berating the Europeans and Americans for their inaction after Hitler's first move: "I am a little man and I have always got along by being acquiescent, conciliatory, anxious to please—that has been my way of avoiding trouble—but before I'd submit to these bullies I'd be shot and so help me I'd take as many of them with me as I could."[12] The next day, when Hubble went down to settle accounts before leaving for California, he was presented with a bill totaling seven cents; as in England, everything else had been paid for by friends, including some very expensive claret.

Back in Pasadena, Maria confided to Grace that Aldous "is very rich now," earning as much as $27,000 a screenplay, about the same amount their mutual friend John Balderston got for similar work on *The Lives of a Bengal Lancer, The Prisoner of Zenda, Gone with the Wind,* and *Gaslight.* The Huxleys' new Cadillac impressed the Hubbles, who traded in the La Salle for one of their own. They also purchased kitchen appliances, and Aldous quipped that the new stove had everything on it but a radio. More interesting to Hubble was the Frigidaire, which was not delivered until three in the afternoon yet produced ice in plenty of time for drinks. He kept going over and opening the door in an attempt to glimpse the magic, a reminder of just how far technology had come since the summer he and Bill had worked the ice wagon in a sweltering Wheaton.

Aldous liked to stroll about the shops of Chinatown and buy things. The Hubbles were with him and Maria one evening when they came to a sign that read: "Astrologer and Fortune Teller, George Chung." They decided to go in at the writer's urging. In the small, bare room sat the self-proclaimed seer, fat and impassive as an idol. He told Grace, Maria, and Aldous several things about themselves that were fair approximations of the truth. Turning at last to Edwin, Chung hesitated for the first time. He picked up an abacus and fumbled absently with the bead counters, then exchanged it for a pack of cards that were shuffled and dealt out. Finally he exclaimed, "Oh, big figures, many big figures, all big . . . 200 horsepower mind, 200 horsepower mind— I can't tell more!" Edwin bowed and thanked him. Outside under the lanterns, Aldous thought it curious that Chung had connected his friend with numbers. And could the 200 horsepower have something to do with the 200-inch telescope? "They sometimes come close although they miss the fact."[13]

Edwin explained that Chung had cleverly played for time by picking up the abacus. He was feeling them out for leads and they had looked impressed, or glanced at one another. If you want your fortunes told, you should keep straight faces.

When Aldous had arrived at Anita Loos's for dinner the night Paris fell, his face was dead white. Anita later wrote that he bore the expression of someone who was peering into hell. The Hubbles were present, but nobody mentioned Paris. Yet discussion of the war was not something that could be avoided at social gatherings. Whenever the subject was broached, Aldous retreated into silence, as did Maria for her husband's sake. Instead, she informed Grace that she lay awake every night examining her conscience. Pacifism seemed such a terrible abstraction in light of what was happening to England. And when it was over, how could they possibly go home—wealthy and secure—without becoming the objects of enmity?[14]

Late in January 1941, Hubble and John Balderston agreed to debate two members of the local chapter of the America First Committee on the NBC radio program *I Disagree*. The question: "Is Britain our first line of defense?" Grace wrote of the marked contrast between the participants—Edwin and John "so beautiful to look at, gentle, honest, outright in expressing what they believed, putting their cards on the table." The other two, one the president of America First, the other a financial columnist, "common, cheap, nasty, hinting at vast stores of material which they refused to submit." Strongly anti-British, "they resorted to school primer statements, claiming the U.S. could do business with Hitler."[15] Summarizing the debate, moderator Lewis Browne said it was the difference between a man who had vision—and who used a telescope—and minds who looked through a microscope and saw a lot of bacteria.

Feeling triumphant, the Hubbles drove to Perino's for a late dinner. At the studio Marian Balderston had given them a copy of *London Front*, a book of pro-British letters edited by their English friends the playright H. M. "Tottie" Harwood and his wife, Fryniwyd "Fryn" Tennyson Jesse, a novelist and popular crime reporter. Several of Grace's letters, along with others written by Marian and the flamboyant drama critic Alexander Woollcott, with whom Grace regularly corresponded, appeared in the volume. The Hubbles read Grace's letters to Fryn and Tottie aloud to each other as they dined: "The Germans are incredible.

Germans here have rung up American acquaintances to ask 'Aren't our armies wonderful?' And have been painfully shocked at the answers. I shall never speak to another German. . . . Edwin says their appalling guilt will never be forgotten."[16]

At least one who listened to the evening's program was as impressed as Grace by her husband's stand. The Huxleys' only child, a tall, blond, unassuming youth named Matthew, whom Hubble termed "his mother's son," had planned to register as a conscientious objector out of filial loyalty. After next hearing Hubble speak in favor of war at Thacher School, he became very upset and said almost nothing for three days. Matthew finally announced to his parents that Dr. Hubble had made him see things differently: he would somehow make his way to England and join the army.[17] When his draft notice finally came, in March 1943, the UCLA student changed his mind once again. He sought and was granted noncombatant status by the army, which posted him to the Medical Corps, where he almost died after losing forty pounds following a severe attack of measles, which embittered his distraught father all the more.

For Hubble it was never a matter of whether, but of when. While sitting on a park bench in Salt Lake City one night, waiting for the train to Rifle, Colorado, he told Grace that America was going to be in the war soon and that he had made up his mind to get back into the infantry and serve abroad.

Meanwhile, he kept hammering away on behalf of the prowar movement. Two weeks after the radio debate he spoke at the Biltmore's regular Town Hall Lunch before an audience of two hundred wealthy businessmen and professionals. His views were greeted by enthusiastic applause, and his proud father-in-law telephoned to say that Edwin was as good as Churchill, who had spoken on the radio the previous night.

Hubble was at work the next morning when Grace received the first of several anonymous calls attacking her husband. When the mail arrived, it contained more of the same. Some of his critics were complete pacifists and didn't believe in the manufacture of guns, bullets, or powder. Grace dismissed one such hysterical woman with the angry riposte: "How do you feel about burglars, Madame."[18]

Early in July, Hubble began making arrangements for Willkie's upcoming speech at the Hollywood Bowl, sponsored by the Fight for Freedom Committee of California. He also drafted a simple statement of principle titled "The Unity Manifesto," then lunched with the cautious heads of the C.I.O. and A.F.L., who eventually pledged their support. The leaders of the Sunset Club were also corralled, guaranteeing Willkie a capacity crowd.

The Millikans called for the Hubbles at 6:30 on the evening of July 23. When they arrived at the Bowl a half hour later the crowd, composed of workingmen and millionaires, was singing at the top of its lungs. Millikan took the podium and read Hubble's Manifesto with spirit, surprising its author when the Nobel laureate was interrupted three times by applause. Promptly at 8:15, with the radio microphones on, Willkie took the stage to a thunderous ovation. Happy and boyish, he spoke for forty-five minutes, prompting the Hubbles to wonder what might have been.[19]

On the morning of Tuesday, November 11, Grace drove Edwin to the corner of Hudson and Green streets in Pasadena, where he climbed aboard a fire engine next to the chief. The Armistice Day parade wound its way through the downtown streets to the city park, where he was scheduled to speak at noon in the so-called Gold Shell. Seated on a bench next to her sister Max, Grace studied the crowd, which included everyone from "pickaninnies" on roller skates to the mayor and Walter Adams of Mount Wilson. A group of officers helped some tottering veterans of the G.A.R. climb the steps to the guest platform, where to Grace they looked like old men of the tribe, quite Homeric.

The flag was raised, and the invocation was delivered by an army chaplain. Grace kept a nervous eye on several suspicious-looking "old trouts" who she feared might resort to heckling. But from the moment Edwin began speaking, they applauded. In between, they listened with more concentration than she had ever seen in an audience. He didn't speak down to them, but he made a fine rabble-rouser. When, some weeks earlier, he had been introduced at the California Club as being pro-British, he shot back, "Hell, I'm for civilization."[20] Yet to him they were one and the same. In an emotional conclusion, he summoned the vision of Henry V, Queen's College's most illustrious ghost, by quoting the lines given him by Shakespeare at Agincourt on St. Crispin's day five centuries earlier. Never an anniversary will come

> From this day until the ending of the world
> But we in it shall be remembered.
> And gentlemen in England now abed
> Will think themselves accurst they were not here,
> And hold their manhoods cheap whiles any speak
> Who fought with us upon St. Crispin's day.

"This is our war. The final stake is our freedom and the freedom of our children. Let us never forget that this freedom, this liberty, is the heritage of brave men who the world over, since time began, have fought to achieve and maintain it."[21] Among the people who pressed around to congratulate him afterward was "the idiot" who attended all his lectures—a Dostoyevsky figure in worker's overalls with a big head bristling with light sandy hair and friendly blue eyes. He had a terrible impediment and only managed to get out his few words of praise with a fierce effort of will. He beamed when Edwin thanked him. *The Star News* ran the speech in full, and it was soon printed up as a pamphlet under the title *O'er the Ramparts We Watch.*[22]

The America First Committee of Pasadena took a more jaundiced view. At its forty-sixth meeting in the McKinley School auditorium, it unanimously passed a resolution charging Edwin Hubble, "allegedly in the employ of Carnegie Institute at its Mount Wilson observatory," with subversion, a criminal violation of President Roosevelt's neutrality proclamation of September 5, 1939. "For the public safety and in the public interest said Hubble should be apprehended at once and brought to summary account for his alleged violation of law."[23] Copies of the unsigned document were sent to Hubble and to the press. A few nights later, Grace and Edwin went to see a picture of the R.A.F. in action and heard people in the seats behind them murmur, "There's Hubble, we thought he was in jail."[24] The constant ringing of the telephone resumed, Grace answering to nameless voices uttering violent threats. Rather than take the receiver off the hook, she had a second phone installed by her chair in the library, where she battled wits with the harassers until the orchestrated campaign finally died down.

III

Before they left for dinner at the Balderstons' on the night of December 6, 1941, the Hubbles received a phone call from Maria Huxley, who was deeply upset. Her brother-in-law, Julian, had just arrived in New York and was being pursued by Hearst reporters because he had remarked privately that it would be a good thing if the United States went to war with Japan. Grace spent a long time trying to calm her friend and had to dress in a hurry.

The following morning dawned bright and warm. Edwin, looking very handsome in his new robe, lounged about the house while his fellow walkers assembled for the usual Sunday trek to the Huntington grounds. They returned to find a note left by a neighbor: "Turn on your radio." Moments later Max phoned. The first reports were coming in; the Philippines and Hawaii had been attacked, all firemen and policemen were to go to their stations, civilians should not use telephones more than necessary, and should stay at home and keep off the streets. Japanese aliens were being rounded up by the F.B.I., and policemen had already been stationed in the section of Los Angeles known as Little Tokyo. Bulletin followed bulletin throughout the afternoon; on CBS the grim voices of Ed Murrow, William Shirer, Eric Sevareid, Albert Warner, and their friend Elmer Davis spoke deep into the night.

The next morning, while their Filipino houseboy John sang "The Star-Spangled Banner" between threats to cut off the heads of the Japanese, Edwin surveyed the headlines and noted that "if we had declared war a year and a half ago this would not have happened." They brought the radio down by the fire, and Max phoned to tell them of a rumor that Japanese planes were flying toward San Francisco. Minutes later all radios on the Pacific Coast were ordered off the air, and people were once again told to stay in their homes.[25]

By Tuesday, December 9, Caltech was under guard, and Mount Wilson had been turned into an observing post for enemy planes. The Crottys came over in the evening, and Homer reported that the damage at Pearl Harbor was rumored to be appalling. The United States was virtually defenseless in the Pacific, nor were there any planes with guns in Southern California. The following night Hubble received a call from the British Information Office in New York, urging him and his fellow trustees to have the contents of the Huntington removed to a safe place as soon as possible. He later described the agonizing process in humorous terms: "We divided them into three classes—world treasures, national treasures, and just treasures." Another call followed from Washington, D.C. He came back from the telephone without comment, and it was not until the end of the war that he told Grace he had been asked to stand by for further instructions and to make no commitments, because what they wanted him to do had already been decided.[26]

Though Edwin cautioned Grace against speculating and dealing in rumors, he predicted an attack on the mainland, with the aircraft factories and Caltech as the primary targets. Weather reports were no longer given, a development which he told Grace to include in her journal. On the night of the tenth, the radio announced the appearance of unidentified planes, and the first of several blackouts was ordered from Santa Barbara to San Diego, and as far east as Las Vegas and Boulder Dam. They stepped out into the yard as the lights began to blink off to the wails and squeaks of makeshift sirens. Then all was silent except for the occasional howling of a dog and the hooting of a quizzical owl. It was lovely without the lights, the oaks very dark, the stars and wisps of cloud, the house glimmering faintly in the celestial glow. "This is rather an important thing," Edwin whispered as they turned and walked inside to cover the terrace windows with blankets.[27]

As it turned out, the secret call from Washington concerning Hubble's wartime fate was premature by several months, during which conditions on Mount Wilson underwent some dramatic changes. The observatory was closed to the public while the astronomers were issued passport photographs that enabled them to negotiate the checkpoints manned by armed guards. Soldiers on maneuvers marched up the Sturtevant Trail at night and were sometimes surprised at the top by a rejuvenated "Major" Hubble, who took it upon himself to remind them of the regulation against lighting up in the dark. From a blacked-out and burning London, Tottie Harwood wrote of his envy of Edwin's relations with the stars. "What a comfort to be able to turn your back on the world and say with Coriolanus: 'I banish *you*. There is a world elsewhere.' How Hitler, if he were capable of grasping it, would hate astronomy."[28]

Grace enrolled in a first-aid class that met at Mrs. Millikan's, which, she confessed, made her feel strangely undignified. Later in the day, she joined sixty-one other women at the Y.W.C.A. gymnasium for "physical jerks" in what she termed "true New Yorker style." At the final meeting a month later, she prided herself on being only one of three to have completed the vigorous workouts. Urged on by her husband, Grace also tried her hand at belles lettres. Hubble himself mailed off her article to Edward Weeks, the editor of *The Atlantic Monthly*. When the manuscript was returned together with a rejection letter a month later, Grace wrote in her journal of planning Mr. Weeks's "future discomfiture."[29]

In April 1942, John D. Rockefeller III wrote Hubble from the national headquarters of the American Red Cross in Washington asking him to consider taking over operations in either England or Australia. Receiving no answer, Rockefeller wrote a second letter, and when this one went unanswered, a third. Forever apologizing for her husband's procrastination with the pen, an embarrassed Grace finally got him to send a long-overdue letter of regret.[30]

It was also in April that Hubble received F.B.I. clearance for access to classified information. A few weeks later, he received a call from his friend Frank Aydelotte, past president of Swarthmore College and American secretary of the Rhodes Trustees. Could he come east to have a look at Maryland's Aberdeen Proving Grounds, which was desperately in need of a head of exterior ballistics? He would have his choice of remaining a private citizen or reenlisting in the army with the rank of lieutenant colonel.

Hubble asked for a couple of days to think matters over. Grace remembered it as the one and only time he sought outside advice. He went to talk to Millikan, who opposed his going, saying he was much too valuable a man and needed in defense work on the West Coast, preferably at a galvanized Caltech. He had also been contacted at some point by Vannevar Bush, Merriam's successor as president of the Carnegie Institution. As director of the Office of Scientific Research and Development, Bush had taken the initiative in mobilizing the nation's powerful communities of science and engineering for war. He was in the early stages of recruiting scientists for the top secret Manhattan Project, and Hubble was invited to take up residence in the small desert town of Los Alamos, New Mexico. He was one of the few who refused Bush's overtures, citing his belief that he should serve his country as a human being first, not as a scientist.[31]

Two days later, as he was considering another wire to Aydelotte, the sound of the Western Union boy's motorcycle could be heard coming up the street. The telegram read: "Aberdeen job extremely important / affect immediate conduct of war / hope you will accept invitation to come east for short period to look over job."[32] Grace called the Union Pacific and canceled their reservations for Rifle, Colorado, then booked a sleeper on the Santa Fe for the East Coast. As Edwin passed by the camp gear, she overheard him mutter, "There are those damned flies."[33] She remembered him writing a letter to Yellowstone to order the Sterry specials when he couldn't be persuaded to answer Rockefeller's queries. At nine that night yet another urgent wire arrived from Aydelotte.

I V

After the war was over and he had returned to Pasadena, Hubble was asked to speak of his experiences before fellow members of the Sunset Club. They thought he was joking when, after having agreed to visit Aberdeen Proving Ground on the shores of Chesapeake Bay, he immediately headed for his set of encyclopedias to look up the term "ballistics." But it was no joke; the man who had been asked by Army Ordnance to head research and development on the trajectories of bullets, bombs, and rockets had told his curious friend John Balderston the same thing during a farewell dinner in July 1942.[34]

At Aberdeen, Hubble had been escorted around the firing range in a jeep while deafening guns blazed away day and night. He was told by the commanding general, Charles T. Harris, Jr., that he had been selected because, according to his fellow scientists Oswald Veblen and Henry Norris Russell, he had a reputation for controlling theory with observation. He could begin as a consultant to avoid red tape, then apply for a commission if he wanted it.

Persuaded that the job was of the utmost importance, Hubble accepted the offer the following day, agreeing to report on August 1. He returned to Pasadena only long enough to collect Grace before departing for Colorado on what would be their last vacation until the end of the war. The cars ahead of theirs were crowded with grim-faced young sailors who had recently seen the worst of combat half a world away. When the Hubbles got off at Rifle they discovered that each car bore the same chalked message: "U.S.S. *Lexington* Special," a tribute to the lost aircraft carrier at rest at the bottom of the Coral Sea.

Hubble was fingerprinted, photographed, and sworn in on schedule, his temporary consultant's fee of forty dollars a day exceeding his salary at Mount Wilson. But the colonelcy he sought was in jeopardy. There had been a recent clamor in Washington over the granting of too many high commissions for political reasons, leaving legitimate as well as questionable appointments in doubt. He wrote Grace, who remained behind in Pasadena feeling useless, that he might have to settle for civilian status after all, if the salary could be worked out.

His days began early in the morning and did not end until ten-thirty at night. As head of exterior ballistics, he was learning about bombs and shells of all kinds. "The most important items in the War now are submarine detection and accurate firing tables. After them come new weapons, new fuzes, new transports etc. I can't write of the activities here but they are many, and always there is the sound of the guns on the proving range." He found it very odd to be away from home, but he had started a journal of his own, which would allow Grace to catch up once he found housing, hopefully in about six weeks. Any apartment would be small and "servants are very hard to come by, mostly darkies, none too clean, I suppose." His greatest asset was a small, almost stubby fellow named Robert Kent, the best ballistics man in the country.[35]

A typical day found Hubble on the trench mortar range in the morning, the bombing range in the afternoon, and out with the 34-

millimeter gun crews at dusk, all in an effort to acquire firsthand familiarity with the firing techniques. He quickly learned that the performance of a newly designed shell could not be predicted from its drawings and specifications. Typically, it was compared to older ordnance most like it, on the often erroneous assumption that it would behave in much the same way. Thus new tables were required not only for each new gun, shell, or bomb but also for each modification of those already in use, as well as for captured enemy matériel, especially that manufactured by the Germans. Time was of the essence, the pressure enormous.

A special group led by Dr. L. S. Dederick, a granite-faced mathematician and chief of the computing laboratory, was set up to provide several dozen firing tables each month. This was accomplished with the help of about 150 human computers, many of whom were W.A.C.s with degrees in mathematics and the physical sciences, together with two of the four differential analyzers in the country, and a huge collection of I.B.M. punch card machines. Once this unit was in operation, Hubble spent an increasing amount of his time troubleshooting, sometimes in the most literal sense.

Soon after he arrived at the proving grounds a new gun called the bazooka seriously injured one of the soldiers who had fired it, adding to its reputation as a risky weapon. Hubble selected another bazooka and went alone to a remote field where he fired the weapon repeatedly until he found the cause of the malfunction. The design was subsequently changed and the armor-piercing gun proved itself effective at close quarters against tanks and other battlefield vehicles.[36]

Most complex of all were the problems associated with dropping bombs and firing projectiles from rapidly moving aircraft. Until now, no one had seriously studied the second-by-second behavior of a bomb released three miles ahead of its target, or the exact path of machine-gun bullets as they passed through terrific crosswinds of three to four hundred miles per hour, let alone the bending of the gun barrels from which they were fired. The introduction of rockets brought with it a new series of headaches. The first such weapons were launched from a ten-foot tube attached to the tail of a bomber. It was feared that the missile, on leaving the tube, might actually turn around and pursue the plane, with disastrous results.

Hubble's unit became specialists in the design and operation of bombsights and was soon scouring the country for the few high-speed movie cameras in existence. Operating simultaneously from aircraft and the ground, the synchronized cameras enabled technicians to determine the position of the curve gradient of a streaking rocket to about one and one half inches over a trajectory of one half mile, at the same time capturing the behavior of the plane both before and during the launch.

The completion of an indoor precision firing range yielded large quantities of new data on the flight characteristics of model shells and bombs. The hundred-yard range had twenty-five stations, each equipped with special cameras. Shadow photographs of the shell taken in two planes were recorded by sparks lasting about one-millionth of a second, tripped by the electrostatic charge on the projectile as it passed. Each station provided the orientation of the shell, and by simply aligning the photographs the yawing of the projectile could be measured and analyzed. Meanwhile, at Caltech a model supersonic wind tunnel had been successfully tested, with its full-scale counterpart scheduled for installation at Aberdeen sometime in 1944.

In addition to his scientific and technical duties, Hubble spent considerable time in the company of distinguished visitors, guiding them through the facilities. Three senators and two staff members from the Senate Special Committee to Investigate the National Defense Program, popularly known as the Truman Committee after the obscure junior senator from Missouri who thought up the idea, followed him around for the better part of a day. "They were no end impressed with the Lab," he proudly wrote Grace, and "spoke of it as one of the bright spots in the war picture as regards efficient planning and operation directed to a specific major objective."[37]

Three months later, on what he termed "Circus Day," a delegation arrived from the Congressional Committee on Military Affairs. Connecticut representative Clare Boothe Luce, the playwright and former managing editor of *Vanity Fair*, stole the show. With the movie cameras rolling, she fired a carbine and several other weapons "like an old-timer." Hubble was introduced to her as the foremost astronomer in the world, and she went out of her way to converse with him about their mutual friends in Hollywood. As the delegation was about to depart, Luce turned to one of the generals in the party and asked whether any of the ballistic research could be used after the war, to which the general replied that, among other things, "we could use it in preparing for the next war."[38]

Lost without Edwin, Grace cut down on her social engagements and made few entries in her journal. With John the houseboy gone to war and Lydia the maid able to come in only once or twice a week, the long days were spent in a kind of menial limbo: "I paint the porch furniture, the mail box and myself, grassgreen." Edwin's letters meant everything; when he sometimes failed to write because of a heavy work-load she panicked and sent a wire to make certain he was well. "How delightful it is to see him write togather, and when he spells occasional with only one s it puts me off a bit. And how like him his letters are, simple and straightforward."[39]

When he wrote telling her to come east after a two-month separation she was "galvanized into action, as a dog that hears his owner's far-off whistle." Not expecting to return for several months, she took down the curtains, covered the furniture with sheets, packed away the books, and wrapped the silverware in paper that had the pervasive scent of mothballs.[40]

Grace arrived late at night and awoke the next morning in the Co-lonial Hotel, a small yellow stucco building in Havre de Grace, Mary-land, an overcrowded town of 8,000 located a few miles from the laboratory. Unable to find a room in the village of Aberdeen, which had burgeoned from 1,600 to over 4,000 in a matter of months, Edwin, who appeared thinner and more keen, had settled into the Colonial a few days after his arrival. They had barely eaten breakfast before he apologized, saying, "I must leave you, my dear," and headed for his office.

Grace spent the following day house hunting far afield, but, as Edwin had warned her, there was nothing to be had within fifty miles, and government housing was not open to men in advanced positions. Dull explosions shook the hotel windows, causing him to remark proudly, "My guns." Any doubts he had of making a significant contribution had long since vanished. He expected to be at Aberdeen at least an-other year, possibly for the duration of the conflict, whose end, he predicted, would likely come in 1944. Of the 280 in his department, who were about equally divided among civilians, officers, and enlistees, some were brilliant but eccentric, some plodding and accurate, but few had any administrative ability. Grace drove with him in the rain down a forest road to the Ballistic Research Laboratory, a large square struc-ture of red brick with a gray stone entrance. An armed guard stood just inside the door, a reminder that this was as far as she could go. Edwin pointed out his office windows on the second floor, then donned his raincoat and waved goodbye. "So this is it, I thought."[41]

Since house hunting was useless, they stole a few days and went to New York, where they stayed in their old room at the Grosvenor, the green goddess of the harbor a welcome sight against the misty skyline. Grace's spirits plummeted when she thought of what lay ahead. "I can't describe the feeling that comes when I am alone and then I see Edwin and the whole of life changes—from charcoal and oyster-shells to diamonds and pearls." She cried at dinner the evening they made their decision. Without a chance of keeping house, life in a small hotel room would quickly become a nightmare. There was nothing for her to do but return to California. As Edwin had explained it: "These are campaign conditions, as if I were in the army, and can't be helped." Grace spent a restless night and arose toward morning to look out on the city. The lights were few, the buildings shadowy, taxis standing at the corners—all still as a painting.[42]

V

Before Hubble joined the staff at Aberdeen Proving Ground, his old interest in the formation of spiral nebulae had been rekindled. He was particularly hopeful of determining the direction of nebular rotation, which had also preoccupied such prominent astronomers as Vesto Melvin Slipher and Bertil Lindblad. There were only three possible answers to the question: the arms may be trailing behind the central regions in all spirals, as Slipher believed; the arms may be leading the rotation, as Lindblad postulated; or the direction of rotation may vary from nebula to nebula. It was one of the few scientific questions that Hubble discussed at length with Grace, who kept track of her husband's progress in her journals.

The breakthrough had come during a run in early February 1941. Hubble phoned home to tell Grace that he had enjoyed three nights of excellent seeing, enabling him to obtain the spectrum of a critical test nebula, the only one among a thousand such objects whose visual orientation was almost perfect. It was as he had always thought: the spiral arms trail behind the more rapidly rotating nucleus.[43] In the months ahead, he succeeded in establishing the same rotational pattern in three more spirals, and by inference added another eleven to the fold, confirming what Grace described as his "aesthetic" impulse. Perhaps in time this discovery would shed light on the evolution of the universe, as Hubble had hoped to do when he first developed his classification scheme for the nebulae. Working up to the last minute before he departed for Aberdeen, he got the material into shape for publication and mailed it off to the *Astrophysical Journal*, where it appeared while he was occupied with the design of machine-gun bullets and plotting the trajectories of incendiary bombs.[44]

Back at Mount Wilson a number of precautions had been taken to safeguard the invaluable mirrors against enemy planes, but, as Adams was quick to acknowledge, such measures would do little good in the event of a direct hit. He doubted that the observatory would become a primary target, but there was always the chance that a Japanese pilot, driven away from a factory or research facility, would decide to unload on the prominent landmark before heading back out to sea.

As the months passed and it became clear that Los Angeles was in no danger of becoming another Pearl Harbor, precautions were relaxed and the public was readmitted to the facility. At first, the normal Sunday flow of traffic remained high, averaging from 500 to 750 automobiles, which to Adams seemed "all out of reason" given the serious shortage of rubber for tires. When gas rationing went into effect the number decreased significantly for a few weeks, then inexplicably began to rise. When Adams calculated the average round trip at between 50 and 80 miles, coupled with the wastage of rubber on the winding mountain road, he ordered that the observatory be closed to the public out of consideration for the war effort.

With the city below periodically blacked out, the view from the mountain had not been so ideal since Hale's earliest nights there. And with many astronomers engaged in war work, those who remained behind received almost all of the telescope time they could use. Ironically, none was in a better position to take advantage of the situation than the German-born Walter Baade, who had been classified as an enemy alien immediately after the United States entered the war.

Baade had come to Mount Wilson from Hamburg Observatory in the autumn of 1931, having established his reputation as an astronomer of the highest technical skill, an assessment soon confirmed by his new colleagues, who dubbed him "a hot man on the buttons." Small of frame with a hawk nose, meticulously parted hair, and Dumbo-like ears, he hated to write and loved to talk. As one of the "dark men" he was somewhat contemptuous of the spectroscopists, whom he considered dull, once confiding to his friend Jesse Greenstein, "They don't eat, they don't drink, they don't love," only he used a different word for love.[45] Fond of parties, Baade hosted a number of his own at which only martinis were served. Drinking little himself, he would move unobtrusively about the room like an agent provocateur, filling half-empty glasses while inhibitions melted away, and the talk got louder and looser.

The one person with whom Baade did not get along was the immoderate and irrational Fritz Zwicky, who was jealous of Baade's discoveries. Zwicky had a violent temper, and called Baade a Nazi to his face, which so unnerved the astronomer that he feared Zwicky might kill him.[46] Baade had been saved from serving in the German army during World War I by a congenital hip defect, which caused him to limp markedly. However, his brother, a member of the National Socialist German Workers' Party, had chosen a military career and was now, as far as anyone knew, a submarine captain somewhere in the North Atlantic.

A procrastinator of the first order, Baade did not take out his preliminary citizenship papers until 1939, eight years after coming to the States. These he proceeded to lose during a move into another house, and he had not bothered to renew his application before December 1941. When war came, he was notified by the government that his status as an enemy alien required that he remain in his home between the hours of 8 p.m. and 6 a.m. He could visit Mount Wilson by day, but his career as an astronomer was suddenly in limbo.

While Baade signed up to buy war bonds, Adams wrote Bush to ask whether anything could be done. Though he was loath to part with the astrophysicist, perhaps it would be better if Baade could be sent to the recently opened McDonald Observatory in Texas, which, unlike Mount Wilson, had not been declared part of a military district. Bush wrote back telling Adams to sit tight for the moment. While there was no question regarding Baade's loyalty, his dilatory ways could not be overlooked: "he is still a citizen of a country with which we are at war." Bush was confident that the matter would soon be resolved to everyone's satisfaction, especially because no work of a confidential nature was being conducted on the mountain.[47]

Within months Baade's case was reviewed and the curfew lifted on the grounds that he had demonstrated his good faith by initiating the naturalization process before Pearl Harbor. Working with new red-sensitive photographic plates, he homed in on M31, the great Andromeda nebula, and its satellite nebulae M32 and NGC 205. He began photographing the myriads of individual stars in the central mass, as no one before him had been able to do. As gifted a theoretician as he was an observational astronomer, Baade was dedicated to establishing a more accurate distance scale for the universe, and he rested his hopes on the hypothesis that different regions of the nebulae contain different types of stars, each with their own peculiarities.

The observational breakthrough came when two distinct stellar populations began to emerge through the filtered light. One group, composed mostly of hydrogen and helium, the only abundant elements when the universe was young, Baade termed Population II stars. These aged red and yellow giants are found primarily in the nucleus of a nebula and in globular clusters. The other group, comprising younger, metal-rich blue stars, congregate in a nebula's disk or along its spiral arms. These the astronomer categorized as Population I stars.

A sudden thought flashed through Baade's mind: What if the variable stars, including the Cepheids that Shapley and Hubble had used as distance indicators, also came in these same two types? After taking scores of additional photographs, he announced in 1944 that long-term and short-term variables were almost certainly different types of stars in different stages of evolution. Those with longer periods were from Population II, and were intrinsically far brighter than their younger and shorter-term cousins. The fact that they appeared so dim must have meant they were farther away; consequently, the nebulae themselves were much more distant from one another than Hubble's calculations suggested.[48]

Though it was wartime, word of Baade's triumph struck the astronomical community like lightning. How Hubble received the news is not known, but he must have been secretly pleased, for in discovering his error in calibration Baade had resolved one of his lingering doubts about his vaunted theory of expansion. Always before, when his value for the rate of expansion was run backward in time, the cosmos imploded at a point some 1.8 billion years ago. However, geologists estimated the age of Earth and other objects in the solar system at somewhere between 3 and 4 billion years. Based on Baade's revised calculations, the distance between the nebulae was almost exactly double the figure arrived at by Hubble, meaning that the universe had also been expanding double the time. If only Baade's work would hold once the 200-inch went on-line, geologists and astronomers would be waltzing to the same tune at last.

VI

In November 1942, Hubble's status was changed from that of consultant to head ballistician. A promotion quickly followed, elevating him to chief ballistician at a salary of $8,000 per year, the highest civilian ranking outside Washington. His disappointment at not receiving the commission he had been promised was offset by equivalent pay and the knowledge, as he wrote Grace, that he was free to "clear out at any time," an option closed to a man in uniform.[49]

Still, certain of the old mannerisms lingered. When an English lieutenant named Strand was assigned to the ballistics unit, Hubble, to the amusement of many, insisted on addressing him as Leftenant Strand. He was visibly disappointed when the young officer was promoted, for there was nothing to be done with "captain."[50]

In quest of an even more secure site at which to test its top secret weapons, the army had leased a small island, some four miles long and two miles wide, from John Pierpont Morgan, son of the famed industrial magnate and philanthropist, who had bought it as a hunting preserve and summer retreat. Named after Colonel Nathaniel Utie, a colonial who built his manor house on the north shore, Spesutie—"Utie's Hope"—was shaped like a horseshoe around a central lagoon that joined Chesapeake Bay at a place called the Causeway. Distinctly flat and only a few feet above sea level, the island was laced by sluggish, canal-like streams and marshes that filled and drained with the tides. A village of muskrat houses dotted the partially submerged western shore, and in the spring and summer migrating ducks and geese from the Atlantic flyway nested and fed by the thousands, their quacking and honking borne like waves on the following wind. When the bay froze in late December, foxes, both gray and red, crossed over from the mainland a thousand feet away, as did red Virginia deer and other animals, slipping into the dense virgin forest on the island's southern end or burrowing beneath the withered stalks of goldenrod, sunflowers, asters, and rose gentian that covered the abandoned fields in summer.

In May 1943, with the tide of battle gradually shifting in favor of the Allies, the last planks of the wooden bridge linking Spesutie to the Maryland shore were bolted into place, followed by orders that twenty-four-hour sentries were to be posted at either end. Among the island's few structures was a nineteenth-century house previously used by the exclusive New York Rod and Gun Club founded by Morgan and two of his Wall Street cronies, George Whitney and Stephen Birch, known locally as "The Three Bears." Built of wood, it sported a Victorian tower, a veranda enlivened by fretwork, and a hallway with windows of colored glass; at one end of the living room, which contained three large black leather chairs and three heavy Pullman lamps for reading the stock reports after the morning's shooting, was a case filled with stuffed game fowl of many kinds, including a Scotch capercaillie.

Barely visible across a wide expanse of lawn was a small empty cottage formerly used by the Negro servant who assisted in the hunt. To Hubble it was little more than a "dirty shack," but to one who deeply missed his wife a shack that just might do. On the first floor were a

living room, a bath, a minuscule kitchen, and a screened-in porch; the
second floor contained two tiny bedrooms and a single closet housing
a large collection of spiders and insects of various descriptions. He
immediately got in touch with Miss Carter, who had served as Morgan's
housekeeper at the lodge before the war, and asked her to render an
independent judgment. When she spoke diplomatically of "possibilities" it was all the encouragement he needed. A series of wires were
exchanged with the Morgan estate, the owner having recently died, and
Hubble was given free use of the place for as long as he wished by the
executor. He then informed an even more anxious Grace that everything would be ready within two weeks. The house, he assured her,
needed little more than some plumbing, a bit of wiring, an electric
stove, a hot-water heater, a refrigerator, some new screens, a thorough
cleaning, and a good coat of paint.[51]

On the eve of heading east, Grace received a going-away letter from
Maria, who together with Aldous was now living in self-imposed isolation near the village of Llano: "I hope there will be a fairy tale castle
with dwarfs to do the work and glass shoes to twinkle on your feet and
the scent of flowers to welcome you. But I know that a shack and the
housework would not matter, even if they were at the end of the journey." Aldous was almost as happy for her as Maria, except for a "selfish
little corner" of his heart which regretted the departure of his most
trusted proofreader.[52]

Within weeks, Grace, having emerged from the "nightmare" of separation, was on hands and knees contentedly scrubbing and polishing
what she described as "this ridiculous ruin [that] ought to have Faces
staring out of the upstairs windows, and may, for all I know." It was
only after she had arrived that Edwin told her the century-old structure
was known locally as Haunted House, and had been the setting for
several ghost stories collected by a member of the Smith family, whose
dead lay beneath a small pyramid of stones at the turn of the lane. No
shrinking violet, Grace nevertheless stayed clear of the second story and
its resident population of arachnids, sleeping next to Edwin on twin
army cots in the living room while field mice skidded across the freshly
painted linoleum. She cooked on two hot plates while awaiting the
delivery of a range, and fought the chill with a belching coal stove
dubbed "Little Filthy." "It is like Haydn's 'Toy Symphony,'" she wrote
in her journal. When the commanding general dropped by for a visit,
he simply shook his head in disbelief, remarking that a second lieutenant wouldn't live in such a place for fear of losing his social position.

If anything, the primitive conditions only drew her closer to Edwin:

It is very natural to find myself mingling again in the stream of ideas and talk. To have E to answer all questions, or to be silent in a complete sense of companionship. . . . I thought while performing my charwoman's tasks, how much of the structure of one's life, in the long run, is dependent on a very few people who have shared experiences and thoughts. E and I, after these years together, recall things and give them life and colour. A word, spoken by someone, will make us look at each other and we think the same thing. Without this, one becomes more deaf, dumb, blind, amputated.[53]

The island was patrolled day and night by car, and no boat was allowed to approach the ragged shoreline. Hubble had his targets set up at the south end and in the bay beyond. Towers had been built for observation posts, and concrete bunkers for protection. When blockbusters and land mines were detonated, Haunted House rose slightly in the air, then shuddered when engulfed by the trailing waves of sound. Every window was cracked, and repairs were futile. On a snowy day in January, the front door was blown from its hinges and cartwheeled into the living room. Grace waded through the drifts to report the damage to the colonel stationed at the bridgehead. Repairs were soon completed and Edwin noticed nothing unusual when he returned home for dinner. "By the way," he asked, "did you hear anything about three o'clock this afternoon?"[54]

The treasury of Hubble's dubious exploits as a young man was added to by a high-ranking officer with a bent for storytelling who found a credulous listener in Grace. Late in 1942, as submarine warfare in the Atlantic was approaching its peak, a German U-boat laden with high explosives allegedly made its way up Chesapeake Bay, where it was captured, presumably by the navy. On board were written orders from Hitler himself, calling for the destruction of the Proving Grounds and with it "Dr. Hubble." At first Hubble was annoyed when he heard what had been said, but after the war Grace told the story to friends without protest from her husband. On another occasion he was summoned to the Pentagon by none other than Major General William "Wild Bill" Donovan, head of the newly created Office of Strategic Services, which the flamboyant lawyer quickly built into a formidable intelligence agency and forerunner of the C.I.A. According to Grace, Donovan wanted Hubble to fly to Switzerland, where he would be given six secret agents, including the baseball player Moe Berg, who, posing as a French physicist from Nancy, had attended a meeting of German scientists at the Technische Hochschule in Zurich, where Heisenberg spoke on nuclear fission. From Berne the seven spies would be flown into France and dropped behind enemy lines on a mission whose purpose was never revealed. Though this account is somewhat more credible than the U-boat tale, Hubble, who was approaching his fifty-fourth birthday and had never jumped from a plane, received no fateful call from Donovan.[55]

VII

Hubble worked as hard at cultivating his image at Aberdeen as he had at Mount Wilson. In a profile of the chief ballistician in the post news-paper, *Flaming Bomb*, he was celebrated as a star basketball player and boxer, with a hint of a British accent. "He finds relaxation in puffing a pipe, which holds anything . . . just so long as he's got something to puff on." When time permits, he likes to take it easy in "the atmo-sphere of plaid ties and comfortable, though unconventional shirts," which Grace identified as English mesh linen from Austin Reed in Regent Street.[56] He continued an old habit of attempting to impress visitors who happened to arrive while he was sorting through his mail. Opening none of the first-class letters, he ceremoniously dropped them into the wastebasket, one by one. Later on, when it appeared that no one was looking, he could be seen retrieving the missives.

Still, his refusal to engage in correspondence had not changed. In January 1943 the council of the American Academy of Arts and Sci-ences elected him a member, together with Harlow Shapley and Bart J. Bok. A year and three unanswered letters later, the council turned the matter of a formal acceptance over to Shapley, who, after hearing nothing from Hubble, appealed to an embarrassed Adams. The Mount Wilson director cited his own grievance "with the notoriously bad cor-respondent. We built a number of cameras for him in our instrument shop but we have never heard from him whether he received them, to say nothing of their operation."[57] Nor had the blood feud with Shapley been resolved. When Hubble learned that the Harvard astronomer had lobbied in secret against building the 200-inch Hale telescope in South-ern California, he spoke of Shapley's "iniquities" to Fritz Zwicky during lunch at the Athenaeum.[58]

It was during dinner with his colleagues on Mount Wilson that Hub-ble perfected the art of intellectual one-upmanship, which he may have also practiced in the Officers' Club at Aberdeen prior to Grace's arrival. He could usually be seen in the astronomy library reading from the

Encyclopaedia Britannica before the bell rang. A few minutes into the meal, he would "just happen" to mention some obscure person or thing, drawing the others into the conversation. He then issued a subtle challenge to their knowledge of the facts and promptly settled the disagreement by getting up from the table and reaching for the pertinent volume of the Britannica. As with the ritual of the mail, the others, having been played for fools once too often, caught on and refused to be baited any longer. The astronomer Gibson Reaves, who met Hubble while completing his Ph.D., noted that he became the subject of ridicule for such egotistical failings. "Yet they show a tragic side to him," Reaves reflected sadly. "They reveal that Hubble did not feel sure of himself, was not satisfied with himself or his achievements."[59]

With the war's outcome no longer in doubt, security precautions were relaxed. On more than one occasion, Hubble invited the wives of his male colleagues to visit the building housing the newly completed supersonic wind tunnel. He stationed himself at the foot of the lobby stairway and greeted each visitor with a handshake before engaging them in informal conversation. A graying Grace teased her husband about a W.A.C. who was smitten with him, but she was secretly proud of his masculine appeal to a pretty woman easily young enough to be their daughter.

Hubble could be observed in his office on the second floor through a window that overlooked the large chamber and its complicated electronic machinery. In an outer room both uniformed and civilian engineers, chemists, physicists, and mathematicians—some American, others English and French—waited turns to see their superior. He sat behind a large desk with several telephones, all of which seemed to be ringing simultaneously, while a steady stream of post messengers left deliveries stamped "urgent" or "top secret" with the officer on duty in the anteroom.

Equipped with a compressor capable of generating 13,000 horsepower, the first of two closed-circuit tunnels was put in operation in December 1944. Known as the Bomb Tunnel, it was destined primarily for the study of finned projectiles. The dummy explosive was supported from the rear by a strut attached to a balance, then pushed forward into the test chamber. The strut could be tilted at an angle of attack varying through a range of 10° above and below the horizontal. Filtered and dried air was then driven through two nozzles, one at subsonic velocities, the other at almost twice the speed of sound. While ballisticians watched through glass observation windows measuring 18 inches in diameter, the forces acting on the balance were transmitted to dials registering drag, lift, and pitch. The technique, which produced quick and accurate results, virtually replaced the slow and costly practice of manufacturing full-scale projectiles and testing them on the firing range, much to the pleasure of the wildlife and the handful of humans on Spesutie. Weeks later the Ballistic Tunnel, which had to await the completion of a flexible nozzle and balance system, came on-line.[60] Both wind tunnels were used to test models of captured enemy ordnance as well.

For months the Hubbles watched as landing barges, alligators, tanks, guns, and bazookas were loaded on the flatcars of the little A.P.G. railroad, covered with canvas, and transported to the nearby docks. Edwin noted with pride that every type of gun used in the war had been tested at Aberdeen. On the eve of D-Day, June 6, 1944, Grace wrote of how the tanks, many of which were destined for Patton's legendary offensive, had been roaring up the test hill across the inlet for months, filling the air with clouds of dust that settled over Haunted House like a pall. The departing ships, laden with young men in their teens and early twenties, fired melancholy salutes as they headed down the bay and out to sea. When the Hubbles turned on their radio at 6:30 the following morning, Ed Murrow was telling the nation that the invasion had begun at a place called Normandy on the northwest coast of France. A vision of the countless young faces brought Grace to the verge of tears: "I had seen them so often coming in amphibious boats to practice landing manoeuvres on the beach beyond Locust Point; on the mainland crawling under barbed wire; climbing over high obstacles; sparring with bayonets; marching at the double-quick along the rocks."[61] More sickening still was Edwin's account of films taken for Intelligence and Ordnance by the Signal Corps. They were of an obscure central Pacific atoll called Tarawa, where the bloated corpses of many soldiers had washed up on the beach, like flotsam. "I think," he told Grace, "I should not like to remember that I had stayed at home."[62]

Ever since coming to Spesutie, Grace had been engaged in the task of rewriting the papers of her husband's young scientists. Their equations make sense, he told her, but the prose is virtually unreadable. It did not matter that the material was highly classified, for she was able to grasp little of it. More to her liking was the proofreading she was doing for Aldous, who had long since despaired of those he had hired to take her place—"a pitch of ineptitude undreamed of in happier times," as he described it.[63] In a letter to her recently widowed mother, who preferred James Fenimore Cooper to Shakespeare, Grace explained that the title of Huxley's latest work—*Time Must Have a Stop*—was drawn from Hotspur's death speech in *Henry IV*:

> But thought's the slave of life, and life time's fool;
> And time, that takes survey of all the world,
> Must have a stop.

She filled her journals with long, often eloquent paeans to nature, and was especially moved by the seasonal changes unknown to Southern California:

> The colours change from day to day. This afternoon, bronze woods in a blue veil of mist, a dark green bay with white caps, and waves breaking on the shore. . . . The ground is white with frost and iron-hard. The dead grass looks like mummies' hair, but the black tracery of the bare trees is very beautiful. It snowed in the night. There is a fringe of long icicles over the kitchen window, snow on the branches, crotches, trunks of the trees, so white that the trees and dead grasses that were gray are now a soft brown. Through the lighted windows the snowfield is electric blue, and the sky dark slate-colour. The swans sang all night. It is said that swans sing louder just before a wind.[64]

Grace haunted the 3rd Battalion library, where she made her only close friend, Page Ackerman, a witty civilian in charge of the various lending institutions on the post. There were speeches before the Maryland Garden Club and dinners in the homes of the tidewater aristocracy, occasional trips to New York and Washington, and endless letters to be written. But the isolation and battling Little Filthy and Old Faithful, the big living-room stove named for its tendency to spew ashes at regular intervals, took their toll. An introspective Grace grew bored. "I sometimes find it difficult," she wrote her mother, "to think of a person, especially myself, as a continuity, but perhaps it is like a piece of music, allegro, allegretto, andante, adagio, etc. One cannot be equally aware of all movements as they are played, until afterwards in silence."[65]

The conduct of the war had wrought no change in Hubble's political views. He commented to Grace on the irony of having to choose between Roosevelt and Dewey not on the basis of who is the better but of who is the worse. Grace still had hopes for Willkie, who, as it turned out, had only weeks to live, and dismissed both candidates as "awful," although she favored F.D.R. over the governor of New York. In the end, she decided not to vote to keep from canceling Edwin's ballot.

On returning to Aberdeen from Washington shortly after yet another disappointing election, the Hubbles were met at the station by an army captain bearing bad news. During their absence Haunted House had caught fire and was uninhabitable. They drove to the island and found the structure intact, but the living-room walls were charred and water-soaked. A gaping hole behind Old Faithful betrayed the culprit. Even though the firemen had torn out the wiring they decided to spend the night before moving into the Officers' Club, making do with candles and heavy blankets.

Grace quickly succumbed to the luxury of maid service and steam heat while seven men were put to work on Haunted House. Still, the Hubbles missed the privacy of the island and were eager to return once repairs were completed two weeks later. The homecoming was more pleasant than either dared hope: "Haunted House, repaired, wind-proofed, repainted, is probably cleaner than it has been for a hundred years," she wrote. "I wish the fire had happened a year ago."[66]

The biggest news story since Pearl Harbor and the Normandy landing broke at 5:47 p.m., Eastern War Time, April 12, 1945. Franklin Roosevelt had died of a cerebral hemorrhage at the "Little White House" in Warm Springs, Georgia, even as the armies and fleets under his direction as Commander-in-Chief were at the gates of Berlin and the shores of Japan's home islands. Suddenly, the man the Hubbles loved to hate had left a terrible void. "He will be remembered as a man of many achievements," Grace wrote her mother. Whether Edwin's fellow Missourian was up to the task of filling the fallen President's shoes seemed doubtful, but Truman's heart-wrenching plea to members of Congress, which was later broadcast on the radio, touched Grace, per-haps because of its allusion to the cosmos: "Boys, if you ever pray, pray for me now. I don't know whether you fellows ever had a load of hay fall on you, but when they told me yesterday what had happened, I felt like the moon, the stars, and all the planets had fallen on me."[67]

In Berlin, Nazi Propaganda Minister Joseph Goebbels placed an ec-static call to Hitler to proclaim another celestial event, a great turning point written in the stars. But like the messages transmitted at Delphi, those of the astrologers are fraught with ambiguity. The most costly and murderous conflict Europe had ever seen ended three weeks later, when the German High Command surrendered to the Allied armies in a brick schoolhouse at Reims, now the headquarters of General Dwight Eisenhower. The following morning, May 8, was proclaimed V-E Day.

In the coming months there was little for Hubble and his associates to do but mark time. He received an invitation to participate in a radio broadcast during the intermission of a New York Symphony concert. Talks by other scientists had so bored him that he switched them off at once. For $500, he quipped, "I can be switched off," and chose for his topic "The Exploration of Space."[68]

Expecting to return to the West Coast within months, the Hubbles were now faced with a domestic crisis whose resolution neither of them wished to contemplate. The couple had never had a pet because of their long and frequent absences from home. But when Edwin didn't have the heart to trap and kill the mice overrunning Haunted House, Grace insisted that they get a cat. Enter General Patton, a half-wild female whose "bloody slaughter" of the rodent population was punctuated by languid afternoons spent sunning herself on the front porch. Weeks later ennui gave way to rapture with the birth of several very fat kittens, including Spesutia and Colonel Utie, who was known more intimately as Whitefront. A second litter, dubbed the Dead End Kids, followed, the favorites among them having been christened Percy and Mouse. The Hubbles obtained a copy of T. S. Eliot's *Old Possum's Book of Practical Cats*, and Grace drew up a color chart with which to keep track of the nineteen felines borne by the aging General. There were afternoon tea parties and evening get-togethers with special guests. A local family of raccoons was let in and came to Edwin for apples and bread while the cats sat behind him in a semicircle, watching. Only when one of the coons approached the cats' dish of milk, dipped its paws, and began licking its "fingers," did a hissing Colonel Utie step forward, to be grabbed by Edwin.

Grace despaired of finding homes for the cats before leaving. She wrote her mother about the fine time she had provided them, but feared she would have to resort to chloroform. As the moment for a decision approached, Edwin had an idea. He told the aristocracy of Harford County that the cats were a special breed, descended from a race of wild black hunting cats already on Spesutie Island when the first settlers arrived. Soon they had more offers of homes than animals, and they parceled them out, two by two, to owners of country estates. "Now," Edwin said, "we do not have to think of them as poor little ghosts." But with their departure Haunted House was suddenly desolate, the surrounding landscape as dim as the Elysian Fields.[69]

When the sky above the Japanese city of Hiroshima glowed brighter than a thousand suns on the morning of August 7, 1945, Grace recorded nothing of it in her journal, and mentioned it only briefly in a letter home: "We are waiting this morning to hear more about the new bomb, and can only hope that it may bring a speedy end to the war, saving not only our lives, but the Japanese as well."[70] One week later, following the atomic bombing of Nagasaki, the Japanese surrendered on what was a bright summer afternoon in Washington. Harry Truman, who had pulled the country through after all, went for a swim, while the Hubbles, like hundreds of thousands of other Americans, talked of finally going home.

On December 3, Edwin spent the day in his office, busy as usual with meetings and details. Grace cleaned Haunted House, then went for a walk in the cold to collect some winter berries. These she arranged in a small vase which she placed on the strangely empty bookshelf. It was nearing dark when Edwin returned; shortly thereafter a staff car pulled up outside. While their luggage was being loaded in the trunk, the couple checked the stoves, switched off the lights, and walked slowly out the front door, which had never had a lock, for the last time.

CHAPTER FOURTEEN

DARK PASSAGE

I

Hubble, who had never been good at marking time, grew restive during his last months at Aberdeen, and returned to Mount Wilson with what Grace described as a sense of urgency. The lost years could never be regained, and work on the 200-inch had been suspended, frustrating any attempts to expand the boundaries of the universe until its completion, which was still more than two years away. In a March 1946 letter to Nick Mayall, who remained at Lick while continuing to nurse hopes of joining the "dark men" at Pasadena, he wrote of Humason's plans to publish 500 velocities—"he likes the round number"—before closing out the redshift program with the 100-inch. With only twenty more to go, "he is so anxious to finish, that he is taking whatever comes along and with luck will be through this season easily." Then must come the "definitive selection of new fields here at Mt. Wilson," which would likely include problems associated with nebular rotations, close pairs, dispersions in groups and clusters, dwarf nebulae, and a good many others.[1]

Despite Hubble's apparent enthusiasm, all was not well on the mountain. He had been eclipsed by Walter Baade during his absence, and tensions had gradually built up between the German émigré and Humason, who continued to do Hubble's bidding. Tempers flared not long after Hubble returned, when Humason and Baade argued over spectroscopic investigations with the 100-inch. Baade, who hoped to give Mayall telescope time for the purpose of photographing emission nebulae in M31, charged Humason with hogging the show. "[I]t is evidently beyond the mental capacities of the former mule driver," he fumed in a letter to Mayall, to cooperate "unless he has the whole field to himself!! The perfect, conceited ass! Of course this cannot go on since it would ruin any observatory." Baade had marched down the hall to the office of Mount Wilson's new director, Ira Bowen, to make his case, but was disappointed when Bowen frankly told him that "until he found his own way," he did not want to give orders to the older men on the staff. Left unsaid was the well-known fact that the director had gained his position at Hubble's expense, and was unwilling to cross swords with the still smarting astronomer. He told Baade to wait a week until Hubble returned from his vacation in Colorado since he "has the responsibility for the nebular department."[2]

The change of administrators had been set in motion by Bush in an October 1944 letter to Walter Adams, who had announced his intention to retire. Taking Adams into his confidence, Bush confessed that "I have not as yet considered that we have a man who meets all of the essential qualifications available in our own group [Mount Wilson], although I may have overlooked or misjudged at some point." In terms of insiders, Hubble was at the top of the list. "His scientific reputation is excellent, his personality is exceedingly attractive, and I believe that his ideals in regard to the future of the Observatory are excellent." But Bush, who had an opportunity to monitor Hubble's wartime labors at the Ballistic Research Laboratory from nearby Washington, was also "quite convinced that he has no liking for administrative work, and hence no skill along these lines."[3] It was a view echoed by Dr. L. S. Dederick, chief of the computing laboratory, who shared an office with Hubble and found him "mildly ineffective" as well as poorly informed when it came to ballistics.[4]

Bush also anticipated that Shapley's name would surface once the announcement of Adams's impending retirement became public knowledge. Having known the Harvard astronomer for years, Bush conceded him a number of "extraordinary assets and capabilities." Unfortunately, "I think he lacks generosity; his own reputation and advancement have often been too keenly in his mind rather than the welfare of his organization or of his colleagues."[5]

Preferring to keep his cards close to his vest for the moment, Adams praised Hubble in his letter of reply, stating that he might turn out to be a better administrator than Bush imagined. However, he could not bring himself to concur with Bush's view that Hubble was an altruist when it came to his employer. "A more serious question . . . has been Hubble's somewhat lukewarm attitude toward the Institution in the past," which may have stemmed from a "feeling that he was not being adequately compensated." Where he would stand when the 200-inch went into operation under the auspices of Caltech was a matter "requiring considerable thought," for the possibility loomed that he could be hired away. As for Shapley, a man of attractive mental qualities but a personality riven with blind spots: "He left Mount Wilson with little regret on the part of staff members," while his scientific accomplishments at Harvard do not rate "highly." Furthermore, Shapley was furious over the episode of the 200-inch, believing that the Rockefeller Foundation should have considered no institution other than his own Harvard College Observatory.[6]

Adams then suggested that Bush take into consideration the name of forty-six-year-old Ira S. Bowen, a handsome, amiable professor of physics at Caltech and protégé of Robert Millikan. Thanks to his brilliant work on the identification of the "nebulium" lines in the spectra of gaseous nebulae, Bowen had been awarded the Draper Medal and was elected a member of the astronomy section of the National Academy of Sciences. "All our staff members know him well and appreciate his ability," Adams enthused. "He . . . has a pleasant personality, and a high standing among both physicists and astronomers."[7]

When Bush replied two weeks later, he made no reference to Hubble. "I think we see completely eye to eye" on the subject of Shapley. "The mention of Bowen is very interesting indeed."[8]

Bush next communicated with Max Mason, the mathematician and former president of the Rockefeller Foundation, who was largely responsible for managing the completion of the 200-inch. Like Adams, Mason readily endorsed Bowen. Not only was he a good scientist; he was also an excellent leader, "utterly honest from an intellectual standpoint and straightforward in his dealings." By contrast, Mason had long considered Hubble arrogant and self-serving, and felt his appointment would be a disaster: "He has never put his mind on [the] . . . somewhat subtle aspects of human relationships, which are rather at the heart of the whole art of administrative work." As a distinguished scientist, he should be encouraged to go on with his scientific effort. "I rather think this is the correct point of view," Bush confided to Adams, but whether Hubble himself "would agree to it or not is another question."[9]

Still, Bush held back. He had also sought the counsel of Caltech physicist Richard Tolman, Hubble's friend and scientific collaborator. While Tolman had nothing against Bowen, he felt that Hubble's standing among astronomers was such that they would not understand if their colleague was passed over. Nor, knowing Hubble, could he be expected to swallow such a bitter pill without a serious reaction. He "would be so completely disheartened that unfortunate results would follow." Bush, who harbored a deep respect for Tolman, wrote Adams that "I am much disturbed at this point of view."[10]

Fearing that Hubble might slip in by the back door, Adams, who no longer had anything to lose, grew bolder in his criticism of his colleague. As a scientist, Hubble had to be considered past his prime, whereas Bowen, a superior mathematician, was conversant with the rapidly developing field of astrophysics, of which Hubble knew relatively little. And while Bowen, some nine years Hubble's junior, had the liking as well as the respect of the Mount Wilson staff, Hubble had only the respect, which would almost certainly lead to difficulties down the road. Adams continued by reciting a list of complaints against Hubble which had been documented in the director's many letters to Merriam over the years—the failure to fulfill his professional obligations when chairing committees; the resentment caused by his unwillingness to answer important correspondence; the "extreme individualism" that manifested itself in long absences in the name of science while raising Anglophilia to the level of a faith. Adams appreciated Tolman's concern regarding what might happen if Hubble did not get the post, but there was little chance that he would resign out of spite. His work was too important to him, and for this he needed the two finest telescopes in the world.[11]

Bush had all but decided on Bowen when he met with Hubble in Washington in May 1945, two weeks after the German surrender. He wrote Adams later that same day:

> I think that Hubble finds himself exactly in the same position that we do. On the one hand, he has the feeling that he ought to succeed as director because of the prestige and because the Director of Mount Wilson represents the Institution in astronomical affairs, and, on the

other hand, I am quite sure that he knows personally that he would be completely swamped if he attempted to carry on the duties of this post because of his general disinclination to handle managerial affairs. Wherever we went in discussion we always came back to this quandary, and we left it unresolved.

Though Bush had said nothing of Bowen to Hubble, he had attempted to soften the impending blow in advance. Whatever the eventual outcome, Hubble should know that the salary of the most eminent astronomer on the staff and that of the new director would be approximately equal in the future, a matter already cleared with the Carnegie Institution's executive committee.[12]

By now, Bush was close to agreeing with Adams that everything possible should be done to retain Hubble short of offering him the directorship, but he wanted to speak with Tolman once more before announcing his verdict. To his surprise, Tolman had come to the conclusion that Bowen would make a fine director, but he also pointed out that Hubble was very likely to bring pressure on the new man "through tortuous paths" in hopes of promoting his personal agenda. Bush replied that if this were true, and he had no reason to question Tolman's judgment, "it is an excellent argument as to why he should not be Director." Indeed, "I [said] I would rather take the risk of losing Hubble than take the risk of having him as Director."[13]

In mid-July, Bush traveled to the New Mexico desert to witness the culmination of a mammoth program of U.S. scientific research and technological development that began in 1940. The first atomic bomb set fire to the sky above the Trinity site near Alamogordo, while J. Robert Oppenheimer quoted from the *Bhagavad-Gita*. A jubilant Bush, who had shepherded the Manhattan Project through the war, headed for Pasadena following the test, where he joined Adams, Mason, and certain of the Caltech trustees at a conference that lasted from 3 p.m. until midnight. When they emerged, all agreed that Bowen was their man, and a verbal offer was tendered and accepted the next day. This was followed by a celebration dinner, after which the only complaint about the new director was heard. Bowen had not partaken of his fair share of precious wartime scotch, a fault the trustees vowed to correct in the future.

Hubble cast a long and palpable shadow, causing Bush to worry whether the same $10,000 salary he offered Bowen would be enough to placate the astronomer. Then he hit on the idea of forming a Scientific Program Committee to which Hubble would be appointed chairman by the director-elect. Though the committee would serve only in an advisory capacity, its formation would help to heal a wounded ego while giving Bowen, who supported the idea, the opportunity to extend the olive branch. Only Mason demurred, arguing that the selection committee was going too far.[14]

Hubble, along with everyone else outside Bush's inner circle, was unaware of Bowen's appointment. On September 10, a week after the Japanese surrender in Tokyo Bay aboard the battleship *Missouri*, he drafted a long, awkwardly written letter to Bush concerning the directorship. It was his belief that Mount Wilson should be headed by an astronomer of the first order, for only a specialist could create a program worthy of the observatory's long and distinguished tradition. However, the "research man" should be relieved of the paperwork and other mundane irritations by an executive officer, who would be responsible for the day-to-day operations at Santa Barbara Street. "I am fully convinced that a research leader, plus a competent executive officer, is the proper solution of the problem, and I am confident that the consensus of opinion among the senior astronomers of the country will support this view."[15]

Bush doubtless experienced a twang of conscience in light of the fait accompli coupled with Hubble's obvious desire to succeed Adams, who had become his secret nemesis. When he wrote again, stating his willingness to postpone his return to Mount Wilson pending Bush's decision, the announcement of Bowen's appointment could no longer be delayed. Bush wrote to Hubble on October 3, 1945, breaking the news as gently as possible. "We are all anxious that your scientific career should go forward with the best of equipment and opportunity, for we know that in so doing, it will add to the lustre of your own reputation and of your Institution." Clearly, "you could not develop the full potentiality of your research if you were burdened with administrative duties, for I know they are not to your liking." To destroy an effective scientific career for the sake of an administrative post would serve no purpose. Finally, Bush understood that Bowen would be writing soon to ask that Hubble accept the chairmanship of a committee to plan a program of research as Mount Wilson embarked on a new and important phase of its history.[16]

By prearrangement, "Ike" Bowen wrote to Hubble from California the day after Bush sent his letter from Washington. "All of us," he began, "recognize that your studies of the spiral nebulae were without question the greatest single factor in inspiring the construction of the 200-inch telescope." The war had cost astronomy a great deal, but the new director promised him every opportunity to push his program forward with the greatest speed and efficiency. As chair of the Program Committee, Hubble would have a virtually free hand in his own field of cosmology. "If you are willing to serve, may I ask you to inform Dr. Bush directly?"[17]

Hubble admitted to being stunned, and several days passed before he drafted a terse and chilly reply to Bowen. The appointment of a physicist rather than an astronomer as director was quite disturbing, especially "if it involves [the] actual control and direction of research." Still, he was willing to give the new arrangement a try in light of Bowen's promise of a free hand "in the field you call cosmology. Several of us have given much thought to the big problems in that field, and have rather definite notions on the method of attack." With this understanding, he would accept the chairmanship of the Program Committee and planned to be back in Pasadena no later than the first of the year.[18]

Bush, who received a similar letter, disliked Hubble's tone, and wrote Bowen that he was inclined to answer by laying down the law as to who would actually be running the institution. The fact that Bush had chosen well became evident when Bowen wrote a diplomatic reply, stating that Hubble's problem was more one of attitude than actual demands. He would rather Bush delay answering, thus giving Hubble the chance to bring his emotions under control. Indeed, "I would be quite willing to have you ignore his letter and wait and see whether things do not work out satisfactorily, rather than making an issue of it now."[19]

II

War had changed Hubble in a manner that he and those who knew him best never could have imagined. Although an archconservative, he returned to California deeply committed to the eradication of atomic weapons, sharing the concern, if not the dark melancholy, of his anguished friend J. Robert Oppenheimer, whose very physical appearance had been transformed since Hiroshima. In 1946 Hubble began lecturing on the potential horrors of future wars before any group that would have him, including the Sunset Club, the Los Angeles Town Hall Meeting, luncheon organizations, university clubs, even the Daughters of the American Revolution. In his most widely noted address, titled "The War That Must Not Happen," he predicted that rival nations would soon have the capacity "to blast out the greater part of the other's material civilization within a brief period measured in hours" via a lethal combination of long-range rockets, guided missiles, and atomic bombs. Then would follow long dreary years of attrition marked by commando raids and small units, a war of brute survival in the shambles of a ruined world. Finally, what remained of the human race would be reduced to exterminating the last of itself with the clubs and spears of a new Paleolithic age.

Given the failure of governments in the past, this vision of a man-made Armageddon, however terrifying, could not be averted by the old expedients of treaties, alliances, and moral sanctions, "any more than respect for traffic regulations can be left to the decency of motorists." In a conclusion that was greeted more often than not by murmurs of skepticism, Hubble argued for the establishment of a world government backed by a powerful international police force. While "it is true that no element of national independence should ever be sacrificed carelessly or needlessly," the concept of isolation is an illusion, for "competition grows as the nations crowd together in the shrinking world. Under these conditions, complete sovereignty leads to war and mutual suicide." The choice is self-evident: "It is now world-government or no world—one world or none."[20]

Still blaming Germany for the recent conflict, Hubble was startled to receive a letter from Hans-Lothar Gemmingen, with whom he had become friends thirty-seven years earlier at Queen's. Gemmingen had read of Mount Palomar and was pleased to hear that Hubble had become "a big man." "When we were together at Oxford we never thought of such a disastrous development of the policy of this world and that two great wars could separate us for so many years." The steel magnate, who like many top German industrialists had spent time in an Allied prison, often reflected on happier days, "especially when we rowed together and won the Fresher's Four and when the most amusing fruit-eating match took place between you and Popp! Do you remember??"[21] The double question marks were to remain unanswered; an unforgiving Hubble could not bring himself to pen a reply.

In October 1946 the Hubbles boarded the Santa Fe Chief and headed east in response to a summons Edwin had received the previous February. Arriving in Aberdeen after a stop in Washington, they were met by Major General Everett Hughes, the new chief of Army Ordnance. Following the short drive to the strangely quiet proving grounds, Hubble was presented with the Medal for Merit, the highest civilian award granted by the President, while Grace, several of his former staffers, and a reporter for the *Baltimore Sun* looked on. The accompanying citation praised him "for exceptionally meritorious conduct in the performance of outstanding service as Chief of the Exterior Ballistics Branch of the Ballistic Research Laboratory." It was as close as he would come to long-dreamt-of glory on the battlefield, and he proudly wore the circle of white stars and eagle in full wingspread on special occasions. "There are two things I like about the medal," he said. "The citation is worded 'In accordance with the order issued by *General* George Washington at Headquarters, Newburgh, New York, on August 7, 1782.' And it is signed 'Harry Truman, *Commander-in-Chief*.' "[22]

They remained at Aberdeen for a week, during which Edwin attended several meetings in his ongoing capacity as a consultant to the army. The pull of memories carried them back to Spesutie one afternoon, where they were photographed in front of the rickety porch of Haunted House. Grace wore a big-buttoned coat and clutched two packs of cigarettes, while Edwin stood in a tweed suit protectively behind her. The place was just as they had left it; the forlorn little bouquet was still on the living-room bookshelf, and the raccoons' deserted drinking bowl sat empty on the back porch. By the time of their next visit Haunted House would be no more. The army tore it down after determining that no one could live in such a place.

III

Like everyone else associated with the 200-inch telescope, Hubble periodically stopped by the Caltech optical shop to check on the progress of the "Big Eye." He was conspicuous but unrecognized behind a glass partition among tourists and schoolchildren, gazing up at the smock-clad opticians who labored all day in dead air carefully guarded against temperature changes. The workers had lately switched from large polishers to "local" instruments, ranging in diameter from 68 to 8 inches. As the day of completion drew near, increased precautions were taken. Street clothes and shoes were banned from the laboratory, while a vacuum cleaner and a magnetic sweeper designed to suck up vagrant metal particles were passed over floors, walls, and machinery at least twice daily.

The moment everyone had been anticipating for years came on an October afternoon in 1947, when it was announced that the fragile colossus was ready at last for transportation to the top of Palomar Mountain. A delicate operation, it was fraught with a whole new set of logistical problems. One month later, at 3:30 a.m., a hulking sixteen-wheel diesel accompanied by a police escort lumbered through the optical shop archway and headed south into the November night, its strange cargo insured by Lloyd's of London for $600,000. The caravan detoured for narrow bridges, heavy traffic, and other obstacles. Averaging nine miles per hour, it stopped for the night thirteen and one-half hours later.

Because of crank letters attributed to "psychopaths" who feared the unlocking of celestial mysteries, a special six-man protection unit took up stations in the glow of flares which surrounded the huge mirror crate as twilight faded. The following morning, they were relieved by fifteen California Highway Patrol officers, who paced and trailed the ponderous freighter over the final thirty-three miles of its uneventful journey. Once safely inside the dome, the mirror was lowered into the vacuum chamber below the main floor to receive its reflective coating of aluminum, the final ritual before fitting glass to metal.

When combined, telescope and mirror equaled 500 tons of precision instrument refined to the limit of human ingenuity. Fifty-five feet long, the great barrel was set on a modified horseshoe mounting so large that it had to be shipped from its Eastern manufacturer, the Westinghouse Electric Corporation, by way of the Panama Canal, then trucked the remaining distance to the observatory. To overcome the massive friction on the axis bearings, it was first decided to employ the mercury flotation system used on the 100-inch, but the expense of several tons of liquid metal was deemed prohibitive. A solution was provided by an adviser from Westinghouse, who proposed that oil be forced through five bearing pads at a pressure sufficient to float the instrument on a thin film of liquid. Requiring only a few gallons of oil under a pressure of 210 to 385 pounds per square inch, the flotation system produced so little friction that 500 tons of telescope could be moved with a motor similar in size and power to that of an ordinary sewing machine.

The 200-inch mirror was tucked away in the butt end of the telescope, protected by the steel leaves of an iris that folded at the push of a button, exposing it to the heavens. Instead of climbing iron steps to his observation post, the astronomer, freed of the time-consuming task of changing the outmoded cages bearing the Cassegrain and Newtonian mirrors, would don an electrically heated suit and enter an elevator which lifted him to the bullet-shaped observer's cage, where he stepped into the very throat of the steel monster, disappearing into blackness. Once settled, he spoke via an intercom to the night attendant who sat at his console somewhere in the dimness below. Tiny lights blinked as buttons were touched; a low rumbling sound issued from the 1,000-ton revolving dome; the telescope was synchronized with the movement of Earth; the target was locked in; the voyage backward into eternity had begun.

Or so Hubble and his anxious colleagues envisioned it. Unfortunately, there was still much to be done. The formal dedication took place on June 3, 1948, when eight hundred invited guests drove up the steep, curving road past small encampments of Palos Indians preparing for one of their annual rites. Among the passengers were Nobel Prize winners, atom bomb scientists, movie and stage actors, Caltech trustees, photographers, newspaper and radio reporters—the last large gathering of laymen permitted to wander over the observatory floor. Carrying cameras and staring upward, they moved about in silent groups or spoke in hushed tones, as in a cathedral. But when Charles Laughton, the crusty, heavy-jowled actor who had recently played the role of Galileo, was asked by an awed friend whether he considered the sight inspiring, Laughton shot back, "Inspiring, my eye! It's damned frightening. What are they going to do with it? Start a war with Mars?"[23] Turning on his heel, he stalked off, hands behind his back, as though he were pacing the deck as the infamous Captain Bligh.

A few paused in silent homage before the new bronze bust of George Ellery Hale, after whom the 200-inch was to be named in a ceremony commencing at two o'clock. The first to address the audience was Dr. James Rathwell Page. Chairman of Caltech's board of trustees, he was moved, like many of those present, by a sense of the religious. He read the entire canticle "Benedicite, Omnia Opera Domini" from the Book of Common Prayer, which, though centuries old, seemed to have been written especially for the occasion: "O ye Sun and Moon . . . ye Stars of heaven . . . ye Light and Darkness, bless ye the Lord." Max Mason, chairman of the Observatory Council, then rose and attempted to dispel the gloom spread by pessimists who fretted over what the telescope might reveal. He quoted Mark Twain: "I am an old man, and have had many troubles. Most of them never happened." Vannevar Bush turned to Emerson when he next paid homage to Palomar's founder: "An institution is the lengthened shadow of a man." Mrs. Hale, white-haired and wan, was asked to step forward while Lee A. DuBridge, the new president of Caltech, read the resolution naming the telescope for her husband. When he had finished, the widow managed a barely audible "Thank you so much" before returning to her seat, where she touched a handkerchief to her eyes.[24]

Ira Bowen, the last to speak, explained the telescope's mechanics in layman's terms and then pushed a series of buttons, initiating a live demonstration. The telescope and dome had just begun to rotate when a linnet flew into the shutter opening and began to warble, an interloper on a mountain once famed for its band-tailed pigeons (in Spanish, "Palomar" means pigeon roost or dovecote). Then the dome swung around, the telescope moved upward, and the bird disappeared into the brilliant blue glare overhead.

While Hubble seemed to be getting along with Bowen, he had preferred to watch from a distance. A reporter who counted his theory of the expanding universe among the greatest concepts in the history of astronomy spotted him quietly viewing the proceedings from the back row. When he approached the tall, broad-shouldered astronomer at the end of the ceremonies to ask whether the ability to see another 500 million light-years into space would support his theory, Hubble snapped, "I am not worried about theory." He then softened a bit when next asked how he felt when he first peered through the eye of the new giant: "I thought I was blasé, but I got all excited about it."[25]

IV

"Peer" was the operative word, for Hubble and others had seen precious little of the universe through the 200-inch to date. As anticipated, a number of "bugs" developed in the mirror and its gigantic mechanism. Among the most serious was a faint but disturbing vibration in the right ascension controls. Too much friction in the movement of the thirty-six supports interrupted the focus, and the telescope had to be shut down while delicate adjustments were made. Technical problems were compounded by the need to become familiar with the idiosyncrasies inherent in such a finely tuned instrument. The telescope's behavior, even in good seeing, was complicated by air disturbances unknown to the 100-inch Hooker, especially when its resolving powers were concentrated on a single object relatively close to Earth. This, combined with Bowen's penchant for seemingly endless tinkering and testing, gave rise to a rumor that the 200-inch was a failure, the origin of which was traced back to Eastern newspapers and a still bitter Harlow Shapley.[26] Conversely, George Adamski, an amateur astronomer and owner of the Palomar Gardens, a little restaurant located halfway up the mountain, let reporters in on some "uncensored dope." The "big boys," who often dropped by for beers, had recently taken thirty-five pictures, including several of Mars, nineteen of which had turned out perfectly. "And do you know what they saw?" Adamski whispered as he looked around to be sure that no one but the coyotes were listening. "They saw the canals all right. They saw them so distinctly they even saw water running through them. What do you think of that?"[27]

Fittingly, Hubble's first official photograph taken with the 200-inch (PH1H) was of NGC 2261, the cometary nebula on which he had trained Yerkes's 40-inch reflector thirty-four years earlier, and then photographed with the 100-inch during his first run on Mount Wilson, in 1919. The seeing was poor and he was unable to discern any change in the object's configuration.[28] Having paid homage to what Grace called his "Polestar," he began concentrating on more revealing plates, including one made during the period when adjustments were still being undertaken. The region photographed was a random sample of the sky near the pole of the Milky Way designated Area No. 57, which had already been thoroughly studied with the 100-inch. Though the sky conditions were average, a mere six-minute exposure recorded all that had ever been photographed with the Hooker. As exposure times were increased with successive plates, fainter and more distant objects were found in constantly increasing numbers. Finally, an hour's exposure reached the sky background, exhausting the full power of the telescope. Just as predicted, the Hale had reached into space at least twice as far as its Mount Wilson cousin, the intercepted light having traveled for one billion years before falling on the planet. In reporting these findings to the press, the ever cautious astronomer sounded a familiar refrain:

> The photographs looked about as we expected, and the scanty data could be forced into any of the current theories of the universe. There was no convincing evidence either of expansion or non-expansion, nor was there any conspicuous change in the pattern of nebulae in the newly observed regions. Definite conclusions on these questions must await long and very precise investigations of many samples of the sky.[29]

Privately, he was delighted. Everything he had discovered to date had held while the celestial mists were parting to the tune of one billion light-years. Wearing a lumberjack's plaid shirt, flannel pants, and knee boots, the familiar pipe dangling from the side of his mouth, he completed his first four runs on Palomar between January and April 1949, garnering sixty-two plates in mostly poor seeing. Grace became aware of a compulsion that had never surfaced before, as well as an abstraction that caused him to overlook the fact that his Santa Barbara Street office had been redecorated by the staff while he was absent during the war. When a storm, no matter how severe, set in, he headed for the mountain anyway, for there was always a chance the sky might clear. At the beginning of one run he put chains on the car at the bottom of Palomar grade and forced his way through the blinding snow to the top, like a burrowing mole. On reaching the summit he found the steel gate locked, the custodian never thinking that anyone but a madman would brave the slope. Hubble simply climbed one of the huge drifts and stepped across the six-foot fence.[30]

Then, just as he and his colleagues were beginning to develop a real feel for the Hale, the great mirror, which was yet to undergo its final finishing, began giving them problems. On May 1, 1949, the Big Eye was removed from the barrel and lowered into the vacuum chamber for retouching, a process that would delay what the press termed the "Great Campaign" another seven months.

Aware that a Nobel Prize was hanging in the balance if only the selection committee would consider astronomers in the same category as physicists, the Hubbles had engaged the services of a publicity agent, who seemed to be working overtime on Hubble's behalf. Newspapers and magazines trumpeted his achievements beneath such eye-catching headlines as "Trouble, Trouble, Toil and Hubble," "Behold, the Universe!," "Pilgrimage into Eternity," and "Palomar Eye Peers Billion Light-Years." In those few instances where his colleagues were mentioned, Hubble always received the most space as well as the largest photograph. Typical in this regard was a long article that appeared in *Collier's*, titled "The Men of Palomar." Three-by-three-inch photos of Zwicky, Bowen, Baade, and Humason were dwarfed by Hubble's six-by-seven-inch portrait, a jet-black Nicolas Copernicus, the "one-man" cat recently added to the household resting contentedly in the crook of his arm.[31] In many instances the rhetoric was no less florid than the headlines: "While Columbus sailed three thousand miles and discovered one continent and some islands," a writer for *Travel* intoned, "Hubble has roved through infinite space and discovered hundreds of vast new worlds, islands, sub-continents, and constellations not just a few thousand miles away, but trillions of miles out yonder."[32]

Hubble, who was approaching sixty, was beginning to look his age. His deeply etched face, resembling a road map stretched over bone, appeared on the cover of the February 9, 1948, issue of *Time*. Dwarfed by his chiseled head was a background drawing of Palomar's dome, from which the ethereal arm of a goddess, bejeweled with stars, thrust upward into the firmament. In an attempt to pin down its subject, *Time* framed the title of its cover story in the form of a question: "Will Palomar's 200-Inch Eye See an Exploding Universe?"

Casting about for a layman's analogy, Hubble compared the expanding cosmos to a rubber balloon marked by small dots, representing nebulae, spaced equally far apart on its surface. As the balloon expands, each dot moves farther away from every other dot. Place an observer on any one of these, and he will see the same picture: every other dot nebula will be moving away from him at breakneck speeds, the farther away the faster. The Milky Way is not the only center of explosion; every other nebula is equally an explosion center. Relativity theory holds that nothing can travel more swiftly than light, at 186,000 miles per second. Thus far he and Humason had measured nebulae about 250 million light-years away moving off at some 26,000 miles per second, more than one-eighth the speed of light. The 200-inch would permit them to measure the speed of objects at much greater distances. Should the nebulae continue, on and on into space, they would eventually exceed the relativistic speed limit, thus defying the supreme law of modern physics. Yet it was equally possible that the universe, while expanding, was also finite, and that the boundaries of curved space can be plotted with such instruments as the Hale telescope, vindicating Einstein once and for all.

Hubble then addressed an alternative hypothesis embraced by those who found the expansion theory too fantastic. The redshift, it was argued, does not indicate expansion but something quite different. Light starts out from a distant nebula as young, vigorous, and violet. But after millions of years its energy is depleted, its waves elongate, and it turns redder, transforming it into the "tired light" captured on the plates taken at Mount Wilson and Palomar. If this is what happens, the nebulae may be moving very little—or not at all.

While Hubble would not be pushed into a corner, he finally admitted that he did "not expect" to find visual evidence that would undermine the redshift hypothesis, yet he would "welcome it if he finds it. Tired light . . . would be a discovery quite as sensational as the exploding universe."[33]

Aldous Huxley, who was featured in the same issue, wrote a facetious note from his latest residence, a desert house in Wrightwood, California, which was then snowbound: "What a happy coincidence that we should have achieved Fame together in . . . *Time*—and achieved it, which is more remarkable, without having been insulted by those inimitable stylists!"[34]

V

With the 200-inch out of commission, there was nothing for Hubble to do but make the most of his runs on Mount Wilson. He had come to Pasadena when it was little more than a quiet village and Los Angeles a distant, smallish city, far from the great industrial center it had become. A strange, acrid haze, known locally as smog, was slowly enveloping the mountain, often dimming the nebulae as well as the hopes of anxious astronomers. The suburbs below sprawled over nearly the whole of the mountain-ringed plain, their lights a glittering lava flow in reverse, creeping up the heights to fog the sensitive photographic plates. The most striking feature in Humason's recent spectrographs was a bold black band caused by the mercury-lit signs of L.A. Grace wrote her mother that the place "is vulgar, vulgar without charm," and has become a "suitable sink and catchall for the sort of people who like that sort of place."[35]

Professionally, Hubble hated the city lights and dreamed of Palomar, where only the campfires of the nut-gathering Palos Indians flickered in the distance. Yet neither could he resist their mesmerizing charms. He still took visitors—his fellow astronomers included—to the brink of the mountain to wax silently on their beauty. Below stretched a Persian carpet woven of alternating shades of colored glitter, the nearer lights twinkling like fallen stars. Just above were long streaks of luminous fog, reminiscent of the great band of the Milky Way. Far off, blurred by haze and distance, lay the downtown, which he referred to as the dense nucleus of the "Los Angeles nebula." "An astronomer," he once told a reporter, "is like a man up here who can never go down to the valley, and who tries to find out, by observing these lights, all that goes on in Los Angeles."[36]

Fewer stars of the earthbound kind came to visit after the war. Anita Loos had sold her beach house and moved to New York, while Charlie Chaplin, hounded by charges of lechery and political disloyalty, was seldom seen and would soon be forced into permanent exile. Though Hubble had rarely turned down a celebrity or statesman who sought him out, he thought long and hard before agreeing to take Eamon de Valera, Prime Minister of Ireland, up the mountain. During one of their trips to England, Grace had hastened to record Oliver Gogarty's description of the former rebel and architect of Irish independence as "that six-penny Savonarola in a Woolworth World."[37] Only after learning that the New York-born de Valera was a mathematician by training did he agree to serve as the Prime Minister's guide, provided that politics was kept out of the conversation. So compelling a figure was his guest that Hubble was charmed in spite of himself, and never once raised the hackles of his Anglophilia.[38]

In July 1948 the Hubbles boarded the *Nieu Amsterdam* for their first crossing of the Atlantic since the war. They were seen off by Anita, who had just finished adapting *Gentlemen Prefer Blondes* to the Broadway stage. The playwright gave Grace a dozen pairs of scarce nylons and told them to look up Helen Hayes in London, who was about to open in *The Glass Menagerie* at the Haymarket. The actress was little known in England, and had jumped at the chance to play the tragedienne, but now thought she had made a mistake. Anita seconded the verdict: "She is scatter-brained and impulsive, and the play is no good."[39]

The Hubbles' only concern on departing was the welfare of Nicolas Copernicus, whom they had left in the care of some friends. A letter concerning the feline reached them at the last moment, stating that all was well, both physically and mentally. Nor had Nicolas carried in any of the extraneous fauna or flora that had made him a frequent topic of conversation among the Hubbles' friends—a white camellia, a lizard drowned in the water pan, a dead mouse, live birds, mostly towhees, and an unharmed dragonfly deposited in the upstairs bedroom. When Fred Hoyle came to dinner and read Shakespeare's plays, the cat sat very close, gazing intently into the astronomer's face. Then one evening, at the climax in *Macbeth*, "Hoyle's voice rang out with violence," causing Nicolas to leap straight up before rushing into the night through the small living-room door cut for him by a carpenter. Huxley, persona non grata in Nicolas's eyes, fared much worse. The writer had to be protected from being scratched, the prelude to an attack signaled by a dance of rage performed upright on hind legs.[40]

The Hubbles ate their usual hearty ship's breakfast of buttered eggs and bacon, a meal they often repeated at lunch. Then they donned formal dress in preparation for the evening banquet; the Medal for Merit hung conspicuously from Edwin's thickening neck. Yet another round of eggs sometimes followed before they retired in the small hours, with Grace gazing out of a porthole at the ever changing colors and reflecting on the fact that no matter what horrors the human race inflicted upon the fair face of the land, it could do nothing to the sea.

They quickly slipped into the familiar routine of parties, guest dinners, and fishing. Edwin tried his luck on the river Test, where the trout were leisurely, and he had to count to two before setting the hook. "This looks," Grace wrote, "as if it's the best time yet. The closest way to describe our strange entranced condition is that we feel all the time slightly drunken. A sense of great heightened intensity of living, mingled with confusion. E. says, like Alice in Wonderland."[41]

They had not yet seen a ravaged London where walls bearing traces of stairways stood open to the sky, and parking lots had mushroomed in the wake of infernos ignited by firebombs and screaming rockets. Water-filled craters formed small artificial lakes ringed by piles of rubble containing bits of blackened plaster cherubs, molding, and arches. Every other window in the city seemed to be missing, including those of their friends the Gore-Brownes, who had lost their porch and nearly every pane of glass to a bomb that detonated while they were eating breakfast. Stray dogs and cats, rare before the war, were everywhere now, as were displaced beggars and dirty children, who played among the ruins while weary parents scolded them in vain about the dangers of unexploded bombs.

The champagne cork came out with a pleasant sound when they met Helen Hayes for dinner at a mutual friend's. The actress's fears had been allayed on opening night by a quick and responsive audience that had given her a warm welcome. Unlike Americans, who arrived late and tight and considered the theater a place to sit between dinner and a nightclub, the English treated the stage with that special reverence accorded a painting by Rembrandt or a concerto by Mozart. The house was quieter when royalty, however minor, was present, and she thought the play was not going well until John Gielgud, who was directing, explained the subtle change of mood. She was hoping to save Henry James's house at Rye, which had been heavily damaged by a bomb and was now sitting derelict while dry rot attacked the rafters and woodwork.[42]

Hubble had been elected an Honorary Fellow of Queen's College in May, and he stopped in Oxford to express his appreciation before departing with Grace for the Continent. Accompanied by the Gore-Brownes, they headed for Zurich, which Grace had been told was the most interesting place in Europe just now. Hubble, who had been chosen as one of six American delegates to the Seventh General Assembly of the International Astronomical Union, was also chairman of Commission No. 28, on extragalactic nebulae. The meeting produced little in the way of light but generated a great deal of heat, thanks largely to the recently instituted Soviet blockade in Berlin, which severed all land and water communications between the western sector of the city and West Germany. After the first course of the inaugural banquet there was some muttering; then chairs were pushed back as the Russians abruptly got up and marched out of the room. One of their number had suddenly realized that the Russian flag was missing. After some confusion and scurrying, the Swiss flag was taken down and a Russian banner put in its place. As if on cue, the miffed delegates marched back into the room to a smattering of applause. Turning to Hubble, the Scottish astronomer G. C. McVitte asked, "What do we do? Are you going to applaud?"

"Darned right I'm not," Hubble shot back, and the faint clapping soon died away.[43]

VI

Lost in the hoopla surrounding the Hale telescope was the commissioning on Palomar of a second extraordinary instrument about which the public had read little or nothing. It was named for its inventor, the late Estonian optician Bernhard Schmidt, who, on a sultry summer afternoon in 1930, timidly unveiled the crude prototype for a friend at the Hamburg Observatory in Bergedorf. They passed the next few hours amusing themselves by reading the epitaphs on the tombstones of a nearby cemetery, much as the 200-inch disk had first captured a row of pinup girls on the wall of the Caltech optical shop.

Whereas the Hale was capable of reaching out some one billion light-years, the range of the 48-inch Schmidt was only about one-third as great. Its advantage lay not so much in its magnifying power as in the fact that its wide-angle camera, ingeniously fitted with a thin correcting lens that eliminates the troublesome distortion of star images called coma, can take in an area of sky one thousand times as large as its big brother, doing in years what it would take the 200-inch 60,000 pictures and millennia to complete. While the 200-inch mirror was getting its rim polished down, the Schmidt was already on its first major assignment—the making of 2,000 star maps on negatives measuring 14 inches square, a veritable atlas of the heavens within reach of Palomar's latitude, or roughly two-thirds of the night sky. Officially designated the National Geographic Society–Palomar Observatory Sky Survey, the program was to continue for four years with the aid of the best modern photographic equipment and darkroom facilities. Experts from Eastman Kodak Company had designed the millimeter-thick photographic plates at a cost of $40,000, half of them sensitive to blue light, the other half sensitive to red. Not only must they be completely homogeneous; they were expected to last unchanged for as long as a hundred years.

While Hubble was in Colorado enjoying the pleasures of the Rio Blanco Ranch, student astronomer Allan Sandage, who had recently made Hubble's acquaintance, spent the summer of 1949 laboring away over "the plates of Moses" in the stifling basement of the cosmic box on Santa Barbara Street. He had reached this exhilarating but nerve-racking juncture in his career via a circuitous route, pursued at every turn, he later revealed, by the "Hounds of Heaven."

Sandage was born in 1926, the only child of a professor of advertising at Miami University, in Oxford, Ohio, and a mother reared in the Philippines, where, in the spirit of Manifest Destiny, her Mormon father had been dispatched by President William Howard Taft under orders to initiate sweeping educational reform. While the youth was religiously inclined, his parents were not, and he sometimes slipped out of the house on Sunday mornings to attend church while they slept in.

For as long as Sandage could remember, two forces were at war in his nature—"the worldly and the otherworldly"—and he seemed forever to be staring into what he called "the Abyss," alternately glimpsing and losing sight of the fingerprint of God. He soon learned that the only way to keep the demon of self-doubt at bay was to immerse himself in his work. "It was an internal drive," he reflected, "that started very young. I was nine or ten years old at the time that I knew I had to be a scientist, and, in particular, I thought I wanted to be an astronomer." The epiphany had come the first time he looked through a telescope set up in the back yard of a childhood friend: "There was a tremendous excitement. . . . From then on, it was just a thing that had to be done. There were just no questions about it."[44]

Thus driven, he completed his first two years of college at Miami before being drafted into the navy to spend the last eighteen months of the war doing electrical maintenance on radios and radar, first in Gulfport, Louisiana, and then at San Francisco's Treasure Island. After his discharge in 1946, the angular-faced veteran, who had developed a taste for leather bomber jackets, Polynesian restaurants, and stiff tropical drinks, enrolled in the University of Illinois, where his father had gone to teach, enabling him to live in comfort at home on the G.I. Bill. Instead of majoring in astronomy, he chose physics, because "astronomy was very easy, and physics was very hard, and it was the physics I needed." In search of intellectual balance, the future scientist chose to minor in philosophy, counting Spinoza, Nietzsche, and Schopenhauer among his favorites, while Hume and Descartes "seemed just like science."[45] (When he returned to this triumvirate after years of trying to find out "what's actually going on" in the universe, he still found their systems appealing but shaky as a bowl of Jell-O.)

Sandage passed many a night alone in a local cemetery after volunteering to undertake research for his undergraduate professor Robert H. Baker, a member of the star-counting circuit headed by Harvard astronomer Bart J. Bok. Guiding a small "patrol camera" across the heavens, he photographed the midnight sky, then spent his days calibrating the magnitude sequences of the stars and constellations he had captured on plates. Mostly what he remembered about the experience was that it was cold, "incredibly cold," and that the constellation Perseus, his main target, was in the winter sky. Yet by teaching himself to develop plates and measure stellar magnitudes, Sandage, like his heroes Galileo, Newton, Einstein, and Hubble, had set his own course down the road to scientific immortality.

To join the blessed atop Mount Wilson had long been his dream. In 1941, while his father was teaching summer school at Berkeley, the two had driven south to Pasadena, where Allan stood rapt before the instrument Hubble was using to pierce the eye of the universe. They also visited the observatory offices and optical shop, where all the talk was of the 200-inch mirror whose outer surface was being ground to a fine glass powder at nearby Caltech. Hoping his chance might yet come, Sandage applied for admission to Caltech's Ph.D. program in physics after completing his degree in 1948. If nothing else, he would be within sight and sound of hallowed ground, where he could continue to daydream of slipping in through the back door.

Then came the letter that would change his life. A new Ph.D. program in astronomy was to be launched in the fall, and Sandage had been admitted to the select company of four other Caltech graduate students. He had also been granted a tuition scholarship and would be paid an additional $900 a year in return for working fifteen hours a week. When he arrived in Pasadena for the beginning of classes a few months later, the Schmidt was already mapping the skies, while the recently dedicated Hale was in the midst of its initial shakedown. In contrast to Urbana, there were no autumn rains and barren trees; indeed, the change of seasons was hardly noticed by the weather-hardened Midwesterner, who exulted in the fact that Perseus or any other celestial object could be observed without freezing to death in the snow.

In April 1948, as the 200-inch was about to go on-line, a deal had been struck between the Carnegie Institution and Caltech resolving the sticky issue of access to the Mount Wilson and Palomar observatories. The facilities became joint operations, albeit operations with separate budgets. The Mount Wilson astronomers were appointed Caltech professors, though their appearance on the hacienda-style campus was sporadic at best. They received neither offices nor pay from the institution. And while they would teach an occasional course and chair dissertation committees, their main concern, as before, was the operation of the great telescopes. In return, Caltech students and faculty were awarded access to the most prized collection of astronomical plates in the world, not to mention the expertise of its most revered astronomers.

The man with the greatest responsibility for seeing that this makeshift deal worked was Jesse L. Greenstein, a thirty-eight-year-old Harvard Ph.D. with a penchant for amateur theatricals and the poetry of Tennyson. Greenstein's round, impish face was highlighted by a thick mustache and dark, wavy hair, which he combed straight back, exposing a cerebral forehead beneath which a twinkling pair of eyes narrowed like the peep sight of a gun whenever technical matters came under discussion. He had temporarily dropped out of college during the Depression, and saved his once prosperous family from foundering by going into commercial real estate. His specialty, which he plied at Yerkes Observatory after finishing at Harvard in 1937, was stellar composition and stellar atmospheres. Old stars and dying stars fired his imagination, and he commonly referred to himself as a "stellar mortician."

Greenstein had been hired by Caltech to handle the new graduate program in astronomy. Fritz Zwicky, the only other astronomer on the staff, simply couldn't be trusted with students. His latest scheme called for the firing of artillery shells over Mount Palomar to make the air more transparent for better seeing. Moreover, Zwicky was constantly at loggerheads with Hubble and the other members of the Observatory Committee, feeling, like van Maanen, who had died in 1946, that he was being denied his rightful share of time on both mountains. It would be decades before another of his harebrained theories—that 90 percent of the matter in the universe seems to be invisible—would gain credence.

For the time being, Greenstein decided to go it alone by teaching all the graduate courses in astronomy himself—stellar atmospheres, stellar interiors, interstellar medium, practical astronomy, astronomical methods of observation, the works. The backbone of Caltech graduate instruction was in mathematical physics, and astronomy students would be granted no exemptions. "The courses," Greenstein wrote, "used problems of numbing difficulty to train all minds, and, it was hoped, to train the best. The survivors knew an order-of-magnitude more physics than my generation had been required to learn."[46]

Sandage thought Greenstein "a remarkable teacher" and "a great man," but he was also "the most disorganized teacher in the world." He and his fellow students often passed the whole night rewriting Greenstein's helter-skelter lectures, absorbing far more astrophysics in the process than they would have otherwise. Then they raced to the library to ransack the journals in preparation for the next day's class. And this not counting the withering core courses in physics—quantum mechanics under Richard Feynman, analytical mechanics under Leverett Davis, plus the contract labor which paid them an extra few hundred dollars a semester. At times, a self-inflicted bullet to the brain seemed the most logical alternative, yet it was the very pressure for excellence coupled with dreams of the looming colossi, both north and south, that kept them going. "We all understood that we were in at the beginning," Sandage reflected. Failure was never a part of the equation.[47]

He had just completed his first year of course work when Greenstein came into his office one morning in May 1949. Hubble had phoned to ask whether he had a student who would be willing to work for him on a new project, the details of which the astronomer had not disclosed. "Why don't you go up to Santa Barbara Street," Greenstein suggested, "and see what Mr. Hubble wants."[48]

The one area in which Greenstein provided no formal instruction was the new cosmology. Sandage had read *The Realm of the Nebulae* on his own, which, together with some articles in the *Astrophysical Journal*, was about the only technical information available. Unlike the book's author, he never doubted that the perceived expansion of the nebulae was a true expansion; now he was about to meet the living legend who had expounded the theory, as well as Hubble's almost legendary assistant, Milton Humason.

So overwrought was Sandage that his memory failed him when he attempted to recount their first meeting some years later. Yet from the beginning to the end of their four-year association, Hubble ever remained his formal self, never laughing or cracking a joke, never attempting to put an abashed Sandage at ease. And though he worked very hard on his public persona, the great man seemed more an actor than a natural patrician, or so it seemed to the self-described "hick from the Midwest." When Sandage was introduced to Grace Hubble several months later the same formality prevailed, causing him to wonder about their conduct when alone together.

Harris tweeds remained in vogue, as did certain of the Briticisms from the old days at Oxford. Having only seen pictures of a younger, more vigorous Hubble, Sandage was somewhat surprised by the fleshy face and creeping age lines accentuated by flecked hair cut too close to the scalp. Still, he was "very formidable-looking" thanks to a square jaw, thin mouth, and chilly gaze; "he carried himself extremely well," and boasted a cane or walking stick in the affected English manner. "A noble man," Sandage recalled, who deported himself as "I imagine a god might." It was no wonder that Humason, a gentleman through and through, always called him Major, while to Sandage he was always Dr. Hubble.[49]

It was the recently launched sky survey under the direction of Rudolph Minkowski that Hubble wanted to discuss. Up to this point, almost all the clusters of extragalactic nebulae had been chance discoveries on the edges of small-area plates taken with the 100-inch Hooker. The Schmidt telescope on Mount Palomar had suddenly opened up a vast new celestial ballpark, its wide-angle camera capturing untold nebulae that appealed to Hubble as few discoveries could. (Shapley had coined the term "cosmography" to describe this form of stellar mapmaking, which, like the term "galaxies," Hubble steadfastly refused to employ.[50]) With negatives now being supplied by Minkowski, Hubble wanted to redo completely the nebular counts he had initiated in the 1930s, with an eye to more accurately calculating the distribution of matter through space, both in quantity and in distance. When carefully analyzed, these plates would not only provide precise counts of galaxies at different levels of magnitude; the calibrated light curves just might confirm the relativists' contention that space is indeed bent, thus revealing the destiny of the universe once and for all.

Thanks to his work at Illinois, Sandage required no instruction from Hubble, who "wanted to find out what I could do." Ensconced in the bowels of Santa Barbara Street, he began working with plates drawn from four areas of the sky, with instructions to take the counts to the 18th magnitude by actual measurement, not estimation. Far from the telescope of his dreams, the graduate student squinted his way to near-blindness in the oncoming summer heat, comparing stars that appeared in the foreground of the plates with the much fainter background dots representing nebulae.

A young man on a mission, Sandage was so pleased with the progress he was making that in July he asked Minkowski, who periodically visited the basement, when Hubble was expected to return. Minkowski wasn't certain and soon left, only to reappear, white-faced and shaken, a few minutes later. The phone was ringing when the astronomer reached his office, prompting him to hasten back down the stairs after hanging up. "Hubble won't be back for many months," he exclaimed in his heavy German accent. "He's just suffered a major heart attack."[51]

VII

Before they departed for Rio Blanco Ranch early in July 1949, the Hubbles deposited Nicolas Copernicus at the Dude Ranch for Dogs in the San Bernardino Mountains near Great Bear Lake. The precocious feline "dictated" a letter to ranch owner Gladys Shipman shortly after his arrival, assuring his owners that he was having a fine time. Nicolas had befriended Mutt, the wildest cat on the place, who was overheard bragging that milk had never touched his whiskers. "The gals are even asking for my autograph—which I usually pass out with a little catnip." Although other cats had tried to outwit him, he had triumphed over all comers. "I brought my crystal ball with me and all the cats flock around to find out what's going on in the cat-world."[52]

Hubble worked through the morning on the day of their departure, returning home an hour before train time and packing quickly. He was exhausted on boarding the coach but said nothing to Grace for fear of dampening her spirits.

They quickly fell into the vacation routine of the place they had been visiting for twenty years: Edwin filled his creel with writhing trout while Grace rode the ridge trails in welcome solitude. In the evenings they smoked and read before the fire, usually in the company of one or more half-wild cats, who inevitably put them in mind of Nicolas. One evening after dinner, Edwin gave an informal talk on astronomy. As Ida Crotty was leaving the lodge, she passed by a distinguished group of men, which included the president of Standard Oil of New Jersey, conversing on the porch. "Isn't it a shame," someone said, "that he wasted such a brain on astronomy." The young in the audience were more appreciative. They eagerly gathered round the astronomer afterward, and one of their number even volunteered to go to the moon.[53]

Several days into their stay, the Hubbles, along with the Crottys, were invited for cocktails at a neighboring cabin. Edwin arrived a bit late, carrying a beautiful basket of trout, but Ida noticed that he seemed "particularly quiet." Following drinks and dinner, Grace informed her that Edwin was feeling ill and wondered if she had brought a cathartic with her. Ida gave her some medicine and the Crottys went to bed, thinking he would be all right by morning.

They were awakened about six by a frantic Grace rapping on the window. "Edwin is in a great deal of pain!" she exclaimed. "I'm quite worried about it." While the Crottys dressed, Grace hurried over to the lodge, from which she telephoned Paul Starr, the Hubbles' physician in Pasadena. Not wishing to disturb Grace, Edwin had endured severe pain through the night before he finally roused her shortly after dawn. Without mentioning his heart, he told her to describe his symptoms to Starr, who would know exactly what to do.

Fortunately, they had included morphine tablets in their first-aid kit in case of an accident while riding cross-country, and Starr told Grace to give him the pills at once. Neither did he pull any punches: Edwin was having a heart attack in the worst of all possible places. The mountain altitude coupled with the absence of professional medical care made for a bleak prognosis. The physician would begin scouting the area for a decent hospital and call back in two hours. If Edwin was still alive, he would have a good chance of making it.[54]

By the time Starr's return call came through Edwin had shaved and dressed himself. With Harry Jordan, the ranch manager, at the wheel of a guest's car, they began the slow, twisting drive down the mountain, after which they were to head southwest for the town of Grand Junction, where Starr had alerted the Sisters of Charity at little St. Mary's Hospital. The lower they dropped the more the pain eased, but it was not until late afternoon, following a 100-mile drive, that they reached their destination.

The sisters had their best room ready. Located in a corner on the second floor, one of its two large windows faced Grand Mesa, darkly forested in the distance. A short walk down the corridor was a room for Grace, who meant to stay by Edwin's side no matter what.

The doctors on staff came in to administer an electrocardiogram. As they fastened the straps on Hubble's strong wrists, burnt brown from fishing, and looked down into his deeply tanned face, one of them remarked, "You certainly don't look as if anything is wrong with you."[55]

The wavering stylus indicated otherwise. A portion of the heart muscle had died because of an arterial blockage, and it was on hearing this news that Grace grasped the full gravity of the situation. She sat by Edwin's bed through the night in the dimly lit room listening for any changes in his cadenced breathing. The liquid bubbling of the oxygen tank mimicked the sound of the river that they had heard only the night before, flowing at the edge of the forest behind their now deserted cabin. As she drowsed, memories of their twenty-five years together floated by like a surreal tableau. Edwin must not die: "She would not accept it; not at all!"[56]

Neither, fortunately, would Mrs. Clark, "a little thing with a low voice and gentle ways." Unbeknownst to Grace, she hadn't practiced her profession in a long time; indeed, the aging nurse had never seen an oxygen tank in use. One night, when she thought Edwin was asleep from Demerol and the other drugs prescribed to keep him immobilized, she was vainly trying to adjust the tank when he opened his eyes and told her how to do it. Still, the doctors considered her the best nurse available, and Grace quickly learned why. "Her spirit was far more important even than her skill." Because of her diminutive stature and soft brown eyes, Edwin privately named her the "Water Rat," and he listened attentively for her approaching footsteps in the echoing corridor.[57]

The crisis came on the fourth night, when a second coronary struck with even greater force than the first. The pain was excruciating, and while the level of drugs was stepped up, an uncomplaining Edwin remained conscious, his mind perfectly clear. An order from the Mother Superior was repeated at intervals—"Quiet in the corridor, please." The doctors told Grace there was no chance of recovery. Prayers were offered in the hospital chapel by the sisters.

Mrs. Clark came on duty at eleven, her faith unshaken. The hours dragged by and Hubble held on. Paul Starr arrived by plane for consultation the next day. Taking Grace aside, he attempted to reassure her. "These young doctors out of the army won't want to take any chances with him, they'll throw everything at him but the kitchen stove." When Starr walked into the patient's room, Edwin asked, "When can I use the 200-inch?" Starr hesitated, then replied, "We'll see, but no altitudes for a time anyway."[58]

A month's uncertain convalescence followed, with Starr telephoning every day. The endearing Water Rat was joined in Edwin's room in the mornings by "Count Dracula," a pale, red-haired intern assigned the task of drawing blood. Edwin chatted amiably with the young man, who was obviously taken by the famous astronomer, but Grace noticed that her husband always searched out the bed frame with his free hand, grasping it with whitened knuckles until the painful ordeal was over. The sisters looked in daily to remind him that he was still in their prayers, and he welcomed the regular visits of Sister Raphaella, a beautiful young nun in charge of surgery. A friend had sent some French fashion journals, and an approving Grace looked on while the unlikely pair turned the pages and examined the pictures together. To ease the tedium, Grace busied herself by performing small tasks such as making tea in the morning, afternoon, and evening in the kitchen at the end of the corridor.

The citizens of Grand Junction soon learned of the celebrity in their midst and could occasionally be seen staring up at Hubble's second-floor window. The Lions Club ordered a large floral arrangement; small boys sent up bits of paper with requests for autographs. The sisters were asked if Hubble could do a radio broadcast from his bed. When he was allowed to read a little, he asked the Mother Superior for a history of the order, which had been founded by St. Vincent de Paul.

Grace did everything possible to keep his condition a secret from the outside world. The staff at Mount Wilson was told that only Mrs. Hubble should enter into a public discussion of their colleague's health—if and when she chose to do so. Shortly after the attack, Walter Baade wrote Nick Mayall the sad but "confidential" news, twice underlining confidential. "Bowen let me in on it besides Milt [Humason] and Minkowski but does not want it to get around since he feels that any statement should be left to Mrs. Hubble or his doctor. Let's hope that he pulls through." The astronomer blamed the heart attack on Hubble's war work, "which sapped his strength. He was completely exhausted when he returned from Aberdeen. We all feel terribly sorry and miserable."[59]

Bowen, like everyone else, attempted to play the part of the optimist, but fumbled badly in spite of his good intentions. Hubble was still in critical condition when the Mount Wilson director wrote to say how sorry he was that his colleague was "temporarily on the sick list." This was followed by an inadvertently distressing paragraph which likely caused Grace to withhold the letter pending Edwin's improvement:

> We had a very pleasant but informal celebration of the taking of the first pictures of the sky survey with the 48-inch last Tuesday. About 20 or 25 reporters were present. I took Minkowski along in your place and we had several very fine discussions with Dr. Briggs of the National Geographic Society. He was very appreciative of the whole process.

This was followed by the news that Baade had just announced the first discovery with the Schmidt, an unusual asteroid with a period of about thirteen months and a perihelion well within the orbit of Mercury.[60] Hardly the best medicine for a man whose placid exterior masked what Ida Crotty described as a "thorough-driving, competitive nature," which was brought to bear with equal force on everything he did, from trout fishing to plying the stars.[61]

Equally well intentioned but more cheering were letters from Anita

Loos and Aldous Huxley. According to Anita, *Gentlemen Prefer Blondes*, with music, was about to go into rehearsal: "Our Blonde is a new comedienne who got the greatest acclaim any comic has since the early days of Fanny Brice. . . . She is Carol Channing, a great big blonde who behaves as if she were a St. Bernard that thinks it is a Pekinese, a really priceless talent." There was also a bit of gossip for Grace. The paramour of Ali Khan, Rita Hayworth, was likely to have a baby at any moment. Neither she nor Ali wanted to get married, but the aging Aga Khan was worried over the bad publicity for India if his playboy son ruined the reputation of the darling of the U.S.A., so he forced the issue. "Rita is homesick, besides being pregnant, and Ali is flitting about Paris with an international beauty called Margaret Sweeney. Hollywood manners and morals have now broken into Europe and no place is safe from them any longer."[62]

Anita had heard nothing from the Huxleys of late and surmised that they were still engaged in buying homes in every district of Southern California. When the restless author of *Brave New World* wrote Hubble, he confirmed Anita's theory. Their latest acquisition was on North Kings Road in Los Angeles, which meant they would be able to visit much more frequently once Edwin recovered. "I don't imagine that Grand Junction is precisely the Athens of the West." Gerald Heard remembered staying there overnight on his way to Aspen to see Albert Schweitzer at a festival celebrating Goethe. It was held, of all places, in a huge circus tent. The speakers droned on interminably about the great man, prompting Schweitzer to whisper into Heard's ear, "After all, he isn't as great as all *that!*"[63]

When Paul Starr next flew up to Grand Junction it was to bring the Hubbles home. Near midnight on August 6, the trio waited in an ambulance on the tracks next to the silent, darkened railway sheds of the Salt Lake station. Starr told all the funny stories he knew, but his mind, like those of his charges, was on other matters. After the physician got Hubble safely from a stretcher to his bed on the train, he poured out three stiff drinks of whiskey, saying, "We can all use it."[64]

Once home there was another bed and another month of convalescence, with Nicolas at his master's feet. The big Persian mix, much to Grace's misgivings, had been hauled up to the Dude Ranch for Dogs in a station wagon together with several canines, but had come back in fine fettle sporting a red bandana. When he was brought into the bedroom he hid his head under Edwin's arm and trembled a little, a far cry from the biting "familiar" known to the Huxleys. When the doctor came he drew nearer, looking intently into his face. As Edwin grew stronger, Grace and Berta, the new maid, heard scuffling in the upstairs dressing room each morning before they brought up the breakfast tray. Edwin laughed while Nicolas uttered cries of simulated horror and menace. After their sparring bout, Edwin ate and read the paper while Nicolas tucked himself close by his side. The familiar pipe, which on doctor's orders was never lit again, still hung from his mouth or drew circles in the air when making a point. Finally came the day when he was allowed to take his first and only step down the stairs. The next day there were two steps; the day following there were three, until he was counting the days by the stairs before they brought him back and made him start all over. Well into his recovery a curious article appeared in the pages of the *Los Angeles Times*. "Reports that Dr. Edwin P. Hubble, renowned Mt. Wilson-Palomar astronomer, is seriously ill were refuted yesterday." Inquiries made at his Pasadena home disclosed that the scientist experienced a "light heart attack six weeks ago while fishing in Colorado, but he is now feeling much better and hopes to be back at work before many weeks."[65]

CHAPTER FIFTEEN

HOME IS THE SAILOR

I

It was well into August, two months after he was stricken, before Hubble finally came downstairs. His first visitor was Sir Oliver Franks, British Ambassador to the United States. Both men had been made Honorary Fellows of Queen's College the same day in May 1948; they drank tea together in the living room, Hubble dismissing his brush with death as "nothing, just an accident, and it [is] over now." After Franks had left, Edwin remarked to Grace, "We had an amusing time, didn't we, in that little hospital?" She may have smiled at the remark, but her thoughts were elsewhere. Hamlet's speech to Horatio had suddenly come to mind: "A man that fortune's buffets and rewards . . ." For the first time the veil had parted and Grace had looked death squarely in the face.[1]

In mid-October, Hubble began showing up in his office for an hour or so at a time. Baade was shocked by his appearance and wrote Mayall that "he looks awful." Hubble confessed that he had become overly excited about a decision made during his absence which, when Baade investigated the matter, seemed inconsequential. Neither would he be able to attend committee meetings for a while, because these, too, threatened his emotional equilibrium. "I think," Baade wrote, "it would be better for him to stay at home at present and to forget about the observatory entirely."[2]

When, if ever, Hubble would be allowed to trek up either mountain was unknown. Even in the best of circumstances he could never again peer into the heavens during the cold winter months, when the seeing was at its best, nor could he ever return to Rio Blanco Ranch to cast for trout, for the swift-running White was a good 2,000 feet higher than either Mount Wilson or Mount Palomar.

Mostly, as Baade had hoped, he remained at home, gazing wistfully at the large collection of straight briar pipes from Dunhill by his chair. A saddened Grace looked on, privately mourning the loss of ritual—knocking the ash out, scraping the inside of the bowl, tamping in the tobacco, lighting it, letting it go out, and relighting it, "all in a deliberate, reflective way." The cost seemed all the greater, since Edwin had no other habits and was "without mannerisms." He never whistled or hummed or sang, or drummed with his fingers or made gestures. "Unless he had something to say he preserved a peaceful silence and a relaxed immobility."[3]

While his Los Angeles tobacconist had lost a preferred customer, Hubble kept an ample supply of pipe cleaners on hand. Nicolas disdained his rubber mouse, ball, and other toys, but delighted in the pliant, tufted rods kept by his master in the bottom drawer of the stand next to his chair. He would either sit by the stand and beg or, when Hubble reached for the handle, dash over and claw out a new one, which eventually turned up floating in the pan of drinking water, or lying beside the food dish. When Hubble was at his study desk Nicolas solemnly sprawled over as many pages as his considerable body and heavy velvet paws could cover. If Grace attempted to shoo him away, her efforts were protested, "He is helping me." In the evenings, while Hubble read or drowsed in his chair dreaming of things still to be done, Nicolas settled into his lap and began purring a slow lionlike purr while staring into the fire with contented eyes.[4]

Grace was kept busy seeing to her husband's needs while answering the dozens of letters from well-wishers spanning the social and astronomical world, as well as other important correspondence. Of the latter, none was more welcome than a filigree announcement from the French Academy of Sciences, naming Hubble a correspondent in the astronomical section. Relying on Aldous Huxley's discretion and mastery of French, he apologized for not replying sooner, but he had only just recovered from *une grave maladie*. Neither he nor Grace ever used the words "heart attack," as though to invoke them would invite divine reprisal. Still, there was no point in taking chances. Shortly after marking his sixtieth birthday, Hubble wrote out a simple will by hand on one side of a sheet of Mount Wilson Observatory stationery, leaving everything he had to Grace.[5]

II

The spring of 1950 found Hubble sufficiently recovered to begin traveling once again, although Paul Starr's moratorium on visiting the 200-inch, which was back in operation, remained in effect. Before departing for the East Coast with Grace, he fulfilled his responsibilities as chairman of the Barnard Medal Committee by nominating physicist Enrico Fermi for the award, which Hubble himself had won in 1935. The committee's decision was unanimous, and Fermi was cited for his "central role in the discovery and elucidation of the principles of quantum mechanics, and their applications to atomic and nuclear phenomena."[6]

Anita and the Huxleys were waiting for them when they reached New York. Paying little attention to his diet, Hubble ordered fish and chips for dinner in the Raleigh Room of the Warwick, together with giant asparagus and hollandaise. Toward the end of the meal there was a splintering crash caused by a waiter dropping a tray of green salad and boats of French dressing down Aldous's back. The horrified headwaiter wiped him off, but the chilled goo penetrated to his skin, and there was no time to change before they walked across the street to the Ziegfeld Theatre.

The couples took their aisle seats while Anita headed backstage. As the lights went down, the orchestra gave a blast that fairly lifted the audience from its chairs. Then the curtain rose on *Gentlemen Prefer Blondes* and Carol Channing took center stage. Described by Grace as "outsize and looking like a creature not quite human," the voluble comedienne somehow "carried the thing, managing to make one feel that her gold digging was a very sensible business after all." "A good musical, as they go," Edwin remarked in the elevator they took up to the star's dressing room following the last curtain call. They stood around awkwardly while Channing, wearing a blue Turkish towel, vigorously rubbed off her greasy makeup. They walked Anita back to her suite at the Fairfax and then departed for their room in the Shelton Hotel, but not before promising to write to everyone from Europe.[7]

The Dordogne Valley and the pretty village of Lascaux, sprinkled with soft creamy stone buildings and steep roofs of brown-pink tile, seemed to be suspended in time. Up a hill and beyond, the Hubbles could see the tall tree which, they were told, marked the original hole of the great cave discovered in 1940 by a curious dog and four equally curious youths. Accompanied by Monsieur Blanc, the regional director of archaeology, they approached the opening, which had been enlarged and faced with stone, then sealed with a heavy bronze door covered by a soft green patina. Only recently had the grotto been closed to the public for fear that the collective respiration of summer visitors by the thousands would destroy the Paleolithic masterpieces. The door was swung open by their guides, two of the very boys, now grown to manhood, who discovered Lascaux amid towering pines and wild bracken.

After descending stairs of stone, the small party was led into the Great Hall, a long, wide, gallery whose walls were painted with enormous frescoes in black and red of giant bulls. On the left a sort of procession of various animals, including ponies and cows, was seemingly driven forward by a strange, rather sinister beast some-how not innocent of volition. Blanc remarked that regardless of any-thing else, the artistry was of the first order, in imagination and technical feeling and power, as if a master had covered a canvas with drawings, unhampered by sharply irregular surfaces and primitive tools and materials.

Another long passage led them to a wall on which was painted a bird-faced man with four splayed fingers, and a bird-headed wand. Nearby was a wounded bison, with a rhinoceros to the left, trotting away. Blanc said the Abbé Breuil interpreted this to mean that the man had speared the bison which in turn killed the man and had itself been disemboweled by the rhinoceros. Everywhere were other animals, the heads of deer, as if swimming; black, fat, high-crested ponies; huge cows pierced by arrows and signs of unknown significance; a fallen horse; a bison with its eye and nostril sunk in the rock, making it seem more lifelike than any other creature. "The general impression," Grace wrote in her diary that night, "was far beyond expectation." Both she and Edwin went to bed with their minds awhirl with great spotted bulls and the human beings who painted them, and the idea of that far-off life still hovering over the forested hills and lush valleys.[8]

Despite the fact that he knew full well that Hubble's long trips to England and the Continent were little more than junkets, Ira Bowen was too smart to play Walter Adams's game of cat and mouse with the celebrated astronomer. He had given Hubble a free hand after the war, stepping in only if there was a conflict of interest between two programs dependent on the same telescope or a question of priority in the construction of some new instrument. After finding Bowen to be as good as his word, Hubble quickly lost interest in the chairmanship of the Program Committee, which, to Bowen's recollection, met only once for about a half hour, and was soon discontinued, its responsibilities parceled out among other groups. "Hubble was rather notorious for not caring about administrative details, and he just never called the committee."[9]

Maintaining pretenses, Hubble spent part of an evening with a small group of astronomers at Burlington House, headquarters of the Royal Astronomical Society. He showed slides of Palomar, after which the party retired to a dinner club for drinks. Mostly they reminisced about Hubble's early days of discovery, but there was also some discussion of new methods and techniques, such as the growing interest in radio waves emanating from "hot spots" in the sky, one in the direction of the Swan, a constellation better known to astronomers as Cygnus. Displaying his usual caution, Hubble termed it an interesting mystery that might come to a great deal, or to nothing at all.[10]

Once back in England from the Continent, he accepted a long-coveted invitation wangled for him by a close friend. He and Grace were to be the guests of the pioneer aviator and industrialist Sir Richard Fairey at Bossington, five thousand pristine acres of Hampshire downlands in the valley of the Test. Starting, at age twenty-eight, with £10,000 of borrowed money, the future Sir Richard became famous for the many aircraft produced by Fairey Aviation—Swordfish, Fox, Flycatcher, the long-range monoplane, the Battle bomber, Firefly Fighter, Barracuda, the Gannet antisubmarine detector, and Delta II, the first plane to break the 1,000-mile-per-hour barrier. A champion yachtsman, crack shot, and first-class fly-fisherman, Fairey loved what he called the "sea-silences" and the sight of "white-horses running." Standing well over six feet with a granite head, graying mane, and deep-set blue eyes, he presided over Bossington like a patriarchal squire of Elizabethan times; no old cottage was ever pulled down; no ancient bridge swept away. The sprawling close-clipped lawns were populated by enormous oaks, cedars, beeches, and chestnuts; the formal garden sported a dozen varieties of flowers. A slender Lady Fairey, looking youngish and very smart in her checkered cotton dress and blue short coat, led her guests through the vegetable gardens, burgeoning with ripe strawberries, peas, and large pots of climbing tomatoes, as well as glassed-in espaliers dangling peaches, nectarines, and orange pippins. In the distance was the remnant of a curving maze, a small chapel, and a few old gravestones in the grass. Unlike the Elizabethans, Fairey abhorred cruelty to animals and was worshipped by his keepers, whose comings and goings were monitored from a sunny window at which the lord of the manor tied flies, including his well-known Father's Irresistible, with infinite patience, dropping now and then a philosophical insight discharged in a kindly but booming voice.[11]

Flowing over a chalk bottom and sharp against the dawn sky, the willow-lined Test had long been considered the Holy Grail of English troutdom. Coventry Patmore, who avidly fished its swirling green pools in the late nineteenth century, wrote his wife that "the river is the loveliest thing in water I ever saw." Together with fourteen other gentlemen, Patmore was a member of an exclusive angling club that celebrated its triumphs and near misses in white ties and swallowtails during champagne dinners at the Boot Inn in the village of Houghton. For individual dues of £25 per year, the members purchased exclusive fishing rights to eight miles of whispering river in which thirty-three trout, ranging from the lawful weight of from $1\frac{1}{2}$ to $8\frac{1}{2}$ pounds, lazed in territorial eddies waiting for the next meal to turn up. An accomplished fisherman knew each on sight by name—Sir William Harcourt, Tom Paine, Dr. Manning, Shaftesbury, Gladstone, Bismarck, and Sir William Temple, the only one never known to have risen to an artificial fly. Each fish was always to be found in the same place, unless one of their number was caught; then all moved up a step but were still identifiable by name. Everyone for ten miles around knew when a fish was taken; if it weighed more than $3\frac{1}{2}$ pounds, the news was telegraphed to all the London papers, and to Portsmouth, where a salute was fired, one gun for every half pound.[12]

Standing on a bridge beneath a silver-white moon and transparent sky, Grace watched a silhouetted Edwin downstream in his waders make his first cast as a large owl flapped overhead in the breaking dawn. The line sang when a big trout rose under the thick turf of the bank and the battle was joined. Five minutes later, he removed a $2\frac{1}{2}$-

pounder from the barbless hook, the first of ten fish taken in the first four days. While there were no salutes by naval cannon, every fish weighing 4 pounds or more was rewarded with a free bottle of fine sherry from Willams & Humbert Ltd., enabling Hubble to restock his liquor cabinet at no expense. The servants, who made the Hubbles feel like "lazy dogs," served the catch poached for breakfast, which the couple ate in bed on trays in the privacy of the largest, most luxurious room they had ever occupied.

Before their departure they visited the head of the Test Valley, once a tidal estuary controlled by marauding Danes who came in their long ships and dug dikes and embankments for a little harbor and village later called Longstock, from which they raided and burned the surrounding country. "We decided," Grace wrote of Bossington when it was time to leave, "that we liked this place best of all; partly for the sense of spacious landscape and the way the sky touches it, and that it is natural, unspoilt meadow and great trees, protected but not turned into someone's idea."[13]

III

While Hubble fished and daydreamed on the Test, Allan Sandage began living the great fantasy he had nurtured since long before he arrived in Pasadena. Together with fellow graduate student Halton "Chip" Arp, he was taken up Mount Wilson by the fastidious Walter Baade, who, at Jesse Greenstein's urging, had agreed to show the neophyte astronomers the ropes.

In the beginning, the two were allowed nowhere near the 100-inch, and were lectured repeatedly on the sins of the "pig spectroscopists," who slopped up the darkroom while developing their plates. Baade shuddered on entering the office of his fast friend Rudolph Minkowski, whose desk was always piled so high with papers that any object placed on it was in danger of sliding off and crashing to the floor. Arp, an accomplished fencer with keen reflexes, remembered standing by self-consciously while a freshly delivered plate began its downward trajectory, then reaching out to avert disaster while feigning nonchalance.[14] Nor could Baade abide Minkowski's clumsiness. "When Minkowski tightens a screw, it's 50 percent that he has the screw head in his hand—broken off." To Baade's horror, his fellow German held the singular record of driving the prime focus platform into the north pier of the 100-inch so that it could not be dislodged, other than by an engineer. An equally appalled Humason privately voiced the opinion that Minkowski should not be permitted to use the instrument; it was too dangerous for him and too dangerous for the telescope.[15]

Duly cautioned, Sandage and Arp began their apprenticeship on the antique 60-inch, perhaps the most complicated of all the Mount Wilson instruments to operate. Because the telescope is constantly in motion, the observer must synchronize the movement of the dome and the Newtonian platform at the same time he is attempting to take a long-exposure photograph. Virtually suspended in air, he must also remain conscious of the fate awaiting the slightest misstep. Sandage, like many before him, found that he was more comfortable in the dark, when he could not see the concrete floor, than he was in daylight climbing about to adjust the hardware. Complicating all this was a piece of information Baade intentionally held back. A periodic error in the drive screw, amounting to 1½ seconds of arc in amplitude, required additional compensation on the part of the astronomer. "Baade wanted to see whether we would discover that, whether we'd push the east button, then the west button, then the east button. He could tell by listening to the relay click and the time period whether we were guiding well or not."[16]

Donning coats and ties, they took subordinate positions at the linen-draped dinner table, their napkins held by plain clothespins rather than the engraved wooden rings signifying high professional achievement. Conversation came easy for Arp, a graduate of prep school and Harvard, but it was all Sandage could do to keep up his end of the wide-ranging discussions. The most terrifying moments came when Baade and Zwicky, old enemies rarely on speaking terms, were on the mountain at the same time. The antagonists talked through Sandage, who was so unnerved that he "sweat through shirts at the speed of light."[17]

A deeper source of anguish was a miscalculation he had made while Hubble was away recuperating from his heart attack. In computing the apparent magnitudes of extragalactic nebulae as he had been ordered to do, the graduate student ran into some inexplicable differences in the transfers he was making to selected areas on the Schmidt plates. Though these amounted to only three-tenths of a magnitude, he considered this discrepancy serious enough to set the project aside. Only when a displeased Hubble was well enough to meet with him did he learn that such a divergence counted for little in the grand scheme of things.

Unbeknownst to a crestfallen Sandage, Hubble was not as displeased as he had let on. The young man had mastered the 60-inch in the span of seven exhilarating nights, puzzling out the idiosyncrasy in the drive screw in short order. Still barred from the mountain by his doctor, Hubble was in need of assistance. After first speaking with Baade, he summoned the still smarting but eager student to his office.

In the years to come, Sandage would look back on this moment as among the most crucial of his career, for Hubble had designed a second test whose stakes were of the highest order. He handed Sandage his stack of plates taken of variable stars in M31 and M33 during the previous thirty years. Known as superluminous blue variables, they are the brightest stars in the universe, having entered the final stage of their stable lifetimes. "All he said was, 'Examine these five variable stars and get their light curves.' "[18]

This time there would be no second-guessing. Doing just as he was told, Sandage rode the observatory truck back up the mountain and went to work with the 60-inch. He first transferred magnitude scales into the nebulae; then, by eye estimates, which he had already been trained to do, got the light curves and delivered the results to an undemonstrative Hubble. "This was, I guess, impressive enough," Sandage reflected. "It was on that basis that he recommended that I take over as his observing assistant." Though there had been no metaphorical laying on of hands, for this was not Hubble's way, he subsequently described Allan Sandage in a letter to a friend as "a young astronomer of great promise and ability, who is working with me."[19] The coming years would prove that his confidence was not misplaced.

IV

In October 1950 a chafing Hubble was finally cleared by Paul Starr to undertake his first run on Mount Palomar in eighteen months. It was a step fraught with the greatest trepidation for Grace, who strained mightily to keep her feelings in check. "She didn't hover over him," Ida Crotty remembered. "She was too smart for that."[20] Nevertheless, Grace insisted on accompanying Edwin to the mountain, where she stayed in one of the guest cottages while he took a room in the all-male Monastery. Instead of being housed on the second floor with the other astronomers, he was quartered near the main entrance, where a button connected to an alarm system had been installed within reach of the bed. Sandage, who was becoming a familiar face in the Hubble home, had earned Grace's trust and from now on would always be on the mountain whenever her husband was on a run, along with Milt Humason, who had tutored Sandage in the use of the 200-inch.

The run lasted three nights. The seeing, due to a hazy sky and partial moon, proved far from optimum. Still, it felt good to be back. Once darkness settled in, Hubble entered the massive dome and, following doctor's orders, wriggled into a pair of electrically heated coveralls before ascending the heights via a series of stairs and elevators. After passing the darkrooms, the night kitchen, the control room, and various offices that ringed the vast inner space, he negotiated a final catwalk and ramp, then climbed into a basket similar to the cherry pickers familar to Californians at harvesttime. The mariner manipulated the controls and, to an eerie siren whine that filled the echoing chamber, began moving out across empty space. Ahead lay the skeletal outline of the giant tube aimed skyward toward the dome. At its upper end, suspended near the center of the vault, hung a mesh cage not unlike the gondola of a hot-air balloon. Within seconds the cherry picker came to a stop; Hubble climbed over a railing and entered the prime focus cage, his face softly illuminated by the red lights highlighting the instruments. As he was rusty from his involuntary layoff, it took some time before he recaptured the seeing eye necessary to establish the "true" focus. Once this was done, he took his first plate, a 30-minute exposure of NGC 7217. Eight others followed, the last clocked in at 4:34 a.m.[21]

With a total of eighteen plates to his credit, Hubble, a relieved Grace at his side, headed for home a few days later. His heart seemed fine, and he was suffering no ill effects from a loss of sleep caused by resident woodpeckers driving holes in the Monastery's copper roof, into which they deposited acorns. He was bothered only by the lack of protocol at the dinner table, which others, especially the younger generation, were happy to be rid of. An awkward moment had taken place on Mount Wilson shortly before Hubble was stricken, when two visiting students from Oxford showed up in T-shirts and blue jeans. A seething Hubble remarked that they would never think of dressing so informally at High Table, sparking the impertinent reply, "This isn't Oxford."[22]

Barred from both mountains in the winter, Hubble was forced to rely on surrogate observers. "Sandage for Hubble" became a frequent notation in the logbooks as the young astronomer, together with Humason, commuted between Pasadena and Palomar as often as three times a month. Sandage couldn't recall which was the most frightening to him in the early days of their association—Hubble or the universe. His was the formal conversation of a diplomat; "there was no easiness to it; he was warm, but there was always a wall." It also seemed to him that Hubble had never failed at anything, never seemed to struggle. "His insight into the problems that he should tackle was infallible. Every paper he wrote was a classic." Always in quest of the big picture, his projects were long-term; "there [were] never any one- or two-night-stand operations that could be made into a little note." Hubble had made up stories about his past, and those stories became his past. Now such stories were no longer needed; as Sandage kept reminding himself, he was apprentice to the greatest astronomer in four centuries.[23]

There was considerable ferment at the observatories, caused mainly by Baade's wartime discovery of two kinds of Cepheids. To Hubble this meant only one thing; he must return to the long and difficult recon-naissance work begun in the 1930s for the purpose of recalibrating the distance scale from scratch. Either he or Sandage would have to pho-tograph the nebulae anew, adding those giant clusters only recently made visible by Palomar's gaping eye. But the question of taking neb-ular counts to fainter magnitudes soon raised the old and touchy issue of observing time. When combined with Humason's ongoing work on redshifts, the program Hubble laid out required roughly half of all the nights available, particularly those occurring during the precious dark of the moon.

Princeton astrophysicist Martin Schwarzchild, a regular visitor to Mount Wilson, long remembered "an extremely difficult" meeting at Hubble's home one afternoon. Others in attendance included Bowen, Baade, Humason, and Richard Tolman, Hubble's longtime friend and sometime collaborator from Caltech. The decision had been made by Bowen that the seeing time assigned to counts of fainter magnitudes would have to be curtailed, while the search for greater redshifts should be pushed as hard as possible with the 200-inch. "I think," Schwarz-child mused, "some sense of personal tragedy was very much in the minds of all of us," but there was also no question that Bowen's de-cision rested on firm scientific ground. Determining the magnitude of extragalactic nebulae on the basis of photographic plates alone was simply fraught with too many variables to achieve an objective analysis. Tolman, who may have been briefed in advance, was "totally magnif-icent" and smoothed the troubled waters with the help of an unusually solicitous Baade, who stood to gain more by this decision than anyone else. Never at ease in the company of the distant Hubble, Schwarzchild nevertheless considered this to be his finest hour. The astronomer, who, after all, was in his own home, acted "very much like the gentleman that he was," displaying "no emotion during the meeting."[24] When

Sandage came aboard some time later, Hubble continued to pursue his dream on a more limited basis, but the writing on the wall was crystal clear: "To find Cepheids beyond the local group was itself a very important first step," Sandage remarked, "but to work them up and analyze them—I think he knew that he couldn't follow that through."[25] Yet while walking with Grace and Ida Crotty on the grounds of the Huntington Library after nearly dying, Hubble remarked to the statues of the resident gods as much as to anyone, "You know, if I can have two more years, I feel I can accomplish as much as I can expect to during my lifetime."[26]

Humason, who was nearing sixty, had never lost a step in pursuing what Hubble, in a poetic turn of phrase, termed his "adventures among the clusters." Having suffered a lingering illness in 1939, he complained of a chronic loss of memory in a recent letter to Nick Mayall, but no one else seemed to notice.[27] As secretary of Mount Wilson Observatory, he came into contact with a steady stream of visiting VIPs. "Milt," Mayall wrote, "could unerringly pick out the phonies and the self-inflated types, and he had little liking for visiting scientists who acted as elite members of society," though he treated everyone with uniform courtesy.[28] Years of successful observing and a long string of publications in astronomy's most respected journals had helped him overcome feelings of inferiority stemming from a grammar school education. Then, in 1950, he received his greatest honor; Sweden's University of Lund did what no American university would do by awarding him an honorary Doctor of Philosophy degree.

After dealing with Hubble, who reminded Sandage of General Douglas MacArthur and whose hatred for Harry Truman was expressed frequently and at length, Humason was as amiable and compelling as a morning in May. A "true gentleman" but earthy, he told ribald jokes, swore when upset, and loved to organize fishing expeditions whose gregarious participants were liberally supplied with whiskey. The two-dollar window at Santa Anita remained one of his favorite pastimes. Like Mayall, Sandage admired his tutor's gift for "catching somebody out, some stuffed shirt."[29] Yet Humason was almost as conservative politically as Hubble and was not above some election year chicanery. When assigning telescope time on the mountain, he would make sure that Democrats were given runs over election day so they wouldn't be able to vote. Besides acting as Hubble's eyes, Sandage noted that Humason was liaison to Hubble's fellow astronomers, some of whom had never been introduced to Grace.[30]

With the closing of the first phase of Hubble's radial-velocity program in 1936, which Humason had marked by reaching the limits of the 100-inch, redshifts approaching 40,000 kilometers per second had been recorded in the Boötes cluster and in Ursa Major No. 11. Palomar promised more, much more; primed by Hubble's normally guarded enthusiasm, Humason often drove to the mountain from Pasadena thinking that he might reach the limit or, as Hubble termed it, "the horizon" of the universe. As with everyone else, his initial reaction had been one of awe. First scanning the firmament with his unaided eyes, he stood transfixed by the spectacle of the streaming Milky Way, "a beautiful sight" that put him in mind of his old friend Lord Byron. The smog and light pollution that turned the heavens a dull gray above Mount Wilson were nowhere to be seen. "It was clear 99 percent of the time."[31] He peered into the eyepiece and discovered a black field like a blackness he had never known. Sounding more like a neophyte than a seasoned observer, Humason caught himself reaching out to touch nebulae only dimly visible on Mount Wilson, as if they were his children approaching home after an incredibly long journey.

As soon as the sky survey of the 48-inch Schmidt was launched, lists of nebular candidates were made for further study. These would then be photographed by Hubble or Sandage with the Hale and given to Humason, who homed in on his assigned targets with his newly fitted spectrographic equipment. Slowly and with great pains, the Hubble diagram of redshifts was fleshed out by adding hundreds of nebulae to the original clusters that defined the linearity of the velocity-distance relation. Hubble was interested not only in extending the rate of expansion but also in establishing that redshifts occurred in nebulae spread across the whole of the sky, a concept fundamental to his long-standing principle of isotropy.

During the first season that a nebular spectrograph could be used on the 200-inch (1950–51), Humason obtained spectra in three clusters well beyond the limit reached with the Hooker. The greatest velocity was 61,000 kilometers per second, or about one-fifth the speed of light, and was recorded in the constellation Hydra situated in the equatorial region of the southern sky near Cancer, Libra, and Centaurus. His appetite whetted, Humason felt confident that he would be able to reach out to nebulae receding up to one-quarter the velocity of light as soon as the sky survey supplied him with the proper candidates.[32]

That confidence was suddenly shattered when Humason hit an invisible wall. He spent the better part of two dark runs on a cluster whose redshifts Hubble and Sandage calculated at some 120,000 kilometers a second, four-tenths the speed of light. On plate after plate the H and K lines were simply too dim to be read, a fault not of the theory or the men applying it, but of the limitations of the equipment at Humason's disposal. At such distances, which were later confirmed by Sandage after the aging mariner had passed from the scene, the sky brightness was swamping the nebular spectra. Only with electronic spectrographs equipped to perform what astronomers refer to as "sky subtraction techniques" would Humason's seven-league boots, donned by another, press onward once again.[33] Gazing back from a retirement during which he passed his days fishing with "a Chinaman" whose grandfather had allegedly crossed the Pacific in a sampan, he remarked wistfully, "Well, there is apparently no horizon, at least as far as the 200-inch goes—it just continues—the number of stars continues to build up and the nebulae go on, they get fainter and fainter, and so we did not accomplish anything like that."[34]

v

The steady flow of honors which had been interrupted by a lengthy convalescence quickly resumed once word of Hubble's recovery began to circulate in the intellectual world. The night before he was to receive an honorary doctorate from the University of California at Berkeley, the Hubbles stayed with Donald and Mary Shane in the director's house at Lick Observatory. Thirty years earlier, Shane, then a Ph.D. candidate in astronomy, had ridden the autostage up Mount Hamilton with a tall uniformed gentleman, who spoke with a British accent and introduced himself as Major Hubble. The two men had remained close during the coming years, and on this occasion they reminisced deep into the night. Shane remained mystified at his friend's continuing reluctance to accept the redshift as a literal expansion, though he was too polite to press him as to why.[35] When it began to snow at two in the morning, Shane, fearing they might not get off the mountain in time for the afternoon's ceremonies, moved his car so that it would not be plowed in by the trucks clearing the road. The couples began their treacherous descent in a blinding storm a few hours later and finally reached Berkeley in full sunshine. The ceremony, which had a decidedly English flavor, was held in the Greek Theatre. Hubble and his fellow honoree Lord Alexander, the Governor-General of Canada, were near the head of the academic procession, which filed slowly by to Handel's "March of the Priests of Judas Maccabeus." President Robert Sproul, an old friend and a fellow member of the Sunset Club, smiled broadly as Hubble spoke, and Hubble smiled broadly back.[36]

Though Grace reveled in pomp and circumstance, even she had her limits and was highly amused over an incident that had recently occurred during the inauguration of a new president of U.C.L.A. Oxford University asked Edwin to represent the institution, which he happily agreed to do. "Oxford is the oldest of all the universities," he chortled as he was leaving the house, "so I shall lead the procession." When he returned a few hours later, a sheepish grin had displaced his glee. "The University of Paris, of course, is the oldest; I had forgot!" He thus had to endure the ignominy of trailing "a very diminutive Frenchman," who basked in the glory of precedence.[37]

Hubble was honored in ways even he never expected. In late 1951, Robert S. Labonge, the assistant editor of *The Tidings*, the official newspaper of the archdiocese of Los Angeles, wrote to him and included the text of a recent speech given by the Holy Father, Pius XII. The Pope had mentioned Hubble's name and scientific writing in connection with a discussion on the beginning of time. The text had also been released to the newspapers, and a jocular Elmer Davis, who shared his old friend's skepticism on matters of religion, was quick to respond: "I am used to seeing you earn new and ever higher distinctions; but till I read this morning's paper I had not dreamed that the Pope would have to fall back on you for proof of the existence of God. This ought to qualify you, in due course, for sainthood."[38]

Seeking to get at her husband through Grace, Cass Canfield, chairman of the board of Harper & Row, wrote to ask if she might find the time to produce a book on some literary subject. Her talents had come highly recommended by their mutual friend Susan Ertz, whom the Hubbles often visited at Pooks Hill. "While I am at it, may I ask you to 'needle' your husband from time to time about a book on the telescopes concerning which I have been writing him. There is certainly material here for a fascinating work." Though he had heard nothing from Hubble, Canfield was not easily dissuaded. "It was quite a while ago," he noted two years later, "[that] I wrote to you suggesting you describe your experience at Mount Wilson Observatory and give the layman an idea of the workings of the telescope as well as the discoveries made as the result of its use. Your memories could include all this as well as touching on various other phases of your life and experience."[39] Again Hubble chose not to reply; nor was there any evidence that he was actively pursuing a project apparently conceived before the war—an atlas of the spiral nebulae based on more than two decades of photographing the giant clusters from atop Mount Wilson. As early as 1942, Shapley, who mourned its absence, had written Adams of "this colossal contribution" in a letter asking where matters stood. "Apparently," Adams replied, "the publication has not gone farther than the stage of discussion, although Hubble had gained the approval and financial support of Vannevar Bush "when the time comes."[40]

In February 1950, with his reputation at its zenith, Hubble wrote H. H. Plaskett at Oxford to ask a special favor. He was to have his portrait painted for an unnamed institution, and Grace wanted him to pose in the scarlet broadcloth and silver taffeta robe in which Oxford had granted him his honorary Doctor of Science degree. "Would you do me the great kindness to call at the proper place and arrange for a new robe. I will pay in dollars, of course, and neither the price nor the shipping and other charges are of great importance." Treating Plaskett more like his tailor than a fellow astronomer, he concluded, "You will recall that I am about 6'2", and weigh about 190 pounds."[41]

The idea for the portrait was Ida Crotty's. She persuaded her father, Ralph B. Lloyd, a Caltech trustee, to underwrite its cost, with the understanding that the completed work would hang in the Athenaeum. Only after he ordered his scarlet robe did Hubble meet with artist Arthur Cahill, who flew down from San Francisco and ate lunch with Grace and Edwin at their home. The work required several sittings, but neither the Hubbles nor Ida thought the artist had done his best work. Hubble remarked, "He attempted the portrait of an astronomer, and he tried too hard." The work was temporarily hung in the meeting room of the Sunset Club. So many complained of the stilted pose that Cahill retouched the mouth, but to little effect. When the portrait was then moved to what was supposed to be its permanent home, it caused "a furor," in Ida's words. In the first place, Hubble was not a regular member of the Caltech faculty; second, it was argued that Hale's portrait should have been done first; after all, he was the force behind the founding of Caltech and the construction of the Athenaeum. Hubble, looking overdone in his formal gown, became a running joke until, one day, the portrait quietly disappeared. It later turned up in the Astrophysics Library, a facility little visited then, as now, by anyone but preoccupied students and their professors.[42]

Hubble, who had many teaching as well as administrative offers from universities, including the presidency of Stanford, entered the classroom for the first and only time a year after his recovery. Like Baade, Minkowski, and others at Mount Wilson, he took his turn with the graduate students at Caltech by offering a seminar in his specialty. His appearance was preceded by great anticipation, which quickly turned to disappointment, and, on the part of some, outright embarrassment. The man who had confirmed the existence of other galaxies, determined that the universe was expanding at breakneck speed, and elegantly classified the nebulae was a naif among the new generation of aspiring astrophysicists, who, like Sandage and Arp, were writing their dissertations under Baade. Both they and their professors, who often sat in on the lectures, looked forward to a high-powered seminar laced with challenging equations and advanced physics; instead, they were treated to a somewhat updated rehash of *The Realm of the Nebulae*, which had been published in 1936 and was well known to each of them. What is more, Sandage recalled, Hubble never broached the question of expansion because that would have raised the issue of cosmological models and a discussion of theory, a quagmire to be avoided at all costs.[43]

Hubble was no more forthcoming about his innermost thoughts when he and Sandage were alone. "We never did talk about cosmological models. It was almost always on a technical level that we talked. He was of a poetic nature, he was an intellectual of a most profound type, but he didn't really open up in these ways, in philosophical discussions, at least with me."[44]

Had Hubble been so inclined, he had plenty of reason to share his thinking. Caltech had become something of a second home to Fred Hoyle, his longtime friend, who regaled the young astronomers with his latest venture into cosmology, dubbed the Steady State theory. It had been born of skepticism about the Big Bang during late-night conversations among Hoyle, Thomas Gold, and Herman Bondi that usually took place in the latter's spartan digs at Trinity College. While the trio had no argument with the idea that the universe is expanding, they were skeptical of the idea that everything came into being as the result of a primordial singularity. Rather, they argued, as the galaxies move farther and farther apart, new galaxies are formed in the empty space left behind, and at a rate such that the universe would seem to be unchanging. Inside a massive volume of space, you would find roughly the same number of stars and galaxies 10 billion or even 50 billion years ago as now. The Big Bang universe is evolving; the Steady State universe is not; its overall properties remain unchanged with time.

It was clear from the outset that the idea of a continuous creation violated a fundamental law of thermodynamics, which states that the total amount of energy in a closed system never changes. How could new atoms, the building blocks of galaxies, be created out of the void?

Yet it could also be argued, and was, that the same is true of the fireball cosmology that postulated an instantaneous creation reaching back in time 15 billion or 20 billion years.

Sandage, who was present at all of Hoyle's lectures, wasn't buying, and neither, as he recalled, were any of the other California astronomers. "I think really the Steady State theory was dead as soon as it was formulated." For one thing, where were the young galaxies needed to confirm Hoyle's hypothesis? "Hoyle, I think, was . . . more interested in all the worlds that could be, instead of the world that is," while Hubble took "what the universe gives you."[45] While he never expressed himself directly on the matter, Grace provided a large clue as to her husband's position following a visit to Los Angeles's Griffith Observatory, where Edwin addressed an enthusiastic group of amateur telescope makers. Afterward, the couple was swarmed by adoring youths whose intent gaze reminded her of Piero della Francesca's singing angels. As Edwin was signing autographs, one of the boys remarked, "How proud you must be to be married to a man like that." He also volunteered that he had been reading Eddington's *The Expanding Universe*, but "was doubtful about Hoyle's [latest] book and his assumptions."[46]

VI

Berta, the Hubbles' German housekeeper, spoke a language full of mystery laced with unintentional humor. There was always a long pause between hearing what she said and puzzling out what she meant. "Twenty thousand whoop-caps are stolen every month," she announced one evening after listening to the radio. Someone had "upheld" a liquor store. "Icing-dogs are needed" for the blind. "The Americans hold the topknots of the Russians in Germany," the meaning of which neither of the Hubbles ever resolved. When delivering a message taken over the phone, she rarely got the caller's name right, so that deciding who it was became a tricky game all its own. "How," for example, "did one know that Lucy Booth was Clare Boothe Luce?" To be the subject of a malapropism was considered an honor among the Hubbles' acquaintances, who loved to recall their favorite slippages during cocktails sotto voce.[47]

One day there was a tinkling crash in the kitchen and Berta was heard to exclaim in a plaintive tone, "*His* Cup!" She proudly assembled all the Hubbles' supplies for their trips to Palomar, compiling a list of everything from linen to matches, then packing cartons with roast chicken and interesting things in big glass jars and covered enamel containers, each item a favorite of Edwin's. "Will you want your devil-bag?" she asked while he was assembling his clothes. "When I travel I do not wish to look suspikious." She came outside with Nicolas in her arms and watched as they drove away. "He is like Columbus," she once remarked to an unsuspecting Grace, who suddenly became aware that Berta had a deeper sense of the implications of his work than many of their friends.[48]

The Huxleys, who other than the Crottys and Max were the Hubbles' most frequent guests, announced during one of their visits that Aldous was beginning research for a long article on Edwin in which the humanist would take precedence over the man of science. The writer had already gone through books and materials at the Huntington Library and was looking for more. He did his work on a small Corona, which to a visiting Grace sounded like a very active woodpecker. Lately, the couples had been driving out of the city to explore such geographical sites as the San Andreas Canyon and its sun-cooked rocks—strange, broken, tumbled masses topped by a thin jagged crest against the skyline. Aldous found the topography Wagnerian: "One would expect Melchior to appear on it at any moment." As they got into the car, Grace spied a black, crested bird perched on the branch of a cotton-wood. "I think that's a phainopepla," she exclaimed, though she had never before seen a member of the rare species. They stayed to watch it and it remained fearlessly close.[49]

Edwin suggested that they drive over to Palm Springs and look it up in the museum. They arrived at dusk near closing time, but Aldous, who had recently spoken at the institution, was warmly greeted by the curator. Edwin found a book about birds and looked up the phaino-pepla. Sure enough, Grace's eye had not deceived her. The staff was as delighted as the birdwatchers and said they would send someone over to shoot it for the collection.

There was a sudden stillness. Edwin put the book away and walked across the room. "Where," he asked abruptly, "did you get that stuffed wildcat?" It had appeared nearby, someone said, and had been shot because it was feared that it might harm some of the children. "Since there are so many children," he shot back, "and so few wildcats, I can only deplore your decision." At this point, Aldous intervened and got Edwin back into the car as quickly as he could. No one spoke of the phainopepla on the way home, for they all felt they had killed it.[50]

Among the Huxleys' friends were the Sitwells—Edith, Osbert, and Sacheverell, or "Sachie"—perhaps the most clannishly eccentric and committed family of show-offs in the history of English letters. The Hubbles were introduced to Edith and Osbert in January 1953, during their visit to Hollywood to consult on a movie script on the life of Anne Boleyn. Dame Edith wore a black turban with a border that reached her shoulders, so that only her perfectly framed face could be seen. Though in her mid-sixties, she was described by Grace as fresh and fair, with a transparent complexion, pale, slanting blue eyes, gold lashes, an aquiline nose, and a witty, thin-lipped, smiling mouth. Her dress, fashioned from masses of black drapery that reached to the floor, concealed her figure completely. Her only jewelry was a ring set with two aquamarine stones atop one another, both as large as walnuts. Osbert, five years younger than his sister, combed his thinning blond hair straight back but otherwise bore a striking remembalance to her.

Together with Aldous, who was in fine form, they attacked the "monstrous Victorians," who became increasingly alien and monumental, seeming more and more like characters in *Alice in Wonderland* and *Through the Looking Glass*. Osbert told of Thomas Carlyle, leaning over the balustrade and shouting to W. H. Mallock, author of *The New Republic*, "Don't come here again, I disliked you very much." Aldous repeated Tennyson's riposte to Benjamin Jowett, who warned him against publishing "Dora": "When it comes to that, the port you gave us at luncheon was filthy." Others who fell to the ax were Mrs. Browning, Christopher Fry, Arthur Miller, and Ford Madox Ford and his "dreary" wife.[51]

At tea the following day Edwin took Osbert into his study to look at some recent Palomar plates. He said his sister must see them, and left to fetch her. When she came out she inscribed her *Song of the Cold*: "After one of the greatest experiences in my life for those who gave it to me and their greatness." Before leaving, both made the Hubbles promise they would call during their upcoming visit to England, though they might have to take potluck. The old cook had died and Osbert was having a difficult time finding a couple to replace her at the inflated postwar wage of £7 a week.

With the Rio Blanco Ranch now a poignant memory because of the altitude, Hubble discussed rivers with fellow trout fisherman Robert Cleland of the Huntington Library. Montana's Blackfoot was Cleland's favorite, and he recommended the E Bar L Ranch outside Missoula, where he always stayed. The elevation was between 5,000 and 6,000 feet. About 500 feet below the cabins flowed the Blackfoot, wide, swift, and pristine.

Hubble fished up and down the river and found it tough going in his heavy waders. Two fishermen from camp were swept from the rocks and carried downriver before they managed to fight their way up a steep bank. Grace took some long rides with one of the wranglers; in the evenings the couple sat on the porch of their cabin watching the sunsets flame and fade into darkness as the first stars came out. Sometimes Cleland joined them and talked of Lewis and Clark, who had come down Nine Mile Prairie across the river. They slept with their door open; in the gray of early morning a grouse led her brood inside while chipmunks wandered in and out. Though neither spoke of it, they missed the high country. Grace wistfully gazed upward at the forest and bushes and undergrowth they both knew so well. Sharing her thoughts, Edwin remarked only that he enjoyed the fishing and that the Blackfoot was good, but he liked the White best of all.[52]

The Hubbles sailed aboard the *Mauretania* in April 1953. They settled into their large comfortable cabin on A Deck, where they lingered, looking at the passing ships and fading shoreline through field glasses as Edwin fought to keep the equilibrium that inevitably failed him on the ocean.

For the first time in years he was committed to more than a token schedule of scientific appearances, a measure of both his and Grace's confidence that he had recovered from the heart attacks suffered almost three years earlier. On the evening of May 8, he delivered the prestigious George Darwin Lecture at the Royal Astronomical Society. While he held the floor, it seemed to Grace that time had turned back upon itself, recasting its magical spell. The only disappointment came when it was announced that there would be no questions from the audience, as the lecturer's wife had forbidden it.

A week later, following a lecture at the Royal Society of Edinburgh, the Hubbles were back in London, where they drove to the opening ceremony of Cicely House on Cochrane Street. There was a marquee in the court, a red-carpeted dais with a big chair, and banks of flowers, including masses of orchids. A bitterly cold wind chilled the audience, most of whom, including the small children, were clad in summer frocks. Minutes later the curtains at the back of the tent parted and the Queen entered, walking slowly down the strip of red carpet followed at a respectful distance by the Bishop of London, whom Grace characterized as a portly figure in scarlet sleeves and black gaiters, and the Archdeacon carrying a tall silver crozier. Elizabeth II, only twenty-six and in the second year of her reign, moved in a graceful, unbroken rhythm onto the platform and everyone stood while "God Save the Queen" was played. Grace was among those selected to present the Queen with a "purse," which was actually a small silk envelope, and could hear an upset Edwin muttering something from the side of his mouth as they moved forward. He was forced to repeat himself before she realized that she was supposed to leave her bag on a table before entering the royal presence. She described the scene as almost hypnotic—"there must be no slips, nothing to break it. So she becomes a symbol." After Elizabeth delivered her speech she was taken to a room for champagne. Maintaining decorum to the last, she drank it well away from the window so, she said, "the populace can't see me."⁵³

It was tails, white tie, and pearl studs for Edwin, and for Grace a full black heavy taffeta with black lace, a low neck, and rolled collar. She walked into the circular lecture hall of the Royal Institute on the arm of Lord Brabazon at a measured pace, like a dance in the age of Louis XIV. He delivered her to her black leather chair, bowed low to her after she was seated, and she nodded in recognition. At the first stroke of nine, two large wooden doors swung open and Hubble, "handsome and self-possessed," strolled into the room to the desk. Hoyle, whom Brabazon claimed to have discovered, had spoken here, as had Jeans, of whom he was afraid, and Eddington, who afterward accompanied him to the Embassy, a London nightclub, with two pretty girls. Hubble began his speech, titled "The Observational Evidence for an Expanding Universe," without a formal introduction. Again, he was the object of rapt attention. As the negative of the farthest nebula taken with the 200-inch was put on the screen, he remarked quietly, "This is the last horizon," and the audience broke into applause.⁵⁴

Three days later the Hubbles took the train to Winchester, arriving at 5:50, in a fine rain. They were met by Keyes, Sir Richard Fairey's driver, who ushered them into a vintage 1940 Cadillac limousine. The rain spread a veil like evening over the green meadows and trees of Test Valley. Half an hour later they were shown to their "old room" on the second floor of Bossington House.

The next morning, after a fortifying breakfast of brown eggs, yellow butter from the resident Guernseys, hot milk, coffee, toast, and marmalade, Hubble entered the lists. His first foe was a 4½-pound rainbow, followed by a 4-pound brown and another half as large. Returning in the afternoon, he netted a 5¼-pound rainbow, which, unlike the others, was not released, because of the gamekeeper's belief that any fish over 5 pounds turns cannibal. His luck held the following day. Casting near Lower Horse Bridge, he took three giants weighing a total of 11¼ pounds. The inveterate fisher of rivers had never experienced anything like it; when accounts were drawn up, C. J Humbert wrote that he was happily shipping seven bottles of sack, one for each of Hubble's "Bottle Fish" over 4 pounds.[55]

When Hubble addressed the French Academy on the subject of "Science and Human Values" in early June, he was asked to answer a list of questions for the newspapers concerning his background, his major accomplishments, his opinion of Einstein, his favorite maxim, and what he would choose for his last words. He filled out everything but the last two. At the dinner that followed, Grace was seated next to Gaston Berget, the youngish Directeur de l'Enseignement Supérieur. They had scarcely spoken a word before Berget turned abruptly to her and said, "The two things that matter in the world are love and hate. What have you to say to that?"

"I have never been much interested in hate," Grace replied. "I have thought that anyone I loved would love me."

"So you thought," he repeated, "that anyone you loved would love you."

"Do not misunderstand. I didn't mean 'Come on Wednesdays or Fridays.'" Berget looked startled. "To love is to try to remove, as far as possible, the orange peelings from the pavements of another's life."

"What quotation is that?" he asked.

"It isn't a quotation. Nobody said it—or rather, just now, I did."

"Oh," he said, "Do you mind if I use it?"[56]

Before leaving Paris the Hubbles took a long walk along the Seine and through the winding back streets of the Left Bank. Rounding a corner, they encountered a cluster of what were apparently young students, the men sporting outrageous clothing and ponytails tied back with rubber bands, the vanguard of the Beat generation.

After a nostalgic trip to Cambridge, where they paid final tribute to Newton while revisiting Edwin's days as a young officer, they sailed for New York in early July. They reached Pasadena on the twentieth, and a deeply contented Nicolas, who had been sleeping under Berta's bed, returned to his normal place at the foot of theirs. "We say it has been a perfect trip," Grace wrote in her benedictory entry, "no regrets, no omissions, not much that we would have had happen differently. And we should be grateful for having had it so."[57]

VII

The Hubbles left Pasadena for Palomar in the observatory truck on the morning of September 1 at 8:30, Edwin's first run on the mountain in eleven months. This time they decided to stay together in a cabin outside the gate, just to the left of the road. In front of it was the forest of white firs and incense cedars, rising like a great fragrant screen.

After an early dinner, Hubble slung his rucksack over his shoulder and drove the truck to the dome of the 200-inch. Grace spent the evening reading and was distracted only by a mouse that had bounded with high leaps into the living room, as if pursued by some predator. She picked up a flashlight and went out to see if there might be a snake on the porch, but found nothing.

Edwin returned in buoyant spirits as the night sky was paling. Although the log reveals that he had taken only five plates with a total exposure time of three hours and ten minutes, he remarked that he was "back in his old form" and not the slightest bit fatigued. "If you see a mouse in the living room," a drowsy Grace murmured, "treat it with every consideration."[58]

They stayed two more nights while he took another thirteen plates, the last a 40-minute exposure of NGC 520, his 176th on the mountain. Before leaving at the end of the run, he took Grace into the darkroom and showed her the nebulae on the freshly developed plates. "In two years," he said, "I shall have determined the redshift. But the big cosmological program I shall not live to see." Then he suggested they do something they had not done before: "Let's go up and have a look at the telescope." As they finally turned away, he remarked, "I'm rather proud of the part I had in designing that."[59] On the way out he unlocked the stellar treasury where all the plates were kept, reminding her that his next run would be for four nights, from October 2 to October 6.

On September 27, the Sunday before his next scheduled run, he spent the afternoon and evening in his book-lined study. Nicolas, batting at his moving pencil, lay sprawled across a pile of papers containing long columns of small numbers, curves, and diagrams. Edwin had returned from England looking bronzed, strong, and vigorous. A routine checkup by Paul Starr indicated that he had recovered wonderfully well for a man who had weathered such a violent storm. At the office the next morning he received a call from the Los Angeles agent of Williams & Humbert Ltd. informing him that his bottles of sherry were ready for delivery. "Bring the sherry to my house this evening," he replied, "we'll crack a bottle together." Then he asked Milt Humason to come to his office, where they talked of a new plan of attack Hubble had thought of for the redshift. "He spoke," Humason recalled, "rapidly and, somehow, urgently as he explained what he had in mind." Then he got up and left to walk home for lunch. "We noticed his ease and energy and how well he seemed."[60]

Grace was on her way home after running some errands and spied Edwin striding along California Street, swinging his stick. She picked him up, and he asked her, as he always did, "What sort of morning did you have?" It was almost noon and they were about a mile from home. Grace began telling some story of no importance when, as she was about to turn into the driveway, something made her stop the car and look at him. He was staring straight ahead, with a rather puzzled, thoughtful expression, and breathing in an odd way, though with little effort, through his parted lips. She was not alarmed but curious and asked, "What's the matter."[61]

"Don't stop, drive in," he replied in a quiet voice, and Grace suddenly became alarmed. She wheeled into the courtyard, got out, and ran around to his side of the car, screaming for Berta. In the few seconds it took he seemed to have fainted, for when she reached him he was relaxed and pale and gave the impression he was about to say something. But he failed to respond to her voice and touch, and Berta, who had come at once, felt for his pulse but there was none. Grace ran inside and telephoned Paul Starr, who came immediately. There was nothing she could have done, the physician assured her; the cerebral thrombosis had been almost instantaneous and without pain. "It could happen to anyone at any time."[62]

Hubble was carried upstairs and placed on the bed he had shared with Grace for almost thirty years. He looked to her as he always did when asleep, like an athlete at rest. Nicolas took his accustomed place at the foot of the bed, where he stayed all night, for Grace refused to let the undertakers bear Edwin away, legalities notwithstanding. Years earlier he had said that when the time comes, "I want to disappear quietly," and she was determined to carry out his wishes.[63] The next day, the body was driven to nearby Monrovia for cremation. Grace, Max, her husband, Clarendon, and the Crottys followed the hearse to Live Oak Memorial Park. When the scheduled cremation took place some while later, Clarendon returned with Grace to stand vigil while waiting for the ashes. There was no funeral, no memorial service, no grave at which stunned mourners such as Humason and Sandage could pay their final respects.

The copper box containing the ashes was buried in a secret place by the same five who accompanied Hubble on his final journey, only one of whom survives to this day.[64] Had he chosen his own epitaph, he would have doubtless selected lines penned by some English writer, perhaps the following from Robert Louis Stevenson's *Requiem*:

> *Under the wide and starry sky,*
> *Dig the grave and let me lie*
> *Glad did I live and gladly die,*
>   *And I laid me down with a will.*
>
> *This be the verse you grave for me:*
> *Here he lies where he longed to be;*
> *Home is the sailor, home from the sea,*
>   *And the hunter home from the hill.*

Letters of regret poured in. Aldous, whose article on Edwin, "Stars and the Man," found no publisher, wrote, "Love remains—for it is what 'moves the sun and the other stars,' which is something that Edwin knew even better than Dante."[65] On hearing the news, Anita Loos thought of "the England that Edwin loved so much and which appreciated him so much more than his own country does. The U.S. will take a long time to understand, really, his place in history and science."[66] In seeking to comfort, Fryn Harwood put her finger on the secret of the Hubbles' enduring relationship. While Edwin was independent and self-confident in his work, "that did not make him so in his emotional life. The two things are quite different; he rested in you, if you know what I mean. That, I think, is what I am trying to say— he rested in you so completely."[67] Finally, Edith Sitwell recalled that not very long ago afternoon in Hubble's study, when he showed her plates of "universes in the heavens" millions of light-years away. "How terrifying!" she had remarked. "Only at first," he replied. "When you are not used to them. Afterwards, they give one comfort. For then you know that there is nothing to worry about—nothing at all!"[68]

For months after Hubble was gone, Nicolas sat, twice a day, before noon and before six, at the window anxiously looking up Woodstock Road. He had never paid any attention to the telephone, but now, whenever it rang, he hurried over and stood on his hind feet, wanting to hear the voice on the line. Grace would put the receiver to his ear, but he walked away as soon as he heard it wasn't Hubble. Finally, he did not come any longer. The fur on his head and shoulders had turned gray. He is growing older, Grace thought to herself; in the spring he will be eight. But when spring came the cathartic metamorphosis was completed when his new coat grew in as black and as thick as ever. During the long evenings while she read he sat beside her in a large chair, sometimes speaking in a rambling soliloquy, very like an imitation of human language. Gradually, the broad oak planks beneath his door were worn to a little hollow by his soft tread. He lived another eight years and died, at age sixteen, on Christmas Eve 1962. Grace buried him beneath the great Engelmann oak he had loved to climb.[1]

In February 1955, Maria Huxley succumbed after a long and painful bout with breast cancer. Aldous, who had begun experimenting with drugs, an activity that Grace disapproved of, grew more distant, and they saw little of one another after his marriage to Laura Archera in 1956. Anita Loos never returned to Los Angeles, except for an occasional visit, and, try as she might, was unable to convince Grace to spend time with her and their mutual friends in New York. Harlow Shapley, that venerable scientist and warhorse, outlived his old foe by nineteen years and was still going strong when he died in October 1972. During an extensive series of interviews which became the basis of his autobiography, he remarked, offhandedly, "Hubble, by the way, was an excellent observer, better than I. He was patient."[2] For one of the few times Milt Humason, whose death preceded Shapley's by months, would have agreed.

Grace heard that Enrico Fermi and Subrahmanyan Chandrasekhar, both members of the Nobel Committee, had joined their colleagues in unanimously voting Hubble the prize in physics, a rumor later confirmed by the astronomers Geoffrey and Margaret Burbidge after speaking with "Chandra."[3] But the Nobel Prize is not awarded posthumously, and death had intervened at the critical moment, thus denying the century's greatest astronomer his due. It was at this point that his sixty-three-year-old widow decided to commit what remained of her life, another twenty-seven years as it turned out, to preparing the way for Hubble's future biographer by sorting through his papers and putting her journals in order, which involved much pruning.[4] Like everything else having to do with Edwin and his work, she had very specific ideas on the subject, which she set down in a 1954 letter to Leslie Bliss of the Huntington Library, where Hubble's papers are on deposit:

> As to the length of time before the material can be used, I agree with your suggestion of about 20 years—unless a very good biographer should happen to turn up. But I am sure you also will agree with me that,
>
> (1) The writer must be a scholar with sufficient scientific background and not a writer of popular biography.
>
> (2) The writer must be a man, not a woman.[5]

Though a number of individuals, both male and female, set out to write Hubble's life, Grace refused to correspond with or to be interviewed by any of them, and their projects, for one reason or another, never reached fruition.

Before Hubble died, a still nervous Allan Sandage had become a regular visitor at Woodstock Road, where he came to know the Huxleys and such other members of the intelligentsia as Igor Stravinsky. Still speaking as a god might, Hubble addressed him on a number of subjects ranging from art and philosophy to music, literature, and religion. Feeling woefully inadequate, Sandage added novels, recordings, and classics in the humanities to his already impossible schedule. Then, suddenly, the king of the mountain was gone and he, Allan Sandage, the self-described "hick" from nowhere, became the newest dark man on the observatory staff, despite the fact that his dissertation was not yet finished. Depending on the day and the circumstances, it seemed to him at once an incredible opportunity and a life sentence. Hubble had died too young; it simply wasn't fair. And yet there his program was, all laid out by him. "If you were assistant to Dante and Dante died," Sandage once lamented to an interviewer, "and you had in your possession the whole of the *Divine Comedy*, what would you do? What would you actually do?"[6]

Grace now turned to the young man they had "treated like a son," asking him to complete Edwin's most ambitious publishing project. The magnificent collection of plates garnered between 1919 and 1948 with the 60-inch and 100-inch telescopes was to have been the basis of Hubble's revised classification of the nebulae. When, in 1961, *The Hubble Atlas of Galaxies* was finally published, Sandage wrote, "I have acted mainly as an editor, not as an editor of a manuscript but rather an editor of a set of ideas and conclusions that were implied in the notes."[7]

Just as when Edwin was away from home, Grace saw no reason to keep a journal after his passing, and most of the rest of her years were spent in quiet obscurity reading and rereading the works of English literature that had captivated her since high school. She continued to walk the grounds of the Huntington on Sunday mornings and could be seen dining at the Athenaeum on occasion, but her circle of friends gradually narrowed until only a handful of intimates remained. ·

Still, she was never heard to complain about her life after Edwin died. "There was a lot of stoic in Grace," Page Ackerman, her librarian friend since the war days, noted. "She coped."[8] Her health remained good until she reached her late eighties, when infirmity set in, requiring that she have someone in the house full-time. She began experiencing trouble walking. Over her protestations, a handrail was installed on the steps leading from the front hallway down to the sunken dining room, but she refused to use it and suffered a fall not long thereafter. She had barely recovered when she fell again and had to be moved to nearby

Las Encinas Hospital, a private facility located on Pasadena's East Del Mar Boulevard. There her closest friends, Page Ackerman, Florence Nixon, a Chaucer scholar whose brains she loved to pick, and Ida Crotty, visited her every Sunday. Grace turned ninety in August 1980 and, though mostly bedridden, was still dreaming of the day when she could return to Woodstock Road. Like Edwin, she died quietly of a cerebral thrombosis seven months later. Her final journey was a repeat of his, and when the ashes were ready the copper box containing them was carried to the same secret place as before. Under the heading "Primary Occupation" on the death certificate, someone entered "Housewife." Under "Years at This Occupation," the number "60" appears.[9] But in the short obituary announcing that she had requested no service, Grace was called a writer, as she was at heart.

Far above the purple rim of the horizon, a great scientific instrument the size of a railroad boxcar circles this planet. Below, dozens of anxious astronomers who never pass a night in a freezing observatory sit by their computers waiting to decipher billions of electronic whispers from the abyss. Free from the distortion caused by the atmosphere of Earth, the Hubble Space Telescope's unblinking eye probes the cosmos with a sensitivity to the 28th magnitude, twenty-five times better than the finest ground-based instrument. Hubble could spot a firefly 10,000 miles distant or home in on a small battery-powered flashlight held by an astronomer standing on the surface of the moon. Its pinpoint accuracy would enable a golfer of mythic attributes to sink a putt from 1,500 miles away, the distance from Washington, D.C., to Dallas. Before its batteries fade and its gyroscopes are stilled, it may gather light from as far as 14 billion light-years into space, perhaps enabling humankind to see almost to the beginning of time.

Thirty-seven years before the telescope's launch in April 1990, the Hubbles were in the middle of what would be their last visit abroad. After Edwin delivered the Cormac Lecture before the Royal Society of Edinburgh, he and Grace paused briefly to chat with the physicist Max Born, who was about to share the Nobel Prize with Walter Bothe. Afterward, as the couple crossed the street to their hotel, they were pursued by a rosy-faced young man, who blurted out the question: "Sir, what about telescopes on artificial satellites?"[10]

Unfortunately, Grace did not record her husband's reply.

Given the data now being collected by his mechanical namesake from the distant nebulae, Hubble would have declared the idea well worth pursuing, no matter what its consequences for his own findings. Yet remarkably, his discoveries, like the theoretical equations from the pen of the gnomish Albert Einstein, have not been eclipsed. The universe in which we live continues to expand, but at a much faster rate than Hubble himself projected—speeds that seem to increase with each new study of hitherto invisible Cepheids conducted via the space telescope. The uncountable galaxies beyond the Milky Way are more uncountable still, as they swarm and multiply by the millions on graphic after computer graphic. And in all directions, great swaths of the universe resemble other great slices of creation as far as the lens can probe. Finally, with only rare exceptions, the new nebulae can be fit into the classification scheme developed by Hubble atop Mount Wilson during the 1920s. Achievements such as these remind one of a line from the verses the great astronomer Edmond Halley composed and prefixed to Isaac Newton's incomparable *Principia*:

Nearer to the gods no mortal may approach.

NOTES AND ABBREVIATIONS OF FREQUENTLY USED SOURCES

Grace Burke Hubble's journals and other writings cover many hundreds of pages, which frequently bear no numbers. In most instances, citations are by date. However, she occasionally began by numbering pages in a document, and then, at a certain point, started numbering again, sometimes repeating this process three or four times. Thus when citing page 2, for example, it might be from her second, third, or fourth series of numbers. In a few instances, she provided neither dates nor page numbers: when this occurs, the citation "unp." (unpaged) has been employed.

Unless otherwise acknowledged in the citations, all interviews were conducted by the author.

Manuscript Collections

AIP   American Institute of Physics, College Park, Maryland

CIW   Carnegie Institution of Washington, Washington, D.C.

HCOA   Director's Correspondence, Harvard College Observatory Archives, Harvard University, Cambridge, Massachusetts

HUB   Edwin P. Hubble Manuscript Collection, Henry Huntington Library, San Marino, California

LOA   Lowell Observatory Archives, Flagstaff, Arizona

MWOA   Mount Wilson Observatory Archives, Henry Huntington Library, San Marino, California

MWODF   Mount Wilson Observatory Archives Director's Files, Henry Huntington Library, San Marino, California

PUL   Seeley G. Mudd Manuscript Library, Princeton University Library, Princeton, New Jersey

SALO   Mary Lea Shane Archives of the Lick Observatory, University Library, University of California, Santa Cruz

UCA   University of Chicago Archives, The Joseph Regenstein Library, Chicago, Illinois

YOA   Director's Papers, Yerkes Observatory Archives, Yerkes Observatory, Williams Bay, Wisconsin

Periodicals

APJ   *Astrophysical Journal*

CG   *Cap and Gown* (University of Chicago student annual)

DM   *The Daily Maroon*

LAT   *Los Angeles Times*

MC   *The Marshfield Chronicle*

MM   *The Marshfield Mail*

NYT   *The New York Times*

PASP   *Publications of the Astronomical Society of the Pacific*

WI   *Wheaton Illinoian*

Individuals

A.A.C.   Albert A. Colvin

A.S.   Allan Sandage

C.A.W.   Charles A. Whitney

B.S.   Bert Shapiro

E.B.F.   Edwin B. Frost

E.H.   Elizabeth "Betsy" Hubble

E.P.H.   Edwin P. Hubble

G.B.H.   Grace Burke Hubble

G.E.H.   George Ellery Hale

H.H.L.   Helen Hubble Lane

H.S.   Harlow Shapley

I.B.   Ira Bowen

I.C.   Ida Crotty

J.C.M.   John C. Merriam

J.F.L.   John F. Lane

J.P.H.   John Powell Hubble

M.L.H.   Milton L. Humason

N.U.M.   Nicholas U. Mayall

S.W.   Spencer Weart

V.B.   Vannevar Bush

V.H.   Virginia "Jennie" Hubble

V.M.S.   Vesto Melvin Slipher

W.S.A.   Walter S. Adams

CHAPTER 1: MARSHFIELD

1. *Personal Reminiscences and Fragments of the Early History of Springfield and Greene County, Missouri, Related by Pioneers and Their Descendants*, Springfield Sesquicentennial Edition (Springfield: Museum of the Ozarks, 1979), 3–4.

2. *The Springfield Leader*, Mar. 31, 1914, 1.

3. G.B.H., "Family History," HUB 82(1), Box 7, 7.

4. Ibid., 1–4; also see *History and Genealogy of the Hubble Family*, ed. by Harold Berresford Hubbell, Jr., and Donald Sidney Hubbell, 3rd ed. (New York: Theo. Gaus, Ltd., 1980).

5. G.B.H., "Family History," 3.

6. *History of Greene County, Missouri* (St. Louis: Western Historical Society, 1883), 274, 456, 475, 477, 426.

7. Drury College Archives, *Second Annual Catalogue of Drury College at Springfield, Greene County, Missouri, for the Year 1874–75*, 34–35.

8. Jonathan Fairbanks and Clyde Edwin Tuck, *Past and Present of Greene County, Missouri*, 2 vols. (Indianapolis: A. W. Bowen & Company, 1915), 2, 419.

9. *Second Annual Catalogue of Drury College*, 23–24.

10. Washington University Archives, *Washington University Annual Catalogue, 1878–79*, 18.

11. Ibid., 143, 147–48.

12. *History of Greene County, Missouri*, 507; *The Springfield Leader*, Feb. 28, 1920, 1; *The Springfield Republican*, Jan. 15, 1886, unp.

13. *Springfield City Directories, 1899–1920*.

14. "Mortgages of Deeds of Trust," 51, 231–25, 637, Webster County Courthouse, Marshfield, MO.

15. *MC*, June 26, 1890, 3.

16. *History of Laclede, Camden, Dallas, Webster, Wright, Texas, Pulaski, Phelps and Dent Counties, Missouri* (Chicago: The Goodspeed Publishing Company, 1889), 239; *History of Webster County, 1855–1955*, compiled and ed. by Floy Watters George (Marshfield, MO: Webster County Historical Society, undated), 171, 201.

17. Interview with E.H. and H.H.L. by J.F.L., Sept. 1991 (courtesy of J.F.L.).

18. G.B.H., "Family History," 4.

19. Interview with E.H. and H.H.L. by J.F.L.; *History of Laclede, Camden, Dallas . . . and Dent Counties, Missouri*, 854.

20. *MM*, Nov. 2, 1893, 3.

21. "Register of Attorneys," March Term, 1882, unp., Webster County Courthouse, Marshfield, MO.

22. *History of Greene County, Missouri*, 761.

23. H.H.L., "Edwin Powell Hubble," HUB 1097, Box 8, 1.

24. *MC*, Nov. 28, 1889, 3; Dec. 12, 1889, 3.

25. Ibid., June 26, 1890, 3.

26. Interview with E.H. and H.H.L. by J.F.L.

27. *MC*, Jan. 12, 1893, 3; *MM*, Mar. 29, 1893, 1; June 28, 1893, 1; Aug. 9, 1893, 1; Mar. 22, 1894, 3; May 9, 1895, 3.

28. G.B.H., "Family History," 6.

29. H.H.L., "Edwin Powell Hubble," 2.

30. Interview with E.H. and H.H.L. by J.F.L.

31. H.H.L., "Edwin Powell Hubble," 1.

32. Ibid., 5.

33. Ibid., 2.

34. This letter was supposedly written between E.P.H.'s eighth and twelfth birthdays. A search of the microfilmed newspapers has yielded no trace of it.

35. G.B.H., "Family History," 4.

36. Interview with Lena James Jump, Aug. 19, 1991.

37. G.B.H., "Family History," 7.

38. *MM*, May 3, 1893, 1.

39. G.B.H., "Family History," 6.

40. Sam J. Shelton to C.A.W., Nov. 12, 1971 (courtesy of C.A.W.).

41. Ibid.

42. *MM*, Aug. 25, 1898, 3.

43. See, for example, "Warranty Deed Record," 39, 496; "Deed Record," 5, 583–84, Webster County Courthouse, Marshfield, MO.

44. *MM*, Oct. 12, 1899, 3; Nov. 9, 1899, 3; Nov. 16, 1899, 3.

CHAPTER 2: "AN AWFUL MOMENT"

1. *MC*, July 19, 1900, 3; Sept. 6, 1900, 3.

2. Graham Burnham, *Wheaton and Its Homes* (Chicago, 1892), unp.

3. Ibid.

4. *WI*, Feb. 8, 1904, 1.

5. Burnham, *Wheaton and Its Homes*, unp.

6. J.P.H., *Fire Insurance: What It Is* (Chicago: The Greenwich Insurance Company, 1900–1), 7–8.

7. Central School Transcript, 1901–2, Wheaton Public Schools, Wheaton, IL.

8. A.A.C. to C.A.W., June 14, 1971 (courtesy of C.A.W.).

9. Interview with E.H. and H.H.L. by J.F.L., Sept. 1991 (courtesy of J.F.L.).

10. A.A.C. to C.A.W., Aug. 28, 1971 (courtesy of C.A.W.); interview with E.H., Mar. 18, 1992; H.H.L., "Edwin Powell Hubble," HUB 1097, Box 8, 2.

11. Central School Transcripts, 1902–3, 1903–4.

12. G.B.H., "Family History," HUB 82(1) Box 7, 9; Grote Reber to Donald Osterbrock, Oct. 24, 1988 (courtesy of Donald Osterbrock).

13. CIW, Mt. Wilson Observatory, f. Archival File.

14. Interview with E.H.

15. Interview with Herman A. Fischer, Jr., by C.A.W., Oct. 5, 1971 (courtesy of C.A.W.).

16. *Great Events of the 20th Century*, ed. by Richard Marshall (Pleasantville, New York: The Reader's Digest Association, Inc., 1977), 49.

17. Burnham, *Wheaton and Its Homes*, unp.

18. H.H.L., "Edwin Powell Hubble," 8.

19. Interview with E.H.

20. Interview with Herman A. Fischer, Jr., by C.A.W.

21. H.H.L., "Edwin Powell Hubble," 8.

22. The Colvin home, now beautifully restored, houses the Wheaton Historical Center.

23. H.H.L., "Edwin Powell Hubble," 3–5, 8.

24. Ibid., 5.

25. Interview with E.H.

26. *WI*, Jan. 4, 1901, 3.

27. Interview with Lora Fox Conley by C.A.W., Oct. 5, 1971 (courtesy of C.A.W.).

28. H.H.L., "Edwin Powell Hubble," 4.

29. Ibid., 1.

30. Interview with Lora Fox Conley by C.A.W.

31. A.A.C. to C.A.W., June 14, 1971; interview with A.A.C. by C.A.W., Oct. 5, 1971 (courtesy of C.A.W.).

32. Interview with E.H. and H.H.L. by J.F.L.

33. Central School Transcript, 1904–5.

34. Interview with Olive Stark Grove by C.A.W., Oct. 4, 1971 (courtesy of C.A.W.).

35. *The 1909 Wheatonia*, 29; HUB 43, Box 2, undated clipping; *WI*, June 1, 1906, 5.

36. Interview with A.A.C. by C.A.W.; Joan L. Anderson to C.A.W. (courtesy of C.A.W.); *WI*, Oct. 6, 1905, 3.

37. HUB 43, Box 2. A clipping states that the winning jump was 5 feet 10 inches, one and one-half inches higher than the figure recorded in E.P.H.'s notebook.

38. G.B.H., "Family History," 8.

39. H.H.L., "Edwin Powell Hubble," 9.

40. Ibid.

41. Interview with Lora Fox Conley by C.A.W.

42. Wheaton High School Senior Class Report, HUB Box 25, f. 1.

43. *WI*, June 22, 1906, 1.

44. Interview with A.A.C. by C.A.W.

45. G.B.H., "Family History," 9.

CHAPTER 3: "A THING SO OUTLANDISH"

1. *The Chicago Tribune*, May 2, 1893, 1.
2. Henry Adams, *The Education of Henry Adams* (New York: Houghton Mifflin Company, 1974), 340.
3. "Building the City Grey," *The University of Chicago Magazine* (October 1991), 26.
4. Ibid., 29.
5. Ibid., 16–17. For additional background, see Richard J. Stoor, A *History of the University of Chicago: Harper's University, the Beginnings* (Chicago and London: The University of Chicago Press, 1966); Thomas Wakefield Goodspeed, A *History of the University of Chicago: The First Quarter-Century* (Chicago: The University of Chicago Press, 1916); *The Idea of the University of Chicago: Selections from the Papers of the First Eight Chief Executives of the University of Chicago from 1891–1975*, ed. by William Lucas Murphy and D. J. R. Bruckner (Chicago and London: The University of Chicago Press, 1976).
6. *CG, 1909*, 150; *CG, 1910*, 246.
7. *CG, 1910*, 246.
8. Ibid., 247.
9. G.B.H., "University of Chicago, 1906–1910, 1914–1917," HUB 82(2), Box 7, 4.
10. *DM*, Oct. 15, 1908, 2; *CG, 1910*, 43.
11. G.B.H., "University of Chicago, 1906–1910, 1914–1917," 1.
12. J. Miller to E.P.H., July 12, 1906, HUB, Box 20, f. 981; *Register of the University of Chicago: 1907–1908*, 75.
13. Interview with E.H. and H.H.L. by J.F.L., Sept. 1991 (courtesy of J.F.L.).
14. G.B.H., "University of Chicago, 1906–1910, 1914–1917," 8.
15. John J. Schommer to G.B.H., May 15, 1958, HUB, Box 19, f. 928.
16. G.B.H., "University of Chicago, 1906–1910, 1914–1917," 4.
17. Henry S. Tropp, "Forest Ray Moulton, *Dictionary of Scientific Biography*, 9 (New York: Charles Scribner's Sons, 1974), 552.
18. *Carnegie Institution of Washington Yearbook*, no. 3 (1904), 255–56.
19. G.B.H., "University of Chicago, 1906–1910, 1914–1917," 4.
20. UCA, A. A. Stagg Papers, Football Scrapbook, June 3 to Oct. 6, 1906, 75.
21. G.B.H., "University of Chicago, 1906–1910, 1914–1917," 2.
22. H.H.L., "Edwin Powell Hubble," HUB 1097, Box 8, 1.
23. G.B.H., "University of Chicago, 1906–1910, 1914–1917," 2.
24. Ibid., 3.
25. *DM*, Dec. 5, 1906, 4.
26. Ibid., Jan. 15, 1907, 1.
27. HUB, Box 28, Scrapbook, 3; *DM*, Jan. 23, 1907, 1.
28. UCA, A. A. Stagg Papers, Track Notebook, 1906–1910, 16–17, 19, 21–23.
29. Sam J. Shelton to C.A.W., Nov. 12, 1971 (courtesy of C.A.W.).
30. HUB, Box 28, Scrapbook, 2.
31. Interview with E.H., Mar. 18, 1992; Lena Jump, "Orphan Trains," *The Webster County Historical Journal*, no. 12 (October 1979), 12–14.
32. E.P.H. to Martin Jones Hubble, Aug. 24, 1909, HUB unclassified.
33. H.H.L., "Edwin Powell Hubble," 2.
34. Ibid.; H.H.L. to Larry Blakée, July 6, 1985 (courtesy of L.B.).
35. *DM*, Feb. 14, 1908, 1.

36. Ibid., Mar. 31, 1908, 1, 6; Apr. 3, 1908, 3.
37. Ibid., Mar. 10, 1908, 1; CG, 1909, 206, 211.
38. UCA, A. A. Stagg Papers, Track Notebook, 1906–1910, 63; DM, May 5, 1908, 1.
39. CG, 1909, 206.
40. UCA, E.P.H. transcript; R. A. Millikan to Edmund J. James, Jan. 8, 1910, HUB, Box 18, f. 840.
41. A.A.C. to C.A.W., June 16, 1971, July 2, 1971 (courtesy of C.A.W.).
42. Interview with E.H.; interview with E.H. and H.H.L. by J.F.L.
43. G.B.H., "University of Chicago, 1906–1910, 1914–1917," 5–6.
44. Ibid., 6–7.
45. CG, 1909, 116–21; DM, May 5, 1908, 2.
46. HUB, Box 20, f. 982.
47. See Robert A. Millikan, The Autobiography of Robert A. Millikan (New York: Prentice-Hall, 1950), 68–81.
48. DM, Dec. 1, 1908, 1.
49. Ibid., Jan. 14, 1909, 1.
50. Ibid., Jan. 16, 1909, 1.
51. Ibid., Feb. 16, 1909, 1.
52. CG, 1909, 226.
53. CG, 1910, 194–95, 199–202.
54. E.P.H. to Martin Jones Hubble, Aug. 24, 1909, HUB unclassified.
55. Ibid. Astronomers now know that Halley's comet itself heralded the Norman Conquest. The Star of Bethlehem may not have been a comet but a brilliant conjunction of the slow-moving planets Jupiter and Saturn.
56. Godfrey Elton, The First Fifty Years of the Rhodes Trust and the Rhodes Scholarships, 1903–1953 (Oxford: Basil Blackwell, 1955), 1–9.
57. MWOA, W.S.A. Papers, Box 34, f. 34.588.
58. DM, Jan. 27, 1910, 3.
59. Ibid., Dec. 11, 1909, 1; Dec. 16, 1909, 1.
60. Ibid., Jan. 5, 1910, 1.
61. Ibid., Nov. 5, 1909, 1; Dec. 4, 1909, 1, 4.
62. Ibid., Nov. 27, 1909, 1.
63. Ibid., Dec. 16, 1909, 1.
64. H. E. Slaught to Edmund J. James, Jan. 5, 1910, HUB, Box 19, f. 948.
65. R. A. Millikan to Edmund J. James, Jan. 8, 1910, HUB, Box 18, f. 840.
66. UCA, Box 7, f. 3, Department of Physical Education and Athletic Records, 1892–1974; DM, Jan. 8, 1910, 1; Jan. 22, 1910, 1.
67. DM, Jan. 26, 1910; Edmund J. James to E.P.H., Jan. 31, 1910, HUB Box 16, f. 717.
68. Interview with E.H.
69. DM, Jan. 27, 1910, 3.

CHAPTER 4: A SON OF QUEEN'S

1. Interview with E.H., Mar. 18, 1992.
2. H.H.L., "Edwin Powell Hubble," HUB 1097, Box 8, 3.
3. Interview with E.H. and H.H.L. by C.A.W., 1971 (courtesy of C.A.W.).
4. Quoted in Nigel Calder, The Comet Is Coming! (New York: The Viking Press, 1980), 26.

5. Justin Kaplan, *Mr. Clemens and Mark Twain: A Biography* (New York: Simon and Schuster, 1966), 386.
6. Interview with E.H. and H.H.L. by C.A.W.
7. J.P.H. to E.P.H., Aug. 22, 1912 (Oxford correspondence courtesy of J.F.L.).
8. UCA, E.P.H. transcript.
9. Alfred R. Wooley, *The Clarendon Guide to Oxford*, 3rd ed. (Oxford: Oxford University Press, 1972), 54.
10. E.P.H. to V.H., Sept. 30, 1910.
11. Warren D. Ault, "Oxford in 1907 (With a Glimpse of T. E. Lawrence)," *The American Oxonian*, 66, no. 2 (Spring 1979), 122.
12. E.P.H. to V.H., Sept. 30, 1910.
13. Ibid., Oct. 20, 1910.
14. Jakob Larsen to C.A.W., Sept. 28, 1973 (courtesy of C.A.W.).
15. E.P.H. to V.H., Oct. 20, 1910.
16. Ibid., Oct. 31, 1910.
17. Ibid.
18. Ibid., c. Dec. 1910.
19. Ibid.
20. E.P.H. to J.P.H., Dec. 16, 1910.
21. E.P.H. to V.H., Jan. 14, 1911.
22. Ibid., Mar. 3, 1911.
23. Ibid., Mar. 14, 1911. Also see E.P.H. to J.P.H., June 23, 1911.
24. H.H.L. to C.A.W., Apr. 12, 1975 (courtesy of C.A.W.).
25. E.P.H. to J.P.H., June 23, 1911.
26. E.P.H. to William "Bill" Hubble, June 25, 1911.
27. E.P.H. to V.H., Mar. 14, 1911.
28. Ibid., Feb. 4, 1911, Nov. 13, 1911.
29. Ibid., Spring 1911.
30. G.B.H., "E.P.H.: [Oxford University, 1910–1913]," HUB 82(3), Box 7, 13.
31. Ibid., 13–14.
32. E.P.H. to Reed, Aug. 26, 1911.
33. G.B.H., "E.P.H.: [Oxford University, 1910–1913]," 10.
34. E.P.H. to "Dear Folks," Nov. 25, 1911. Also see E.P.H. to V.H., Nov. 13, 1911.
35. G.B.H., "Quotations," HUB 82(13), Box 8, 8.
36. G.B.H., "E.P.H.: [Oxford University, 1910–1913]," 10.
37. E.P.H. to V.H., Feb. 4, 1911.
38. Ibid., Mar. 14, 1911.
39. E.P.H. to Martin Jones Hubble quoted in G.B.H., "E.P.H.: [Oxford University, 1910–1913]," 17.
40. E.P.H. to "Dear Folks," c. Dec. 1911.
41. Daisy Turner to G.B.H., Oct. 3, 1953, HUB Box 20, f. 972.
42. E.P.H. to V.H., May 20, 1913.
43. Ibid., Mar. 3, 1911.
44. G.B.H., "E.P.H.: [Oxford University, 1910–1913]," 7.
45. E.P.H. to V.H., c. June 1911.
46. Quoted in Lois Meek Stolz to C.A.W., Oct. 10, 1971 (courtesy of C.A.W).
47. Hubble Family Scrapbook (courtesy of E.H.).
48. E.P.H. to "Dear Folks," c. Mar. 1912.

49. E.P.H. to V.H., Mar. 3, 1911.
50. E.P.H. to Lucy Hubble, Jan. 9, 1912.
51. G.B.H., "E.P.H.: [Oxford University, 1910–1913]," 7.
52. E.P.H. to "Dear Folks," Nov. 25, 1911.
53. E.P.H. to V.H., Apr. 2, 1912.
54. Ibid., c. June 1911.
55. Ibid., Mar. 14, 1911.
56. E.P.H. to Reed, Aug. 26, 1911.
57. G.B.H., "E.P.H.: [Oxford University, 1910–1913]," 7.
58. E.P.H. to "Dear Folks," July 25, 1911; E.P.H. to Lucy Hubble, July 29, 1911; G.B.H., "E.P.H.: [Oxford University, 1910–1913]," 3.
59. E.P.H. to "Dear Folks," Aug. 1, 1912.
60. Ibid., c. Dec. 1911.
61. E.P.H. to Reed, Aug. 26, 1911.
62. E.P.H. to "Dear Folks," Aug. 1, 1912.
63. Ibid.
64. J.P.H. to E.P.H., Aug. 22, 1912.
65. Ibid., Sept. 16, 1912.
66. H.H.L. to C.A.W., Apr. 12, 1975.
67. E.P.H. to "Dear Folks," c. Oct. 1912.
68. J.P.H. to E.P.H., Sept. 16, 1912.
69. G.B.H., "E.P.H.: [Oxford University, 1910–1913]," 3–4.
70. E.P.H. to Lucy Hubble, Jan. 9, 1912.
71. E.P.H. to "Dear Folks," c. Oct. 1912.
72. E.P.H. to J.P.H., June 17, 1911.
73. E.P.H. to V.H., c. Oct. 1912.
74. E.P.H. to "Dear Folks," Dec. 6, 1912.
75. Interview with E.H. and H.H.L. by J.F.L., Sept. 1991 (courtesy of J.F.L); *The Louisville Courier-Journal*, Jan. 20, 1913, 24.
76. E.P.H. to V.H., Feb. 14, 1913.
77. Ibid., May 20, 1913.
78. Louisa May Hubble Dickerson to Emily Jane Hubble, Aug. 14, 1912, quoted in G.B.H., "E.P.H.: [Oxford University, 1910–1913]," 16–17.

CHAPTER 5: HEAVEN'S GATE

1. Interview with E.H. and H.H.L. by J.F.L., Sept. 1991 (courtesy of J.F.L.).
2. E.P.H., Rhodes Trust Record, Rhodes House, Oxford.
3. Settlement Book 112, 31, Jefferson County Courthouse, Louisville, KY.
4. Interviews with Elizabeth Winlock Howerton and Virginia Barringer Baskett by C.A.W., Nov. 9, 1971 (courtesy of C.A.W.).
5. Interview with E.H. and H.H.L. by J.F.L.
6. E.P.H., Rhodes Trust Record, Rhodes House, Oxford.
7. Ibid.
8. H.H.L., "Edwin Powell Hubble," HUB 1097, Box 8, 7. A search of Kentucky Circuit Court Record Books for Jefferson and contiguous counties as well as Webster County, Missouri, has proved fruitless. Nor does E.P.H.'s name appear in the records

of the Louisville Bar Association. Also see Joel A. Gwinn, "Edwin Hubble in Louisville, 1913–14," *The Filson Club History Quarterly*, 56 (1982), 417.

9. G.B.H., "E.P.H.: The Astronomer," HUB 82(7), Box 7, 1–2.

10. *The Blotter*, 14, no. 1, Oct. 10, 1913. Also see Gwinn, "Edwin Hubble in Louisville, 1913–14," 415–19.

11. G.B.H., "E.P.H.: The Astronomer," 1–2.

12. Interview with John Hale by Joel Gwinn, Mar. 8, 1989 (courtesy of Ron Brashear).

13. Ibid.

14. Interview with E.H. and H.H.L. by J.F.L.

15. Ibid.

16. Interview with John Hale by Joel Gwinn.

17. *The Blotter*, 14, no. 7, Feb. 27, 1914.

18. Ibid., Mar. 20, 1914.

19. *The Senior Blotter*, New Albany High School, 1914.

20. F. R. Moulton to E.B.F., May 27, 1914, YOA, Box 60, f. 10.

21. E.P.H. to E.B.F., May 29, 1914, YOA.

22. F. R. Moulton to E.B.F., May 27, 1914, YOA.

23. E.B.F. to E.P.H., June 6, 1914, YOA.

24. E.P.H. to E.B.F., June 6, 1914, YOA.

25. E.B.F. to E.P.H., June 15, 1914, July 24, 1914, YOA.

26. V.M.S., "Spectrographic Observations of Nebulae," *Popular Astronomy*, 23 (1915), 21–24.

27. John S. Hall, "Vesto Melvin Slipher, 1875–1969," *American Philosophical Yearbook* (Philadelphia: The American Philosophical Society, 1970), 164.

28. V.M.S. to E.P.H., Mar. 20, 1953, HUB, Box 18, f. 951.

29. Robert Smith, *The Expanding Universe: Astronomy's "Great Debate" 1900–1931* (Cambridge, Eng.: Cambridge University Press, 1982), 19.

30. E.B.F., *An Astronomer's Life* (Boston and New York: Houghton Mifflin Company, 1933), 114.

31. Donald Osterbrock, Ronald Brashear, and Joel Gwinn, "Young Edwin Hubble," *Mercury*, 19, no. 1 (January–February 1990), 7; Donald E. Osterbrock, "The California-Wisconsin Axis in American Astronomy," *Sky and Telescope* 51, no. 1 (January 1976), 9–14.

32. *Annual Register of the University of Chicago: 1914–1915*, 277.

33. E.P.H., "Changes in the Form of the Nebula N.G.C. 2261," *Proceedings of the National Academy of Sciences*, 2 (1916), 230.

34. E.P.H., "Recent Changes in Variable Nebula N.G.C. 2261, *APJ*, 45, no. 5 (1917), 352.

35. E.P.H., "Twelve Faint Stars with Sensible Proper-Motions," *Astronomical Journal*, 29, no. 1 (1916), 168.

36. E.P.H. to E.B.F., Apr. 7, 1917, YOA, Box 69, f. 8.

37. G.B.H., "E.P.H.: University of Chicago, 1914–1917," HUB 82(2), 10–11.

38. Interview with E.H., Mar. 18, 1992.

39. Ibid.

40. H.H.L. to C.A.W., Apr. 12, 1975 (courtesy of C.A.W.).

41. Forest R. Moulton to E.B.F., Mar. 9, 1915; E.B.F. to E. H. Moore, Mar. 3, 1916, YOA, Box 64, f. 1; E.B.F. to R. D. Salisbury, Apr. 4, 1916, Box 67, f. 2; Salisbury to E.B.F., Apr. 7, 1916, YOA, Box 68. f. 1.

42. G.B.H. "E.P.H.: University of Chicago, 1914–1917," 11.

43. W.S.A. to G.E.H., Sept. 23, 1916; W.S.A. to G.E.H., Oct. 21, 1916; G.E.H., MWODF: 1904–1923, Box 148, f. W.S.A.

44. G.E.H. to W.S.A., Nov. 1, 1916, G.E.H, MWODF.

45. Henry G. Gale to W.S.A., Apr. 4, 1917, G.E.H., MWODF: 1904–1923, Box 156, f. Henry G. Gale.

46. E.P.H. to E.B.F., Apr. 10, 1917, YOA, Box 69, f. 8.

47. E.B.F. to E.P.H., Apr. 12, 1917, YOA.

48. E.P.H. to G.E.H., Apr. 10, 1917, G.E.H., MWODF: 1904–1923, Box 159, f. E.P.H.

49. G.E.H. to E.P.H., Apr. 19, 1917, G.E.H., MWODF.

50. E.P.H. to E.B.F., May 1, 1917, YOA, Box 69, f. 8.

51. E.B.F. to E.P.H., May 7, 1917, YOA.

52. E.P.H. to E.B.F., Nov. 3, 1919, YOA, Box 71, f. 3.

53. E.P.H., "Photographic Investigations of Faint Nebulae," *Publications of the Yerkes Observatory*, 4, pt. 2 (1920), 69.

54. Ibid., 7.

55. G.B.H., "E.P.H.: World War I," HUB 82(4), Box 7, 1.

56. E.B.F. to F. J. Gurney, Aug. 16, 1917, YOA, Box 71, f. 7.

57. E.P.H. to E.B.F., June 12, 1917, YOA, Box 69, f. 8.

58. E.B.F. to E.P.H., June 15, 1917, YOA.

59. HUB, Box 25, f. 1.

60. Quoted in G.B.H., "E.P.H.: World War I," 19.

61. E.P.H. to Frank Aydelotte, Sept. 17, 1917, N.U.M. Papers, SALO, f. 17, Papers Referring to Hubble.

62. See J. G. Little, *The Official History of the Eighty-sixth Division* (Chicago: Chicago States Publications Society, 1921).

63. The rifle-range story is from an undated newspaper clipping in a Hubble family scrapbook, bearing the headline "Look Out, Kaiser, We Can Shoot"; G.B.H., "E.P.H.: World War I," 3.

64. G.B.H., "E.P.H.: World War I," 3.

65. Little, *The Official History of the Eighty-sixth Division*, 30.

66. Physical Examination Report, July 9, 1918, HUB, Box 28, Scrapbook, 5; Abram Poole to C. R. Howland, Nov. 6, 1918, HUB, Box 25, f. 1.

67. G.B.H., "E.P.H.: World War I," 10.

68. Ibid., 10–11.

69. Ibid., 13.

70. HUB, Box 25, f. 1.

71. E.P.H. to E.B.F., Aug. 14, 1919, YOA, Box 77, f. 1.

72. G.B.H., "E.P.H.: World War I," 11–12.

73. Ibid., 15.

74. E.P.H. to G.E.H., May 2, 1919, G.E.H., MWODF: 1904–1923, Box 159, f. E.P.H.

75. Royal Astronomical Society Club Dinner Lists (744), July 11, 1919.

76. G.E.H. to W.S.A., Dec. 23, 1918, Jan. 14, 1919, California Institute of Technology, G.E.H. Archives, G.E.H. (personal), Box 2, f. W.S.A.

77. G.E.H. to E.P.H., June 9, 1919, G.E.H., MWODF: 1904–1923, Box 159, f. E.P.H.

78. Discharge Papers, HUB Box 25, f. 1.

CHAPTER 6: RECONNAISSANCE

1. See Helen Wright, *Explorer of the Universe: A Biography of George Ellery Hale* (New York: E. P Dutton & Co., 1966), 318–21; W.S.A., "Early Days at Mt. Wilson," *PASP*, 59, pt. 2, no. 351 (December 1947), 300–1; 100-inch Telescope Log Book, no. 1, MWOA, 7.

2. Alfred Noyes, *Watchers of the Skies* (New York: Frederick H. Stokes Company, 1922), 3.

3. W.S.A., "Early Days at Mount Wilson," *PASP*, 59, pt. 1, no. 350 (October 1947), 227.

4. Quoted in W.S.A., "Early Days at Mount Wilson," pt. 2, 296.

5. W.S.A. Papers, MWOA, Supplement, Box 5, f. 5.136, 3.

6. 60-inch Telescope Log Book, no. 2, Box 1, MWOA, Sept. 30, 1915.

7. W.S.A. Papers, MWOA, Supplement, Box 5, f. 5.154, unp.

8. 100-inch Telescope Log Book, no. 1, 7.

9. Noyes, *Watchers of the Skies*, 3–4.

10. Interview with Charles Donald Shane by B.S., AIP, Feb. 11, 1977, 1. Interview with Charles Donald Shane by C.A.W., undated (courtesy of C.A.W.).

11. 60-inch Telescope Log Book, no. 1, inside of front cover.

12. M.L.H., "Edwin Hubble," *Monthly Notices of the Royal Astronomical Society*, 114, no. 3, (1954), 291.

13. 100-inch Telescope Log Book, no. 1, Sept. 22, 1919.

14. 60-inch Telescope Log Book, no. 1, Dec. 25, 1912.

15. Log Book, 100-inch Reflector, HUB, Box 29, 36. This is E.P.H.'s personal log.

16. G.B.H., "E.P.H.: The Astronomer," HUB 82(7), Box 7, 6. This account is based on G.B.H.'s interviews with M.L.H.

17. Ibid., 4.

18. 60-inch Telescope Log Book, no. 1, Mar. 25, 1913.

19. E.P.H. to E.B.F., Nov. 28, 1920, YOA, Box 80, f. F4.

20. W.S.A., "Autobiographical Notes," W.S.A. Papers, MWOA, 3.

21. H.S., "Walter S. Adams: A Master of Stellar Spectra," *Sky and Telescope* (July 1956), 401.

22. H.S., *Through Rugged Ways to the Stars* (New York: Charles Scribner's Sons, 1969), 57.

23. Harold J. Ryan to W.S.A., Nov. 12, 1924, W.S.A. Papers, MWOA, Box 61, f. 61.1076.

24. H.S., *Through Rugged Ways to the Stars*, 51.

25. Ibid., 68.

26. Ibid., 5.

27. Ibid., 13.

28. Ibid., 23.

29. Bart J. Bok, "Harlow Shapley," *Biographical Memoirs of the National Academy of Sciences*, 49 (Washington, D.C.: National Academy of Sciences, 1978), 243.

30. H.S., *Through Rugged Ways to the Stars*, 36.

31. Ibid., 7.

32. Ibid., 57.

33. G.B.H., "The Missing Years, 1919–1926," HUB 82(6), Box 7, 2.

34. G.B.H., "E.P.H.: The Astronomer," 1a.

35. *The Springfield Leader*, Feb. 28, 1920, 8.
36. Interview with E.H., Mar. 18, 1992.
37. E.P.H. to E.B.F., Nov. 3, 1919, YOA, Box 77, f. 1.
38. Conference on 100-inch Program, Apr. 8, 1920, MWOA, W.S.A. Papers, Supplement, Box 5, f. 5.137.
39. G.B.H., "E.P.H.: The Astronomer," 1–2.
40. Log Book, 100-inch Reflector, HUB, Box 29, 36–44.

CHAPTER 7: THE COSMIC ARCHIPELAGO

1. *Discoveries and Opinions of Galileo*, ed. and trans. by Stillman Drake (Garden City, NY: Doubleday and Company, 1957), 49.
2. Kenneth Glyn Jones, "The Observational Basis for Kant's Cosmogony: A Critical Analysis," *Journal for the History of Astronomy*, 2, pt. 1, no. 3 (February 1971), 33.
3. *Kant's Cosmogony*, ed. and trans. by W. Hastie (Glasgow: James Maclehose and Sons, 1900), 135.
4. William Herschel, "On the Construction of the Heavens," *Philosophical Transactions*, 75 (1785), 260.
5. H.S., *Through Rugged Ways to the Stars* (New York: Charles Scribner's Sons, 1969), 77–78; interview with H.S. by Charles Weiner and Helen Wright, June 8, 1966, AIP, 64, 67.
6. Quoted in M. A. Hoskin, "The 'Great Debate': What Really Happened," *Journal for the History of Astronomy*, 7 (1976), 170.
7. Ibid., 173.
8. H.S., *Through Rugged Ways to the Stars*, 82.
9. Quoted in Robert Smith, *The Expanding Universe: Astronomy's "Great Debate" 1900–1931* (Cambridge, Eng.: Cambridge University Press, 1982), 79.
10. Hoskin, "The 'Great Debate,' " 173.
11. H.S., *Through Rugged Ways to the Stars*, 41.
12. Owen Gingerich and Barbara Welther, "Harlow Shapley and Cepheids," *Sky and Telescope*, 70, no. 6 (December 1985), 540; C.A.W., *The Discovery of Our Galaxy* (New York: Alfred A. Knopf, 1971), 205.
13. Gingerich and Welther, "Harlow Shapley and Cepheids," 541.
14. Quoted in Smith, *The Expanding Universe*, 61.
15. Ibid., 65.
16. Quoted in Michael Hoskin, "Shapley's Debate," in *The Harlow Shapley Symposium of Globular Cluster Systems in Galaxies*, ed. by Jonathan F. Grindlay and A. G. Davis Philip (Dordrecht/Boston/London: Kluwer Academic Publishers, 1988), 5.
17. Hoskin, "The 'Great Debate,' " 178. The text of H.S.'s paper and Curtis's slide remarks are appended to this article.
18. Ibid., 179.
19. Smith, *The Expanding Universe*, 86. Also see Otto Struve, "A Historic Debate About the Universe," *Sky and Telescope* (May 1960), 398–401.
20. Quoted in Hoskin, "The 'Great Debate,' " 174.
21. H.S., *Through Rugged Ways to the Stars*, 79.
22. Interview with H.S. by Weiner and Wright, 65.
23. Quoted in Hoskin, "Shapley's Debate," 8.

24. Quoted in Owen Gingerich, "Harlow Shapley," *Dictionary of Scientific Biography*, 12 (New York: Charles Scribner's Sons, 1975), 348.
25. H.S. to G.E.H., Sept. 22, 1920, G.E.H., MWODF: 1904–1923, Box 168, f. H.S.
26. 60-inch Telescope Log Book, no. 5, Box 1, MWOA, Mar. 15, 1921.
27. Interview with A.S., June 11, 1991. M.L.H. first told this story to A.S. in 1956 and stood by it until his death in 1972.

CHAPTER 8: UNCHARTED WATERS

1. William A. Stuart to E.P.H., May 28, 1921 (courtesy of C.A.W.).
2. E.P.H. to Edward E. Barnard, c. Spring 1920, A. J. Dyer Observatory.
3. E.P.H. to William H. Wright, Apr. 11, 1921, SALO, Hubble, Edwin P., 1921–1949, f. 741.
4. Wright to E.P.H., Apr. 23, 1921, W.S.A. Papers, MWOA, Box 34, f. 34.581.
5. E.P.H., "A General Study of Diffuse Galactic Nebulae," *APJ*, 56 (1922), 162–99.
6. E.P.H., "The Source of Luminosity in Galactic Nebulae," *APJ*, 56 (1922), 400–38.
7. H.S. to G.E.H., Oct. 31, 1921, W.S.A. Papers, MWOA, Box 61, f. 61.1075.
8. Knut Lundmark, "The Spiral Nebula Messier 33," *PASP*, 33 (1921), 324–27.
9. H.S. to Adriaan van Maanen, Jan. 14, 1922, HCOA, 1921–1930, LUP-MER, Box 12, UAV 630.22, f. 94.
10. Knut Lundmark, "On the Motions of Spirals," *PASP*, 34 (1922), 108–15.
11. H.S. to Adriaan van Maanen, June 19, 1922, HCOA, 1921–1930, LUP-MER, Box 12, UAV 630.22, f. 94.
12. Ibid., May 16, 1923.
13. *NYT*, Jan. 19, 1921.
14. Joel Stebbins to G.E.H., Oct. 1, 1923, AIP, AAS Correspondence, Box 8, 1923.
15. Log Book, 100-inch Reflector, HUB, Box 29, 156.
16. Ibid.
17. E.P.H. to H.S., Feb. 19, 1924, HUB, Box 15, f. 611.
18. *Cecilia Payne-Gaposchkin: An Autobiography and Other Recollections*, ed. by Katherine Haramundanis (Cambridge, Eng.: Cambridge University Press, 1984), 209.
19. H.S. to E.P.H., Feb. 27, 1923, HCOA, 1921–1930, HEP-I, Box 9, UAV 630.22, f. 71.
20. E.P.H. to H.S., Aug. 25, 1924, HCOA.
21. H.S. to E.P.H., Sept. 5, 1924, HCOA.
22. Ibid., Aug. 3, 1923.
23. Henry Norris Russell to E.P.H., Dec. 12, 1924, HUB, Box 19, f. 919.
24. Joel Stebbins to E.P.H., Feb. 16, 1925, HUB, Box 18, f. 956; Joel Stebbins Memoir, "Edwin Hubble and the A.A.S. Prize," SALO, N.U.M. Papers, f. Hubble Biographical Memoir, NAS.
25. Stebbins, "Edwin Hubble and the A.A.S. Prize."
26. HUB, Box 17, f. 753.
27. Joel Stebbins to W.S.A., Feb. 21, 1925, MWOA, W.S.A. Papers, Box 35, f. 35.605.
28. E.P.H. to Henry Norris Russell, Feb. 19, 1925, Henry Norris Russell Papers, PUL, Box 44, f. 56.
29. E.P.H. to Joel Stebbins, Mar. 6, 1925, SALO, Hubble, Edwin P., 1921–1949, f. 741.
30. Adriaan van Maanen to H.S., Feb. 18, 1925, HCOA, 1921–1930, HEP-I, Box 12, UAV 630.22, f. 94, LUP-MER.

31. H.S. to van Maanen, Mar. 8, 1925, HCOA.
32. Interview with H.S. by Charles Weiner and Helen Wright, June 8, 1966, AIP, 69.
33. Interview with Margaret Harwood by C.A.W., Mar. 26, 1971 (courtesy of C.A.W.).
34. William H. Wright to John P. Burke, c. Sept. 1923, SALO, William Hammond Wright Papers, 1923, f. 5.
35. Interview with Margaret Harwood by C.A.W.
36. "E.H. by Susan Ertz," HUB, Box 1, f. 3.
37. Interview with I.C., July 6, 1991.
38. Standard Certificate of Death, Earl R. Leib, June 15, 1921, no. 21, Amador County Courthouse, Jackson, CA.
39. *San Jose Mercury-Herald,* June 17, 1921, 1.
40. Ibid., June 19, 1921, 9.
41. G.B.H., "Library," HUB 82(12), Box 8, 2.
42. G.B.H., "E.P.H.: The Astronomer," HUB 82(7), Box 7, 10.
43. G.B.H., "Daily Events at Home, 1939–1940," HUB 77, Box 5, June 23, 1939.
44. G.B.H., "Diary of Daily Events at Home in San Marino, California, From 1937–1939," HUB 76, Box 5, May 7, 1938.
45. Ibid., July 1, 1938.
46. Interview with I.C.
47. Standard Certificate of Marriage, Feb. 28, 1924, Los Angeles, CA; G.B.H., "E.P.H.: The Missing Years, 1919–1926," HUB 82(6), Box 7, 5–6; G.B.H. to Mrs. J. D. Burke, Feb. 28, 1924 (private collection).
48. G.B.H. to Mrs. J. D. Burke, Mar. 18, 21, 29, 1924; EPH to Mrs. J. D. Burke, Mar. 29, 1924 (private collection).
49. G.B.H. to Mrs. J. D. Burke, Apr. 5, 1924 (private collection).
50. G.B.H., "E.P.H.: The Missing Years, 1919–1926," 6.
51. Ibid., 7.
52. Ibid., 9.
53. Ibid., 11.
54. G.E.H. to J.C.M., Mar. 29, 1923, CIW, Mount Wilson Papers, f. General 1902–1930, no. 2.
55. J.C.M. to W.S.A., May 8, 1923, CIW.
56. Quoted in Robert Smith, *The Expanding Universe: Astronomy's "Great Debate" 1900–1931* (Cambridge, Eng.: Cambridge University Press, 1982), 77.
57. G.B.H., "E.P.H.: The Missing Years, 1919–1926," 3.
58. A. Vibert Douglas, "Arthur Stanley Eddington," *Dictionary of Scientific Biography,* 4 (New York: Charles Scribner's Sons, 1976), 281.
59. G.B.H., "E.P.H.: Some People," HUB 82(17), Box 8, 5.
60. Ibid., 1.
61. Ibid., 1, 3.
62. E.P.H. to V.M.S., July 24, 1923, HUB, Box 15, f. 620.
63. E.P.H., "The Classification of Nebulae, 1923," SALO, Hubble, Edwin, P., 1921–1949, f. 741, 14.
64. Ibid., 15.
65. Quoted in R. Hart and R. Berendzen, "Hubble's Classification of Non-Galactic Nebulae, 1922–1926," *Journal for the History of Astronomy,* 2, pt. 2, no. 4 (June 1971), 116.
66. E.P.H., *The Realm of the Nebulae* (New Haven: Yale University Press, 1936), 45.

67. E.P.H. to V.M.S., Feb. 9, 1924, LOA, G6.
68. Heber D. Curtis to V.M.S., June 3, 1924, LOA, G29.
69. H. Knox-Shaw to V.M.S., July 18, 1924, LOA, G35.
70. J. L. E. Dreyer to V.M.S., Aug. 11, 1924, LOA, G39.
71. Solon I. Bailey to V.M.S., Oct. 8, 1924, LOA, G42.
72. Ibid.
73. V.M.S.'s annotated copy of E.P.H.'s "The Classification of Nebulae," LOA, G1, 24.
74. H.S. to E.P.H., Oct. 8, 1924, HCOA, 1921–1930, HEP-I, Box 9, UAV 630.22, f. 7. Also see E.P.H. to H.S., Nov. 2, 1928, and H.S. to E.P.H., May 29, 1929, HCOA.
75. *Transactions of the International Astronomical Union*, 2 (1925), 221, 240.
76. Knut Lundmark, "A Preliminary Classification of Nebulae," *Arkiv for Matematik Astronomi och Fysik*, Band 19B, no. 8 (1926).
77. E.P.H. to V.M.S., June 22, 1926, LOA, G7.
78. E.P.H. to Knut Lundmark, Aug. 26, 1926, quoted in Smith, *The Expanding Universe*, 152.
79. E.P.H. to H.S., Apr. 29, 1927, HCOA, 1921–1930, HEP-I, Box 9, UAV 630.22, f. 71.
80. H.S. to Henry Norris Russell, Aug. 15, 1927, quoted in Smith, *The Expanding Universe*, 152.
81. H.S. to W.S.A., Dec. 2, 1927, MWOA, W.S.A. Papers, Box 35, f. 35.608.
82. H.S., "On the Classification of Extra-Galactic Nebulae," *Harvard College Observatory Bulletin*, no. 849 (Aug. 1, 1927).
83. E.P.H., "Extra-Galactic Nebulae," *APJ*, 64 (1926), 324.
84. Ibid., 323.
85. Walter Baade, *Evolution of the Stars and Galaxies* (Cambridge, MA: Harvard University Press, 1963), 18.

CHAPTER 9: MARINER OF THE NEBULAE

1. G.B.H., "E.P.H.: The Missing Years, 1919–1926," HUB 82(6), Box 7, 10–11.
2. W.S.A. Papers, MWOA, Box 10, f. 10.165.
3. I.B. to Paul A. Scherer, Nov. 4, 1953, CIW, Staff Files, f. Hubble, Edwin P.
4. G.B.H., "E.P.H.: The Missing Years, 1919–1926," 11.
5. Interview with I.C., July 6, 1991.
6. William H. Wright to the Hubbles, Feb. 15, 1927, AIP, Microfilm, 37732.
7. G.B.H., "E.P.H.: The Missing Years, 1919–1926," 11.
8. G.B.H., "E.P.H.: The Astronomer," HUB 82(7), Box 7, 2.
9. Ibid., 1.
10. G.B.H., "E.P.H.: 'High Sierra' and 'Grand Canyon,'" HUB 82(10), Box 8, unp.
11. G.B.H., "E.P.H.: The Astronomer," 2–3.
12. Ibid., 3; G.B.H., "E.P.H.: Characteristics," HUB 82(9), Box 7, 23.
13. G.B.H., "E.P.H.: Characteristics," 2.
14. G.B.H., "Daily Events at Home in San Marino, 1939–40," HUB 77, Box 5, Apr. 1, 1939.
15. G.B.H., "E.P.H.: The Astronomer," 4.
16. Ibid., 1.
17. 60-inch Telescope Log Book, no. 4, MWOA, Jan. 19, 1919.
18. *Los Angeles Times Sunday Magazine*, Dec. 3, 1933, 4.

19. Quoted in N.U.M., "Milton L. Humason—Some Personal Recollections," *Mercury*, 2 (January/February 1973), 5; CIW, Mount Wilson Observatory, f. Actuarial File, Jan. 16, 1918; CIW, f. Biographical Data of Staff Members, H–J; MWOA, W.S.A. Papers, Box 94 (W), L.1804.

20. Interview with H.S. by Charles Weiner and Helen Wright, June 8, 1966, AIP, 49, 63.

21. N.U.M., "Milton L. Humason," 6.

22. *Los Angeles Examiner*, Oct. 28, 1928, 2.

23. G.B.H., "E.P.H.: Travel Diary Between 1928 and 1941," HUB 72, Box 4, Feb. 20–21, 1928.

24. David White to E.P.H., Apr. 27, 1927, HUB, Box 18, f. 847; H. H. Turner to E.P.H., June 9, 1928, HUB, Box 19, f. 905.

25. G.B.H., "E.P.H.: The Astronomer," unp.

26. G.B.H., "E.P.H.: Quotations," HUB 82(13), Box 8, 3.

27. Ibid.

28. Interview with M.L.H. by B.S., c. 1965, AIP, 1.

29. N.U.M., "E.P.H.: Observational Cosmologist," *Sky and Telescope* (January 1954), 80.

30. Interview with M.L.H. by B.S., 1–2.

31. E.P.H., "The Law of Red-Shifts," *Monthly Notices of the Royal Astronomical Society*, 113, no. 6 (1953), 659; interview with M.L.H. by B.S., 2.

32. Interview with M.L.H. by B.S., 1–2.

33. Ibid., 1.

34. E.P.H. to H.S., May 15, 1929, HCOA, 1921–1930, HEP-I, Box 9, UAV 630.22, f. 71.

35. Knut Lundmark to W.S.A., Dec. 13, 1928, MWOA, W.S.A. Papers, Box 40, f. 40.704.

36. W.S.A. to Lundmark (cable), Jan. 4, 1929, MWOA.

37. G. J. Whitrow, "Edwin Hubble," *Dictionary of Scientific Biography*, 6 (New York: Charles Scribner's Sons, 1972), 531.

38. E.P.H., "A Relation Between Distance and Radial Velocity Among Extra-Galactic Nebulae," *Proceedings of the National Academy of Sciences*, 15, no. 3 (March 1929), 173.

39. Interview with M.L.H. by B.S., 3–6; Timothy Ferris, *The Red Limit: The Search for the Edge of the Universe*, 2nd ed. (New York: Quill, 1983), 58–59.

40. W.S.A. to J.C.M., May 20, 1929, CIW, Mount Wilson Papers, f. General 1902–1930, no. 1.

41. E.P.H. and M.L.H., "The Velocity-Distance Relation Among Extra-Galactic Nebulae," APJ, 74, no. 1 (1931), 59.

42. Quoted in G.B.H., "E.P.H.: The Astronomer," 5.

43. Ibid., 7–8.

44. E.P.H. and M.L.H., "The Velocity-Distance Relation Among Extra-Galactic Nebulae," 73.

45. *Los Angeles Illustrated Daily News*, July 25, 1927, 3.

46. E.P.H. and M.L.H., "The Velocity-Distance Relation Among Extra-Galactic Nebulae," 77.

47. E.P.H., "A Relation Between Distance and Radial Velocity Among Extra-Galactic Nebulae," 168–73.

48. E.P.H. and M.L.H., "The Velocity-Distance Relation Among Extra-Galactic Nebulae," 76.
49. Ferris, *The Red Limit*, 56.
50. Interview with M.L.H. by B.S., 3–4.
51. G.B.H., "E.P.H.: The Missing Years, 1930–1936," HUB 82(6), Box 7, "The Athenaeum," 1.
52. *The Athenaeum*, a pamphlet, 1.
53. Interview with N.U.M. by Norriss S. Hetherington, June 3, 1976, AIP, 17.
54. Interview with Horace Babcock, July 11, 1991.
55. G.B.H., "E.P.H.: The Astronomer," 1–2; interview with N.U.M. by Hetherington, 22.
56. Interview with N.U.M. by Hetherington, 18, 21–22. Also see H.S., "Note on the Velocities and Magnitudes of External Galaxies," *Proceedings of the National Academy of Sciences*, 15 (1929), 565–70.
57. Adriaan van Maanen to H.S., Feb. 18, 1931, HCOA, 1930–1940, HEP-I, Box 42, UAV 630.22, f. 307.
58. G.B.H., "E.P.H.: The Astronomer," 2.
59. Ibid., 7.
60. Georges Lemaître, "Un univers homogène de masse constante et de rayon croissant, redant compte de la vitesse radiale des nébuleuses extra-galactiques," *Annals de la Societé Scientifique de Bruxelles*, 47 (1927), 49–56.
61. E.P.H., "A Relation Between Distance and Radial Velocity Among Extra-Galactic Nebulae," 173.
62. Quoted in Ferris, *The Red Limit*, 82.
63. Interview with G. C. McVitte by David DeVorkin, AIP, Mar. 21, 1978, 18–19.
64. Interview with N.U.M. by Hetherington, 19.

CHAPTER 10: "YOUR HUSBAND'S WORK IS BEAUTIFUL"

1. *LAT*, Dec. 28, 1930, 1.
2. *NYT*, Dec. 12, 1930, 1. Also see Ronald W. Clark, *Einstein* (New York and Cleveland: The World Publishing Company, 1971), 424–35.
3. Quoted in Clark, *Einstein*, 428.
4. Interview with Horace Babcock, July 11, 1992.
5. Quoted in Clark, *Einstein*, 434.
6. W.S.A., "Professor Einstein's Visit to Pasadena and the Mount Wilson Observatory," MWOA, W.S.A. Papers, Supplement, Box 4, f. 4.87, 4.
7. Interview with Olin C. Wilson by David DeVorkin, AIP, July 11, 1978, 31.
8. W.S.A., "Professor Einstein's Visit to Pasadena and the Mount Wilson Observatory," 5.
9. Ibid., 6–7.
10. *LAT*, Jan. 16, 1931, 21.
11. G.B.H., "E.P.H.: Some People," HUB 82(17), Box 8, 1.
12. John Barrymore to E.P.H., Feb. 14, 1931, AIP, Hubble Microfilm 37732.
13. Ibid., 2.
14. Walter B. Clausen, Press Release, Feb. 4, 1931, HUB, Box 28, f. 25. Also see *NYT*, Jan. 3, 1931, 1.
15. *The Springfield Daily News*, Feb. 5, 1931, 2.

16. *Punch*, May 27, 1931, 579.
17. G.B.H., "E.P.H.: World War II," HUB 82(5), Box 7, Sept. 27, 1941.
18. G.B.H., "E.P.H.: Some People," 2.
19. Quoted in Clark, *Einstein*, 444.
20. Ibid., 452.
21. Ibid., 439.
22. G.B.H., "E.P.H.: Quotations," HUB 82(13), Box 8, f. 1.
23. Ibid., 1–2.
24. George Arliss to E.P.H., June 18, 1932, HUB, Box 9, f. 120.
25. G.B.H., "E.P.H.: The Missing Years, 1919–1933," 1932, HUB 82(6), Box 7, 1.
26. Ibid., 3.
27. Ibid.
28. E.P.H., "The Exploration of Space," *Harper's Magazine*, 158 (May 1929), 732–38.
29. G.B.H., "E.P.H.: Lecturer," HUB 82(11), Box 8, 2–3.
30. G.B.H., "E.P.H.: The Missing Years, 1919–1933," 1930, 1.
31. "Budget for 1931," MWOA, W.S.A. Papers, Box 10, f. 10.166.
32. Henry Norris Russell to E.P.H., June 3, 1931, Henry Norris Russell Papers, PUL, Box 12, f. 69.
33. E.P.H. to Russell, June 13, 1931, Henry Norris Russell Papers, PUL, Box 44, f. 56.
34. J. Duncan Spaeth to Russell, July 1, 1931, Henry Norris Russell Papers, PUL, Box 56, f. 20.
35. G.B.H., "E.P.H.: Lecturer," The Vanneuxum [sic] Lectures, unp.
36. Henry Norris Russell to E.P.H., Oct. 31, 1932, Henry Norris Russell Papers, PUL, Box 12, f. 69.
37. G.B.H., "E.P.H.: Lecturer," 4.
38. W.S.A., "Memorandum on Salaries, 1926," and W.S.A. to J.C.M., Sept. 30, 1926, CIW, Mount Wilson Papers, f. General 1902–1930, no. 1.
39. Samuel Harden Church to E.P.H., Sept. 21, 1934, HUB Box 10, f. 225.
40. E.P.H. to Church, Oct. 1, 1934, AIP, Hubble Microfilm 37732.
41. G.B.H., "Travel Diary: Eastern United States, England, and Europe, Apr. 8 to Aug. 27, 1934," HUB 74, Box 4, May 8, 1934.
42. G.B.H., "E.P.H.: [Oxford University, 1910–1913]," HUB 82(3), Box 7, 15.
43. G.B.H., "Travel Diary," May 25, 1934.
44. "Award of Honorary Degree," HUB, Box 18, f. 869. Mount Wilson was not named in honor of the twenty-eighth President but for Benjamin Davis "Don Benito" Wilson, a famous pioneer, developer, and early mayor of Los Angeles.
45. G.B.H., "Travel Diary," June 10, 1934.
46. Ibid., July 2, 1934.
47. Ibid.
48. W.S.A. to J.C.M., Mar. 19, 1934; J.C.M. to Adams, Mar. 21, 1934; CIW, Staff Files, f. Hubble, Edwin P.
49. E.P.H. to J.C.M., Aug. 30, 1934, HUB, Box 13, f. 423.
50. J.C.M., "Memorandum," Sept. 27, 1934, CIW, Staff Files, f. Hubble, Edwin P., 6–7.
51. G.B.H., "Travel Diary," July 25, 30, 1934.
52. Interview with E.H., Mar. 18, 1992.
53. A.A.C. to C.A.W., June 16, 1971 (courtesy of C.A.W.).
54. Interview with E.H. and H.H.L. by C.A.W., 1971 (courtesy of C.A.W.)

CHAPTER 11: "ALMOST A MIRACLE"

1. G.B.H., "Diary of Daily Events at Home in San Marino, From 1930 to 1940," HUB 77, Box 5, Jan. 3, 1939.

2. G.B.H., "E.P.H.: Characteristics," HUB 82(9), Box 7, unp.

3. Interview with I.C., July 6, 1991.

4. Ibid.

5. Ibid.

6. Interview with A.S., June 11, 1991.

7. E.P.H. to Willem de Sitter, Aug. 21, 1931, HUB Box 15, f. 616.

8. E.P.H. to de Sitter, Sept. 23, 1931, HUB, f. 617.

9. *The Edwin Hubble Papers: Previously Unpublished Manuscripts on the Extragalactic Nature of Spiral Nebulae*, ed. by Norriss S. Hetherington (Tucson: Pachart Publishing House, 1990).

10. E.P.H. to H.S., Aug. 8, 1932, HCOA, 1930–1940, HEP-I, Box 38, UAV 630.22, f. 273.

11. MWOA, W.S.A. Papers, Box 34, f. 34.589.

12. Interview with Olin C. Wilson by David DeVorkin, July 11, 1978, AIP, 30–31.

13. G.B.H., "E.P.H.: The Astronomer," HUB 82(7), Box 7, 1.

14. G.B.H. to Michael A. Hoskin, March 7, 1968, HUB Box 16, f. 670.

15. W.S.A., "Mount Wilson Observatory Confidential Statement for President Merriam," Aug. 15, 1935, MWOA, W.S.A. Papers, Box 34, f. 34.581.

16. Ibid.

17. Frederick Seares to G.E.H., Jan. 24, 1935, in Daniel J. Kevles, ed. *The George Ellery Hale Papers, 1882–1937*, microfilm ed. (The Carnegie Institution of Washington, D.C., 1971), roll 33, frame 119.

18. W.S.A., "Mount Wilson Observatory Confidential Statement for President Merriam."

19. Ibid.

20. E.P.H., "Angular Rotations of Spiral Nebulae," *APJ*, 81 (1935), 334–35.

21. Hetherington, *The Edwin Hubble Papers: Previously Unpublished Manuscripts*, 73.

22. Adriaan van Maanen, "Internal Motions in Spiral Nebulae," *APJ*, 81 (1935), 336–37.

23. W.S.A., "Mount Wilson Observatory Confidential Statement for President Merriam."

24. G.B.H., "E.P.H.: The Astronomer," 1.

25. G.B.H. to Michael A. Hoskin, Mar. 7, 1968, HUB Box 16, f. 670.

26. Interview with Martin Schwarzschild by S.W., July 30, 1975, AIP, 1.

27. Interview with Bart J. Bok by David DeVorkin, May 15, 1978, AIP, 63.

28. *Cecilia Payne-Gaposchkin: An Autobiography and Other Recollections*, ed. by Katherine Haramundanis (Cambridge, Eng.: Cambridge University Press, 1984), 184. ·

29. Interview with H. H. Plaskett by David DeVorkin, Mar. 29, 1978, AIP, 16.

30. E.P.H. to J. H. Moore, Oct. 23, 1932, HUB Box 14, f. 567.

31. Ibid.

32. *There Was Light: Autobiography of a University*, ed. by Irving Stone (New York: Doubleday and Company, 1970), 113.

33. E.P.H., "The Distribution of Extra-Galactic Nebulae," *APJ*, 79, no. 1 (1934), 8–76.

34. "Annual Report of the Director of the Mount Wilson Observatory, 1933–1934" (Carnegie Institution of Washington, D.C., 1934), 130.
35. N.U.M. to E.P.H., Feb. 6, 1934, HUB Box 18, f. 810.
36. E.P.H. to N.U.M., Feb. 23, 1934, HUB, Box 14, f. 516.
37. Interview with I.C., July 8, 1991.
38. G.B.H., "Diary of Daily Events at Home in San Marino: 1940," HUB 77, Box 5, Jan. 18, 1940.
39. M.L.H to N.U.M., Aug. 9, 1935, SALO, N.U.M. Papers, f. Humason, M., 1933–1948.
40. Ibid., Jan. 17, 1936.
41. Ibid. Also see Mar. 26, 1936.
42. Quoted in Henry C. King, *The History of the Telescope* (New York: Dover Publications, 1979), 387. In addition to King (Chapter 18), the subsequent discussion of the making of the 200-inch telescope owes much to "Facts About the Palomar Observatory and 200-inch Telescope, 1950," CIW, Mt. Wilson Papers, General 1949–1968, no. 1, and to G.E.H., "The Astrophysical Observatory of the California Institute of Technology, *APJ*, 82, no. 2 (September 1935), 111–39.
43. Interview with M.L.H. by B.S., c. 1965, AIP, 9.
44. M.L.H. to N.U.M., May 8, 1934, SALO, N.U.M. Papers, f. Humason, M., 1933–1948.
45. "Construction Work on Palomar Mt., Calendar Year 1936," MWOA, W.S.A. Papers, Box 67, f. 67.1197.
46. Interview with I.B. by Charles Weiner, Aug. 9, 1968, 24.
47. G.B.H., "E.P.H.: Lecturer," HUB 82(11), Box 8, unp.; G.B.H., "Diary of Daily Events at Home in San Marino, California, From 1937 to 1939," HUB 76, Box 5, Oct. 8, 1938.
48. W.S.A. "Confidential Report for Dr. Merriam," July 1936, MWOA, W.S.A. Papers, Box 67, f. 67.1199.
49. M.L.H. to N.U.M., Feb. 20, 1937, SALO, N.U.M. Papers, f. Humason, M., 1933–1948.
50. Ibid., June 22, 1937.
51. G.B.H., "E.P.H.: Library," HUB 82(12), Box 8, unp.
52. Interview with Anne Crotty, June 1991.
53. G.B.H., "E.P.H.: The Missing Years, 1919–1933," HUB 82(6), Box 7, 1.
54. HUB Box 20, f. 1027
55. G.B.H. to Mrs. Clarendon Eyer, Oct. 21, 1935 (private collection).
56. G.B.H.: "E.P.H.: The Astronomer," 5.
57. E.P.H. to N.U.M., Mar. 8, 1937, HUB Box 14, f. 523.
58. E.P.H., *The Realm of the Nebulae* (New Haven: Yale University Press, 1936), 56–57.
59. Ibid., 28.
60. Ibid., 120.
61. Ibid., 119–20.
62. Ibid., 182.
63. Ibid., 201–2.

CHAPTER 12: "NOW WHOM DO WE WANT TO MEET?"

1. G.B.H., "E.P.H.: Some People," HUB 82(17), Box 8, f. 6.
2. G.B.H., "E.P.H.: The Astronomer, The Barnard Medal," HUB 82(7), Box 7, 1–2.
3. Quoted in ibid., 2.
4. G.B.H., "Diary of Daily Events at Home in San Marino, From 1929 to 1937," HUB 73, Box 4, June 8, 11, 1937; HUB Box 18, f. 889.
5. G.B.H., "Diary of Daily Events at Home in San Marino, From 1929 to 1937," Feb. 13, 1937.
6. Ibid., "Special Data."
7. G.B.H. to Mrs. John P. Hubble, Oct. 1, 1936 (private collection).
8. G.B.H., "Diary of Daily Events at Home in San Marino, California, From 1937 to 1939," HUB 76, Box 5, Mar. 24, 1938.
9. G.B.H. "Travel Diary for Trips to England and the Eastern United States from 1936 to 1937," HUB 75, Box 5, 52–53, 85.
10. Ibid., 80, 97.
11. Ibid., 98.
12. G.B.H., "Edwin Powell Hubble: Lecturer," HUB 82(11), Box 8, 4.
13. E.P.H. to J.C.M., Sept. 29, 1935, CIW Staff Files, f. Hubble, Edwin P.; W.S.A. to J.C.M., Sept. 30, 1935; J.C.M. to E.P.H., Oct. 2, 1935, MWOA, W.S.A. Papers, Box 34, f. 34.581.
14. E.P.H. to J.C.M., Nov. 7, 1935, MWOA, W.S.A. Papers, Box 34, f. 34.581.
15. W.S.A. to J.C.M., Nov. 22, 1935, MWOA.
16. J.C.M., "Memorandum Relating to Requested Leave of Absence for Edwin Hubble," Jan. 21, 1936, CIW Staff Files, f. Hubble, Edwin P.
17. W.S.A. to J.C.M., Feb. 19, 1936, MWOA, W.S.A. Papers, Box 34, f. 34.582.
18. Ibid., Feb. 27, 1936.
19. Ibid., Mar. 9, 1936.
20. E.P.H. to J.C.M., Sept. 5, 1936, CIW Staff Files, f. Hubble, Edwin P.; J.C.M. to E.P.H., Nov. 6, 1936, MWOA, W.S.A. Papers, Box 34, f. 34.582.
21. G.B.H., "Travel Diary for Trips to England and the Eastern United States from 1936 to 1937," 133.
22. Ibid., 143.
23. Ibid., 142–43.
24. Ibid., 153–54.
25. Ibid., 155–56.
26. Ibid., 188.
27. E.P.H. to J.C.M., Dec. 10, 1936, CIW Staff Files, f. Hubble, Edwin P.
28. G.B.H., "Diary of Daily Events at Home in San Marino, From 1929 to 1937," Mar. 4, 1937.
29. Ibid., Aug. 4–5, 1937.
30. Ibid., Aug. 4, 1937.
31. HUB Box 8, f. 86, Anita Loos.
32. Anita Loos to G.B.H., Apr. 29, 1936, HUB Box 17, f. 759.
33. Helen Hayes with Sanford Doty, On Reflection (New York: M. Evans and Company, 1968), 253.
34. Ibid.

35. Gary Carey, *Anita Loos: A Biography* (London: Bloomsbury Publishing Ltd., 1988), 175.
36. Anita Loos, *Kiss Hollywood Goodbye* (New York: The Viking Press, 1974), 148.
37. G.B.H., "Diary of Daily Events at Home in San Marino, From 1929 to 1937," Oct. 17, 1937.
38. G.B.H., "Diary of Daily Events at Home in San Marino, California, From 1937 to 1939," Mar. 13, 1938.
39. G.B.H., "Diary of Daily Events at Home in San Marino, California, From 1939 to 1940," HUB 77, Box 5, June 7, 1939.
40. G.B.H., "Travel Diary for Trips to England and the Eastern United States from 1936 to 1937," 109.
41. G.B.H., "Diary of Daily Events at Home in San Marino Between 1928 and 1941," HUB 72, Box 4, Dec. 11, 1940.
42. G.B.H., "Diary of Daily Events at Home in San Marino, California, From 1937 to 1939," Mar. 4, 1938.
43. Ibid., May 11, 1938.
44. Ibid., Apr. 1, May 11, 1938.
45. G.B.H., "Diary of Daily Events at Home in San Marino, California, From 1939 to 1940," July 30, 1939.
46. G.B.H., "Diary of Daily Events at Home in San Marino, California, From 1937 to 1939," Nov. 30, 1938.
47. Ibid., Jan. 27, 1938.
48. Ibid., Feb. 23, 1938; Helen Wright, *Explorer of the Universe: A Biography of George Ellery Hale* (New York: E. P. Dutton & Co., 1966), 426–29.
49. *NYT*, Feb. 23, 1938, 22.
50. "E.H.," by Susan Ertz, HUB Box 1, f. 3.
51. G.B.H., "Diary of Daily Events at Home in San Marino Between 1928 and 1941," Oct. 21, 1940.
52. *LAT*, June 5, 1938; HUB Box 28, Scrapbook, 52.
53. G.B.H., "Diary of Daily Events at Home in San Marino, California, From 1937 to 1939," Mar. 22, 1938.
54. Ibid., Mar. 24, 1938.
55. G.B.H., "Diary of Daily Events at Home in San Marino, California, From 1939 to 1940," June 8, 1939.
56. G.B.H., "Diary of Daily Events at Home in San Marino, California, From 1937 to 1939," Mar. 27, Apr. 11, 1938.
57. Ibid., Nov. 6, 1938.
58. G.B.H., "E.P.H.: Characteristics," HUB 82(9), Box 7, 4.
59. G.B.H. "E.P.H.: Some People," 5–6.
60. G.B.H., "Travel Diary for Trips to England and the Eastern United States from 1936 to 1937," Oct. 12, 1936.
61. G.B.H., "Diary of Daily Events at Home in San Marino, California, From 1939 to 1940," Dec. 22, 1939.
62. G.B.H. "Diary of Daily Events at Home in San Marino Between 1928 and 1941," Nov. 19, 1940.
63. Ibid., Nov. 14, 1940.
64. G.B.H., "E.P.H.: Characteristics," 23.

65. G.B.H., "Diary of Daily Events at Home in San Marino, California, From 1939 to 1940," Dec. 21, 1939.
66. G.B.H., "E.P.H.: Characteristics," 22–23.
67. G.B.H., "Diary of Daily Events at Home in San Marino, California, From 1937 to 1939," June 10, 1938.
68. G.B.H., "Diary of Daily Events at Home in San Marino, California, From 1939 to 1940," Feb. 21, 1939.
69. Sybille Bedford, *Aldous Huxley: A Biography* (New York: Alfred A. Knopf/Harper & Row, 1974), 367–68. Also see David King Dunaway, *Huxley in Hollywood* (New York: Harper & Row, 1989).
70. G.B.H., "Diary of Daily Events at Home in San Marino, California, From 1939 to 1940," Jan. 16, 1939.

CHAPTER 13: LANDLOCKED

1. G.B.H., "Diary of Daily Events at Home in San Marino, California, From 1939 to 1940," HUB 77, Box 5, June 27, 1940.
2. Ibid., Aug. 23, 1939.
3. E.P.H., "Aid to Britain," HUB Box 1, f. 9; *LAT*, Oct. 17, 1940, sec. 1, 5; G.B.H., "Diary of Daily Events at Home in San Marino Between 1928 and 1941," HUB 72, Box 4, Oct. 16, 1940.
4. G.B.H., "Diary of Daily Events at Home in San Marino Between 1928 and 1941," Dec. 12, 1940.
5. Ibid., Dec. 11, 12, 1940.
6. *NYT*, May 18, 1939, 52; May 21, 1939, sec. 4, 8.
7. Arthur Eddington to E.P.H., Jan. 17, 1940, HUB Box 11, f. 258.
8. G.B.H., "Diary of Daily Events at Home in San Marino, California, From 1939 to 1940," Mar. 6, 1940.
9. G.B.H., "Diary of Daily Events at Home in San Marino Between 1928 and 1941," Dec. 10, 1940; *NYT*, May 18, 1939, 52.
10. G.B.H., "Biographical Memoir of E.P.H.: World War II," HUB 82(5), Box 7, 4.
11. G.B.H., "Diary of Daily Events at Home in San Marino Between 1928 and 1941," Dec. 16, 1940.
12. Ibid., Dec. 22, 1940.
13. G.B.H., "E.P.H.: The Missing Years, 1940–1941," HUB 82(6), Box 7, 1–2.
14. G.B.H., "Diary of Daily Events at Home in San Marino, California, From 1939 to 1940," July 3, 1940.
15. G.B.H., "Diary of Daily Events at Home in San Marino Between 1928 and 1941," Jan. 25, 1941.
16. *London Front: Letters Written to America, 1939–1940*, ed. by H. M. Harwood and F. Tennyson Jesse (New York: Doubleday, Doran and Company, 1941), 402–3. The Woollcott letters, which Max sent to G.B.H. at her request while she was in Maryland with E.P.H. during the war, have since disappeared.
17. G.B.H., "Diary of Daily Events at Home in San Marino, California, August 8, 1941, to September 15, 1942," HUB 78, Box 6, Aug. 28, 1942.
18. G.B.H., "Diary of Daily Events at Home in San Marino Between 1928 and 1941," Feb. 10–11, 1941.
19. Ibid., July 1, 22–23, 1941.

20. G.B.H., "Diary of Daily Events at Home in San Marino, California, August 8, 1941, to September 15, 1942," Oct. 23, 1941.

21. E.P.H., *The Nature of Science and Other Lectures* (San Marino: The Huntington Library, 1954), 62.

22. G.B.H., "Diary of Daily Events at Home in San Marino, California, August 8, 1941, to September 15, 1942," Nov. 11, 1941.

23. HUB Box 9, f. 104.

24. G.B.H., "Biographical Memoir of E.P.H.: World War II," 8–8a.

25. G.B.H., "Diary of Daily Events at Home in San Marino, California, August 8, 1941, to September 15, 1942," Dec. 7–9, 1941.

26. G.B.H., "Biographical Memoir of E.P.H.: World War II," 9–10.

27. G.B.H., "Diary of Daily Events at Home in San Marino, California, August 8, 1941, to September 15, 1942," Dec. 10, 1941.

28. Quoted in ibid., Mar. 3, 1942.

29. Ibid., June 23, 1942.

30. John D. Rockefeller III to E.P.H., Apr. 20, 1942, HUB Box 19, f. 894.

31. G.B.H., "E.P.H.: Characteristics," HUB 82(9), Box 7, unp.

32. G.B.H., "Diary of Daily Events at Home in San Marino, California, August 8, 1941, to September 15, 1942," June 21, 1942.

33. G.B.H., "Biographical Memoir of E.P.H.: World War II," 14.

34. E.P.H., *The Nature of Science and Other Lectures*, 64; G.B.H., "Biographical Memoir of E.P.H.: World War II," 16.

35. E.P.H. to G.B.H., Aug. 10, 1942, HUB Box 13, f. 463.

36. G.B.H., "Biographical Memoir of E.P.H.: World War II," 7.

37. E.P.H. to G.B.H., Dec. 25, 1942, HUB Box 13, f. 477.

38. Ibid., Mar. 24, 1943, f. 484.

39. G.B.H., "Diary of Daily Events at Home in San Marino, California, August 8, 1941, to September 15, 1942," Aug. 14–15, 1942.

40. Ibid., Sept. 6, 1942.

41. G.B.H., "Biographical Memoir of E.P.H.: World War II," 1–7.

42. Ibid., 4.

43. G.B.H., "E.P.H.: The Missing Years, 1940–1941," Feb. 1, 1941.

44. E.P.H., "The Direction of Rotation in Spiral Nebulae," *APJ*, 97 (1943), 112–18.

45. Interview with Jesse Greenstein by S.W., AIP, Apr. 7, 1977, 92.

46. Ibid., 94.

47. W.S.A. to V.B., Apr. 6, 1942; V.B. to W.S.A., Apr. 8, 1942, CIW, Mount Wilson Papers, f. General 1931–1948, no. 1, CIW Staff Files, f. Baade, Walter.

48. Walter Baade, "The Resolution of Messier 32, NGC 205, and the Central Region of the Andromeda Nebula," *APJ*, 100 (1944), 137–46.

49. HUB Box 25, f. 1; E.P.H. to G.B.H., Aug. 12, 1942, HUB Box 13, f. 463.

50. Jacob Larsen to C.A.W., Sept. 28, 1973 (courtesy of C.A.W.).

51. E.P.H. to G.B.H., May 17, 23, 1943, HUB Box 13, fs. 487, 488.

52. Maria Huxley to G.B.H., May 13, 1943, HUB Box 16, f. 691.

53. G.B.H., "Biographical Memoir of E.P.H.: World War II," July 8, Sept. 2, 1943.

54. Ibid. "The Island I," 3.

55. Ibid., 8, 1.

56. Quoted in ibid., 2.

57. H.S. to W.S.A., Jan. 13, 1944; W.S.A. to H.S., Jan. 19, 1944, MWOA, W.S.A. Papers, Box 61, f. 61.1084.
58. G.B.H., "Diary of Daily Events at Home in San Marino Between 1928 and 1941," Sept. 1, 1938.
59. Gibson Reaves to author, June 29, 1993.
60. Staff of the Ballistic Research Laboratory, "Some Recent Advances in Ballistics," *Journal of Applied Physics*, 10, no. 12 (December 1945), 773–80.
61. G.B.H., "Biographical Memoir of E.P.H.: World War II," June 7, 1944.
62. Ibid., Feb. 8, 1944.
63. Aldous Huxley to G.B.H., May 19, 1944, HUB 16, f. 675.
64. G.B.H., "Biographical Memoir of E.P.H.: World War II," Oct. 23, Nov. 19, 1943; Jan. 16, 31, 1944.
65. G.B.H. to Mrs. John P. Burke, June 12, 1945 (private collection).
66. Ibid., Jan. 19, 1945.
67. Ibid., Apr. 17, 1945.
68. G.B.H., "Biographical Memoir of E.P.H.: World War II," June 19, Nov. 18, 1945.
69. Ibid., Nov. 11, 1945.
70. G.B.H. to Mrs. John P. Burke, Aug. 7, 1943 (private collection).

CHAPTER 14: DARK PASSAGE

1. E.P.H. to N.U.M., Mar. 1, 1946, HUB Box 14, f. 551.
2. Walter Baade to N.U.M., Sept. 18, 1946, SALO, N.U.M. Papers, f. Baade, W., 1936–46.
3. V.B. to W.S.A., Oct. 23, 1944, MWOA, W.S.A. Papers, Box 11, f. 11.196. Also see Donald E. Osterbrock, "The Appointment of a Physicist as Director of the Astronomical Center of the World," *Journal for the History of Astronomy*, 23 (1992), 155–65.
4. Interview with L. S. Dederick by C.A.W., Nov. 6, 1971 (courtesy of C.A.W.).
5. V.B. to W.S.A., Oct. 23, 1944, MWOA, W.S.A. Papers, Box 11, f. 11.196.
6. W.S.A. to V.B., Oct. 28, 1944, MWOA.
7. Ibid.
8. V.B. to W.S.A., Nov. 9, 1944, MWOA.
9. V.B. to W.S.A., Apr. 13, 1945, MWOA, Box 9, f. 9.152.
10. Ibid.
11. W.S.A. to V.B., Apr. 24, 1945, MWOA.
12. V.B. to W.S.A., May 22, 1945, MWOA, Box 11, f. 11.97.
13. V.B. to W.S.A., June 12, 1945, MWOA, Box 9, f. 9.152.
14. V.B. to W.S.A., July 5, 1945; V.B. to I.B., July 18, 1945; W.S.A. to Walter M. Gilbert, July 20, 1945, MWOA.
15. E.P.H. to V.B., Sept. 10, 1945, MWOA.
16. V.B. to E.P.H., Oct. 3, 1945, MWOA.
17. I.B. to E.P.H., Oct. 4, 1945, MWOA, I.B. Papers, Box 1, f. 1.6.
18. E.P.H. to I.B., Oct. 16, 1945, MWOA.
19. I.B. to V.B., Oct. 26, 1945, MWOA.
20. E.P.H., *The Nature of Science and Other Lectures* (San Marino: The Huntington Library, 1954), 77–79.
21. Hans-Lothar Gemmingen to E.P.H., Oct. 20, 1948, HUB Box 11, f. 286.

22. G.B.H., "E.P.H.: The Missing Years, 1945–1946," HUB 82(6), Box 7, 2.
23. Quoted in Emil Corwin, "Palomar's New Look," *Script* (July 1948), 24.
24. Ibid., 25–26.
25. Ibid., 27.
26. Dennis Overbye, *Lonely Hearts of the Cosmos: The Story of the Scientific Quest for the Secret of the Universe* (New York: HarperCollins, 1991), 19.
27. Corwin, "Palomar's New Look," 30.
28. Log Book: 200-inch Reflector, HUB Box 29, Jan. 26–27, 1949.
29. E.P.H., "A Photograph with the 200-inch Hale Telescope," Mar. 2, 1949, HUB 3, f. 56.
30. G.B.H., "E.P.H.: The Astronomer," HUB 82(7), Box 7, 11.
31. John Kord Langemann, "The Men of Palomar," *Collier's* (May 1949), 25, 64.
32. Gordon L'Allemand, "Pilgrimage into Eternity," *Travel*, 93, no. 3 (September 1949), 24.
33. *Time*, Feb. 9, 1948, 62.
34. Aldous Huxley to E.P.H., Feb. 7, 1948, HUB Box 16, f. 673.
35. G.B.H. to Mrs. John P. Burke, Nov. 9, 1947 (private collection).
36. *Time*, 59.
37. G.B.H., "Diary of Daily Events at Home in San Marino, From 1929 to 1937," HUB 73, Box 4, Jan. 22, 1937.
38. Joseph Scott to E.P.H., Mar. 18, 1948, HUB Box 9, f. 930.
39. G.B.H., "Travel Diary of Trips to England and Europe for 1948," HUB 79, Box 6, July 1, 1948.
40. Ibid., July 3, 1948; G.B.H., "E.P.H.: Nicolas Copernicus [cat]," HUB 82(16), Box 8, 5, 8.
41. G.B.H., "Travel Diary of Trips to England and Europe for 1948," July 15, 1948.
42. Ibid., July 27, 1948.
43. Ibid., Aug. 13, 1948.
44. Interview with A.S. by S.W., AIP, May 22, 1978, 3–4.
45. Ibid., 5.
46. Jesse Greenstein, "An Astronomical Life," *Annual Review of Astronomy and Astrophysics*, 22 (1984), 17–18.
47. Interview with A.S. by S.W., 13, 16.
48. Interview with A.S., June 11, 1991.
49. Ibid.; Timothy Ferris, *The Red Limit: The Search for the Edge of the Universe*, 2nd ed. (New York: Quill, 1983), 44, 113.
50. Interview with A.S.
51. Interview with A.S. by B.S., AIP, Feb. 8, 1977, 1.
52. Nicolas Copernicus to Dear Folks, July 9, 1949, HUB Box 16, f. 6.71.
53. Interview with I.C., July 8, 1991.
54. Ibid.
55. G.B.H., "E.P.H.: The Missing Years, 1948–1949," HUB 82(6), Box 7, 3.
56. Ibid., 3; interview with I.C.
57. G.B.H., "E.P.H.: The Missing Years, 1948–1949," 3.
58. Ibid., 4.
59. Walter Baade to N.U.M., July 21, 1949, SALO, N.U.M. Papers, f. Baade, W., 1947–60.
60. I.B. to E.P.H., July 21, 1949, MWOA., I.B. Papers, Box 13, f. 13.191.

61. Interview with I.C.
62. Anita Loos to E.P.H., July 31, 1949, HUB Box 17, f. 756.
63. Aldous Huxley to E.P.H., July 30, 1949, HUB, Box 16, f. 674.
64. G.B.H., "E.P.H.: The Missing Years, 1948–1949," 6.
65. *LAT*, Aug. 24, 1949, 6.

CHAPTER 15: HOME IS THE SAILOR

1. G.B.H., "E.P.H.: The Missing Years, 1948–1952," HUB 82(6), Box 7, 6.
2. Walter Baade to N.U.M., Oct. 22, 1949, SALO, N.U.M., Papers, f. Baade, W., 1947–60.
3. G.B.H., "E.P.H.: Characteristics," HUB 82(9), Box 7, 9.
4. G.B.H., "E.P.H.: Nicolas Copernicus [cat]," HUB 82(16), Box 8, 2, 6.
5. Last Will and Testament of E.P.H., no. P346389, Registrar-Recorder, Los Angeles County Court House, Los Angeles, CA.
6. E.P.H. to F. E. Wright, Apr. 1, 1950, HUB Box 18, f. 854.
7. G.B.H., "Travel Diary of Trips to the Eastern United States, England, and France from April 18 to July 14, 1950," HUB 80, Box 6, Apr. 26, 1950.
8. Ibid., June 4.
9. Interview with I.B. by Charles Weiner, AIP, Aug. 26, 1969, 45, 46.
10. G.B.H., "Travel Diary of Trips . . . from April 18 to July 14, 1950," May 12.
11. Ibid., June 23; HUB Box 25, f. 5.
12. Patmore's letter, dated May 23, 1890, was copied by G.B.H. in her journal, "Travel Diary for Trips to Arizona, the Sierra Nevada, and Arizona and Diary of Daily Events at Home in San Marino from 1929 to 1937," HUB 73, Box 4, Jan. 25, 1937.
13. G.B.H., "Travel Diary of Trips . . . from April to July 14, 1950," June 25.
14. Interview with Halton C. Arp by Paul Wright, AIP, July 29, 1975, 9, 12.
15. Interview with Martin Schwarzchild by S.W., AIP, Mar. 10, 1977, 55.
16. Interview with A.S. by S.W., AIP, May 22, 1978, 20–21.
17. Interview with A.S., June 11, 1991.
18. Ibid.
19. E.P.H. to Hazel Littlefield Smith, Mar. 12, 1953, HUB Box 15, f. 624. Also see E.P.H. and A.S., "The Brightest Variable Stars in Extragalactic Nebulae. I. M31 and M33," *APJ*, 118 (1953), 353–61.
20. Interview with I.C., July 8, 1991.
21. Log Book: 200-inch Reflector, HUB Box 29, Oct. 13, 1950, 10–11.
22. This story has been repeated by many associated with Mount Wilson, including A.S. and Olin Wilson.
23. Interview with A.S. by B.S., AIP, Feb. 8, 1977, 4.
24. Interview with Martin Schwarzchild by S.W., 148–50.
25. Interview with A.S. by B.S., 4.
26. Interview with I.C.
27. M.L.H. to N.U.M., June 20, 1947, SALO, N.U.M. Papers, f. Humason, M., 1933–1948.
28. N.U.M., "Milton L. Humason—Some Personal Recollections," *Mercury*, 2 (January/February 1973), 7.
29. Interview with A.S.
30. Interview with A.S. by S.W., 65.

31. Interview with M.L.H. by B.S., AIP, c. 1965, 8.

32. E.P.H., "The Law of Red-Shifts," *Monthly Notices of the Royal Astronomical Society,* 113, no. 6 (1953), 658–66.

33. Interview with A.S. by B.S., 5–8.

34. Interview with M.L.H. by B.S., 9.

35. Interview with C. Donald Shane by Helen Wright, AIP, July 11, 1967, 72.

36. G.B.H., "E.P.H.: The Missing Years, 1948–1952," 1.

37. Ibid.

38. Robert S. Labonge to E.P.H., Dec. 4, 1951, HUB Box 20, f. 969; Elmer Davis to E.P.H., Nov. 23, 1951, HUB Box 10, f. 238.

39. Cass Canfield to G.B.H., Mar. 1, 1951, HUB Box 10, f. 213; Canfield to E.P.H., Aug. 14, 1953, f. 212.

40. H.S. to W.S.A., Sept. 30, 1942, MWOA, W.S.A. Papers, Box 61, f. 61.1083; W.S.A. to H.S., Oct. 13, 1942.

41. E.P.H. to H. H. Plaskett, Feb. 14, 1950, HUB Box 15, f. 583.

42. Interview with I.C.; G.B.H., "E.P.H.: Characteristics," 1.

43. Interview with A.S. by S.W., 15.

44. Interview with A.S. by B.S., 7–8.

45. Interview with A.S. by S.W., 42, 56.

46. G.B.H., "Diary of Daily Life in San Marino, California, and Diary of Trip to England from January 1 to July 26, 1953," HUB 81, Box 6, Feb. 10.

47. G.B.H., "E.P.H.: Some People (Berta)," HUB 82(17), Box 8, 1–3.

48. Ibid., 3.

49. G.B.H., "E.P.H.: The Missing Years, 1948–1952," Feb. 10, 1949.

50. Ibid.

51. G.B.H., "Diary of Daily Life in San Marino, California, and Diary of Trip to England . . . to July 26, 1953," Jan. 29–30.

52. G.B.H., "E.P.H.: The Missing Years, 1948–1952," Summer 1952, 2.

53. G.B.H., "Diary of Daily Life in San Marino, California, and Diary of Trip to England . . . to July 26, 1953," May 13.

54. Ibid., May 15.

55. Ibid., May 15–23.; C. J. Humbert to E.P.H., June 12, 1953, HUB Box 20, f. 1015.

56. G.B.H., "Diary of Daily Life in San Marino, California, and Diary of Trip to England . . . to July 26, 1953," June 9.

57. Ibid., July 20.

58. G.B.H., "E.P.H.: The Astronomer (Palomar)," HUB 82(7), Box 7, 11–12; G.B.H., "E.P.H.: September, 1953," HUB 82(14), Box 8, 1; Log Book: 200-inch Reflector, Sept. 1, 1953, 18–19.

59. G.B.H., "E.P.H.: September, 1953," 1–2.

60. Ibid., 3–4.

61. Ibid., 4.

62. Ibid., 4–5. Another account of E.P.H.'s death, differing slightly in detail, is contained in HUB Box 12, f. 340.

63. G.B.H., "E.P.H.: September, 1953," 5.

64. Turner and Stevens Co. Funeral Home Records, Sept. 28, 1953, no. 365; interview with I.C., July 8, 1991.

65. Aldous Huxley to G.B.H., Sept. 30, 1953, HUB Box 16, f. 676; "Stars and the Man,"

HUB Box 8, f. 85, 10 pp. Huxley's article was eventually published as part of a novel: Tom Bezzi, *Hubble Time* (San Francisco: Mercury House, 1987), 56–66.

66. Anita Loos to G.B.H., Sept. 29, 1953, HUB Box 17, f. 777.

67. Fryniwyd Tennyson Harwood to G.B.H., Oct. 1, 1953, HUB Box 12, f. 328.

68. G.B.H., "E.P.H.: Quotations," HUB 82(13), Box 8, 1a.

EPILOGUE

1. G.B.H., "E.P.H.: Nicolas Copernicus [cat]," HUB 82(16), Box 8, 7–8.

2. Interview with H.S. by Charles Weiner and Helen Wright, AIP, June 8, 1966, 58.

3. G.B.H, "E.P.H: The Astronomer," HUB 82(7), Box 7, 14; interview with Geoffrey and Margaret Burbidge by C.A.W, undated (courtesy of C.A.W.).

4. Anita Loos to G.B.H., June 16, 1954, HUB Box 17, f. 782.

5. G.B.H. to Leslie Bliss, Mar. 13, 1954, Henry E. Huntington Institutional Archives, f. General Library Correspondence.

6. Dennis Overbye, *Lonely Hearts of the Cosmos: The Story of the Scientific Quest for the Secret of the Universe* (New York: HarperCollins, 1991), 28.

7. Quoted in G. J. Whitrow, "Edwin Powell Hubble," *Dictionary of Scientific Biography*, 6 (New York: Charles Scribner's Sons, 1972), 533.

8. Interview with Page Ackerman, July 13, 1991.

9. G.B.H., Certificate of Death, no. 356, Mar. 22, 1980, Deputy Registrar—Vital Statistics, Pasadena Public Health Department, Pasadena, CA.

10. G.B.H., "Diary of Daily Life in San Marino, California, and Diary of Trip to England from January 1 to July 26, 1953," HUB 81, Box 6, May 11.

BIBLIOGRAPHY

The reader seeking a complete bibliographical treatment of Edwin Hubble's published scholarship is referred to the lengthy annotated article by N. U. Mayall, "Edwin Powell Hubble," *Biographical Memoirs of the National Academy of Sciences*, 41 (1970), 175–214.

Books

Arp, Halton. *Quasars, Redshifts, and Controversies*. Berkeley: Interstellar Media, 1987.

Baade, Walter. *Evolution of the Stars and Galaxies*. Cambridge, MA: Harvard University Press, 1963.

Bedford, Sybille. *Aldous Huxley: A Biography*. New York: Alfred A. Knopf/Harper & Row, 1974.

Carey, Gary. *Anita Loos: A Biography*. London: Bloomsbury Publishing Ltd., 1988.

Clark, Ronald W. *Einstein*. New York and Cleveland: The World Publishing Company, 1971.

Drake, Stillman, ed. and trans. *Discoveries and Opinions of Galileo*. Garden City, NY: Doubleday and Company, 1957.

Dunaway, David King. *Huxley in Hollywood*. New York: Harper & Row, 1989.

Ferris, Timothy. *Coming of Age in the Milky Way*. New York: William Morrow, 1988.

———. *Galaxies*. New York: Stewart, Tabori & Chang, Publishers, 1982.

————. *The Red Limit: The Search for the Edge of the Universe.* 2nd ed. New York: Quill, 1983.

Frost, Edwin Brant. *An Astronomer's Life.* Boston and New York: Houghton Mifflin Company, 1933.

Grindlay, Jonathan F., and A. G. Davis Philip. *The Harlow Shapley Symposium of Globular Cluster Systems in Galaxies.* Dordrecht/Boston/London: Kluwer Academic Publishers, 1988.

Haramundanis, Katherine, ed. *Cecilia Payne-Gaposchkin: An Autobiography and Other Recollections.* Cambridge, Eng.: Cambridge University Press, 1984.

Harwood, H. M., and F. Tennyson Jesse, eds. *London Front: Letters Written to America, 1939–1940.* New York: Doubleday, Doran and Company, 1941.

————. *While London Burns: Letters Written to America, July 1940 to June 1941.* London: Constable and Company Ltd., 1942.

Hastie, W., ed. and trans. *Kant's Cosmogony.* Glasgow: James Maclehose and Sons, 1900.

Hayes, Helen, with Sanford Doty. *On Reflection.* New York: M. Evans and Company, 1968.

Hetherington, Norriss S., ed. *The Edwin Hubble Papers: Previously Unpublished Manuscripts on the Extragalactic Nature of Spiral Nebulae.* Tucson: Pachart Publishing House, 1990.

Hubbell, Harold Berresford, Jr., and Donald Sidney Hubbell, eds. *History and Genealogy of the Hubbell Family.* 3rd ed. New York: Theo. Gaus, Ltd., 1980.

Hubble, Edwin. *The Nature of Science and Other Essays.* San Marino: The Huntington Library, 1954.

————. *The Observational Approach to Cosmology.* Oxford: The Clarendon Press, 1937.

————. *The Realm of the Nebulae.* New Haven: Yale University Press, 1936.

King, Henry C. *The History of the Telescope.* New York: Dover Publications, 1979.

Krisciunas, Kevin. *Astronomical Centers of the World.* Cambridge, Eng.: Cambridge University Press, 1988.

Kron, Richard G. *Evolution of the Universe of Galaxies: Edwin Hubble Centennial Symposium.* San Francisco: Astronomical Society of the Pacific, 1990.

Little, J. G. *The Official History of the Eighty-sixth Division.* Chicago: Chicago States Publications Society, 1921.

Loos, Anita. *Kiss Hollywood Goodbye.* New York: The Viking Press, 1974.

Millikan, Robert A. *The Autobiography of Robert A. Millikan.* New York: Prentice-Hall, 1950.

Neyman, Jerzy, ed. *The Heritage of Copernicus: Theories 'Pleasing to the Mind.'* Cambridge, MA: MIT Press, 1974.

Noyes, Alfred. *Watchers of the Skies.* New York: Frederick H. Stokes Company, 1922.

Overbye, Dennis. *Lonely Hearts of the Cosmos: The Story of the Quest for the Secret of the Universe.* New York: HarperCollins, 1991.

Sandage, Allan, ed. *The Hubble Atlas of Galaxies.* Washington, D.C.: Carnegie Institution of Washington, 1961.

Shapley, Harlow. *Through Rugged Ways to the Stars.* New York: Charles Scribner's Sons, 1969.

Sharov, A. S., and Egor Novikov. *Edwin Hubble: The Discoverer of the Big Bang Universe.* Cambridge, Eng.: Cambridge University Press, 1993.

Silk, Joseph. *The Big Bang: The Creation and Evolution of the Universe.* San Francisco: W. H. Freeman and Company, 1980.

Smith, Robert W. *The Expanding Universe: Astronomy's "Great Debate" 1900–1931.* Cambridge, Eng.: Cambridge University Press, 1982.

Trefil, James S. *The Moment of Creation: Big Bang Physics from Before the First Millisecond to the Present Universe.* New York: Charles Scribner's Sons, 1983.

Whitney, Charles A. *The Discovery of Our Galaxy.* New York: Alfred A. Knopf, 1971.

Wright, Helen. *Explorer of the Universe: A Biography of George Ellery Hale.* New York: E. P. Dutton & Co., 1966.

Articles

Adams, Walter S. "Dr. Edwin P. Hubble." *The Observatory,* 74 (1954), 32–35.

————. "Early Days at Mount Wilson." *Publications of the Astronomical Society of the Pacific,* 59, pt. 1, no. 350 (October 1947), 213–31.

————. "Early Days at Mount Wilson." *Publications of the Astronomical Society of the Pacific,* 59, pt. 2, no. 351 (December 1947), 285–304.

Ault, Warren D. "Oxford in 1907 (With a Glimpse of T. E. Lawrence)." *The American Oxonian,* 66, no. 2 (Spring 1979), 121–28.

Baade, Walter. "The Resolution of Messier 32, NGC 205, and the Central Region of the Andromeda Nebula." *Astrophysical Journal,* 100 (1944), 137–46.

Bok, Bart J. "Harlow Shapley." *Biographical Memoirs of the National Academy of Sciences,* 49 (Washington, D.C: National Academy of Sciences, 1978), 241–91.

Gingerich, Owen. "Harlow Shapley." *Dictionary of Scientific Biography,* 12 (New York: Charles Scribner's Sons, 1975), 345–52.

————. "Harlow Shapley and Mount Wilson." *American Academy of Arts and Sciences Bulletin,* 26 (1973), 10–24.

———— and Barbara Welther. "Harlow Shapley and Cepheids." *Sky and Telescope,* 70, no. 6 (December 1985), 540–42.

Greenstein, Jesse. "An Astronomical Life." *Annual Review of Astronomy and Astrophysics,* 22 (1984), 1–35.

Gwinn, Joel A. "Edwin Hubble in Louisville, 1913–14." *The Filson Club History Quarterly,* 56 (1982), 415–19.

Hale, George Ellery. "The Astrophysical Observatory of the California Institute of Technology." *Astrophysical Journal,* 82 (September 1935), 111–39.

Hart, R., and R. Berendzen. "Hubble's Classification of Non-Galactic Nebulae, 1922–1926." *Journal for the History of Astronomy,* 5, pt. 2, no. 4 (June 1971), 109–19.

Hetherington, Norriss S. "Edwin Hubble: Legal Eagle." *Nature,* 319 (Jan. 16, 1986), 189–90.

————. "Hubble's Cosmology." *American Scientist,* 78 (March/April 1990), 142–51.

Hoskin, M. A. "The 'Great Debate': What Really Happened." *Journal for the History of Astronomy,* 7 (1976), 169–82.

Hubble, Edwin. "Angular Rotations of Spiral Nebulae." *Astrophysical Journal,* 8 (1935), 334–35.

————. "Changes in the form of the Nebula N.G.C. 2261." *Proceedings of the National Academy of Sciences,* 2 (1916), 230–31.

————. "The Direction of Rotation in Spiral Nebulae." *Astrophysical Journal,* 97 (1943), 112–18.

————. "The Distribution of Extra-Galactic Nebulae." *Astrophysical Journal,* 79 (1934), 8–76.

———. "The Exploration of Space." *Harper's Magazine*, 158 (May 1929), 732–38.

———. "A General Study of Diffuse Galactic Nebulae." *Astrophysical Journal*, 56 (1922), 162–99.

———. "The Law of Red-Shifts." *Monthly Notices of the Royal Astronomical Society*, 113, no. 6 (1953), 658–66.

———. "Photographic Investigations of Faint Nebulae." *Publications of the Yerkes Observatory*, 4, pt. 2 (1920), 69–85.

———. "Recent Changes in the Variable Nebula N.G.C. 2261." *Astrophysical Journal*, 45 (1917), 351–53.

———. "A Relation Between Distance and Radial Velocity Among Extra-Galactic Nebulae," *Proceedings of the National Academy of Sciences*, 15, no. 3 (March 1929), 168–73.

———. "The Source of Luminosity in Galactic Nebulae." *Astrophysical Journal*, 56 (1922), 400–38.

———. "Twelve Faint Stars with Sensible Proper-Motions." *Astronomical Journal*, 29, no. 1 (1916), 168–69.

——— and Milton L. Humason. "The Velocity-Distance Relation Among Extra-Galactic Nebulae." *Astrophysical Journal*, 74 (1931), 43–80.

——— and Allan Sandage. "The Brightest Variable Stars in Extragalactic Nebulae. I. M31 and M33." *Astrophysical Journal*, 118 (1953), 353–61.

Humason, Milton L. "Edwin Hubble." *Monthly Notices of the Royal Astronomical Society*, 114, no. 3 (1954), 291–95.

Jones, Brian. "The Legacy of Edwin Hubble." *Astronomy*, 17, no. 12 (December 1989), 38–44.

Jones, Kenneth Glyn. "The Observational Basis for Kant's *Cosmogony*: A Critical Analysis." *Journal for the History of Astronomy*, 2 (1971), 29–34.

Lemaître, Georges. "Un univers homogène de masse constante et de rayon croissant, redant compte de la vitesse radiale des nebuleuses extra-galactiques." *Annals de la Société Scientifique de Bruxelles*, 47 (1927), 49–56.

Lundmark, Knut. "On the Motions of Spirals." *Publications of the Astronomical Society of the Pacific*, 34 (1922), 108–15.

———. "The Spiral Nebula Messier 33." *Publications of the Astronomical Society of the Pacific*, 33 (1921), 324–27.

Maanen, Adriaan van. "Internal Motions in Spiral Nebulae." *Astrophysical Journal*, 81 (1935), 336–37.

Mayall, N. U. "Edwin Hubble: Observational Cosmologist." *Sky and Telescope* (January 1954), 78–80, 85.

———. "Milton L. Humason—Some Personal Recollections." *Mercury*, 2 (January/February 1973), 3–8.

Osterbrock, Donald. "The Appointment of a Physicist as Director of the Astronomical Center of the World." *Journal for the History of Astronomy*, 23 (1992), 155–65.

———. "The California-Wisconsin Axis in American Astronomy." *Sky and Telescope*, 51, pt. 1, no. 1 (January 1976), 9–14.

———, Ronald Brashear, and Joel Gwinn. "Young Edwin Hubble." *Mercury*, 19, no. 1 (January/February 1990), 2–15.

Robertson, H. P. "Edwin Powell Hubble, 1889–1953." *Publications of the Astronomical Society of the Pacific*, 66 (1954), 120–25.

Sandage, Allan. "Edwin Hubble, 1889–1953." *Journal of the Royal Astronomical Society of Canada*, 83, no. 6 (December 1989), 351–62.

Slipher, V. M. "Spectrographic Observations of Nebulae." *Popular Astronomy*, 23 (1915), 21–24.

Smith, Robert W. "Edwin P. Hubble and the Transformation of Cosmology." *Physics Today*, 43, no. 4 (April 1990), 52–58.

Staff of the Ballistic Research Laboratory. "Some Recent Advances in Ballistics." *Journal of Applied Physics*, 10, no. 12 (December 1945), 773–80.

Struve, Otto. "A Historic Debate About the Universe." *Sky and Telescope*, 19 (May 1960), 398–401.

Whitrow, G. J. "Edwin Powell Hubble." *Dictionary of Scientific Biography*, 6 (New York: Charles Scribner's Sons, 1972), 528–33.

ACKNOWLEDGMENTS

The largest collection of the Edwin Hubble papers by far is located at the Henry E. Huntington Library in San Marino, California. These, together with the invaluable journals and other writings of Grace Burke Hubble, are the foundation on which this biography rests. If there is a more congenial place—for both inner and outer weather—at which to pursue the scholar's task, it must be on some distant planet. The staff archivists, librarians, and other specialists have been helpful in countless ways, generous with their time, instructive, patient, and always willing to offer sound advice when asked for it. I am forever indebted to them all, but for their particular help and continuing friendship I wish to express my utmost gratitude to President Robert Skotheim and to Martin Ridge, the now retired director of research, to Ronald Brashear, curator in the history of science, to Virginia Renner, reader services librarian, and to Pat Parrish, secretary to the president. I am also indebted to the Huntington for financial support in the form of a Fletcher Jones Foundation Fellowship.

During my time in California, I had the privilege of interviewing some of the individuals who worked with Edwin Hubble at Mount Wilson and

Mount Palomar observatories, as well as others who were neighbors and personal friends of the astronomer and his wife. Preeminent in this regard is the astronomer Allan Sandage, who inherited Hubble's mantle when the master mariner of the nebulae was suddenly felled by a stroke in 1953. Through his reminiscences, Sandage helped bring Hubble to life, and his advice on technical matters and library sources has helped me avert any number of errors and oversights. The retired solar astronomer Horace Babcock approaches the status of a living archive; his extensive recollections of Mount Wilson in the "old days" supplied information available nowhere else; nor will I ever forget the wonderful day we spent at Mount Palomar, where I was given the Grand Tour by the ultimate guide. Larry Webster accorded me the same treatment atop Mount Wilson, where the ghost of Edwin Hubble still circles the 100-inch telescope on the darkest of dark nights. The staff of the Carnegie Observatories on Pasadena's Santa Barbara Street granted me free access to their indispensable collection of papers and journals, not to mention use of the photocopying machine. The California Institute of Technology Archives, the Pasadena Public Library, and the Pasadena Historical Society also placed their collections and staff members at my disposal. The Los Angeles County Courthouse made available their records of marriages, deaths, and wills. Astronomer Gibson Reaves kindly committed his encounters with Hubble to paper for my use.

In terms of their private lives, no one knew Grace and Edwin Hubble better than their longtime neighbor and steadfast friend Ida Crotty, who opened the album of memory to me and let me leaf through those wonderful pages at will. Her faith and support give warmth to this below-zero winter's day. I am also grateful to Anne Crotty for providing much additional background information. Retired university librarian Page Ackerman recalled the war years at Aberdeen Proving Ground, while Joyce Eyer and her son granted me access to certain of Grace Hubble's correspondence with her family. Linda and Robert Mollno, now the owners of the former Hubble home, graciously showed me around, answered all sorts of questions, and then allowed this stranger to take photographs of Hubble's study and several other rooms. Thanks are also due astronomer Jesse Greenstein, who spent an afternoon with the tape recorder running, and to the staff at Mount Palomar, who made my time there not only a pleasant venture but an aesthetic experience. Donald Osterbrock of Lick Observatory, both an astronomer and a historian of science, was giving of his time, source materials, and insights into Hubble's work and personality. He also paved the way for my research in the Mary Lea Shane Archives of Lick Observatory at the University of California, Santa Cruz, where I was entrusted to the capable hands of librarian and archivist Dorothy E. Schaumberg. The collections of the San Jose Public Library were most helpful in compiling background on the early life and first marriage of Grace Hubble, as were Sara Timby of the Stanford University Archives, Sheldon D. Johnson, the Amador County Recorder of Jackson, California, and Leslie Martin of the Amador County Library.

My quest for Edwin Hubble in Marshfield, Missouri, where it all began, was aided by Patricia Holik of Marshfield's Webster County Library, the Webster County Historical Association, the staff of the Recorder's Office of the Webster County Courthouse, Larry Blakée, and Lena James Jump, whose Hubble and James family photographs and recollections breathed new life into the days of Hubble's youth a century ago. Microfilms of the local newspapers were loaned to me by the State Historical Society of Missouri at Columbus, and much additional information on the Hubbles of Springfield was obtained through the Springfield Public Library, the Drury College Archives, and the Washington University Archives at St. Louis. Clifton H. Hubbell and Harold B. Hubbell, Jr., of the Hubble Family Historical Society also rendered much welcome assistance.

As can be seen by a perusal of the end notes, the name Charles A. Whitney figures prominently in my research. The Harvard astronomer and historian of science once contemplated writing a biography of Hubble, and conducted extensive interviews with several who knew him during his childhood and adolescence in Wheaton, Illinois, as well as in his later years. Virtually all have long passed from the scene, but because of Whitney's generosity, they live again in these pages. I am also indebted to the Wheaton History Center and to Diane Batson of the Wheaton Public Schools for providing transcripts and background material on Hubble's student days.

Richard Popp and the congenial staff of the University of Chicago Archives made available their excellent collection of background material dating from Hubble's undergraduate and graduate years. Farther north, at Yerkes Observatory in Williams Bay, Wisconsin, I spent a beautiful week in May rummaging the archives by day and observing the sky by night through the 40-inch telescope, thanks to the assistance of archivist Judy Bausch and astronomer-technician John Briggs. The archives of the University of Wisconsin, Madison, also provided me with correspondence between Hubble and the astronomer Joel Stebbins.

When I undertook this project, two of Edwin Hubble's sisters survived, Helen Hubble Lane and Elizabeth "Betsy" Hubble. A week before Helen passed away at age ninety-two, her son, John F. Lane, graciously interviewed the sisters on my behalf, employing questions I had submitted in advance. Subsequent to Helen's death, I interviewed Betsy myself. Needless to say, the sisters' recollections are among the most important to this work, for which I am indebted to them beyond measure. Moreover, Helen's 1987 memoir, drafted at the behest of Caltech professor Kip S. Thorne and simply titled "Edwin Powell Hubble," is among the most evocative of all the Huntington Library documents. As a further measure of his generosity, John Lane made available to me the surviving letters from Hubble's Rhodes Scholar years, without which this chapter in his life would not only be much thinner but much the poorer. Indeed, these missives, more than any other documents, define Edwin Hubble's personality, for he was ever and always a spiritual Englishman—a true son of Queen's. Additional background on the Oxford years was provided by the Rhodes Trust.

For the time Hubble spent in Louisville, Kentucky, after returning from England the author is indebted to the labors of Joel Gwinn of the University of Louisville and to various staff members of the Floyd, Clark, and Harrison county courthouses in Indiana, and the Shelby, Oldham, Jefferson, and Bullitt county courthouses in Kentucky. Memorabilia from Hubble's teaching and coaching days was supplied by the New Albany Senior High School through the offices of superintendent Dr. Tracy Dust and librarian Barbara Beury.

Background on Hubble's military service in World War I was provided by the National Personnel Records Center in St. Louis, Missouri, and the U.S. Military History Institute, Carlisle Barracks, Pennsylvania.

Much of the scientific correspondence is widely scattered. Among the important collections beyond the Huntington is the Hubble-Shapley material in the Harvard College Observatory Archives of Harvard University. This collection was made available to me through the office of Irwin Shapiro, director of the Harvard-Smithsonian Center for Astrophysics, and the aid of Robin McElheny, assistant curator for University and Public Service. Owen Gingerich, professor of astronomy and the history of science, also assisted my research efforts during my stay in Cambridge.

Princeton University Archives houses a small but important collection of Hubble letters, mostly written to the astronomer Henry Norris Russell. University archivist Ben Primer first brought this material to my attention as a result of my query in *The New York Times Book Review*, and Curator of Manuscripts Don. C. Skemer provided assistance once I arrived in Princeton. It was through this same query that I was contacted by L. Ann Silirie, public affairs officer at Aberdeen Proving Ground, who made available materials relating to Hubble's work in the Exterior Ballistics Program during World War II.

Although Mount Wilson is located in California, the observatory is under the auspices of the Carnegie Institution of Washington, D.C. Its important files and photocopying facilities were opened to me by staff member Susan Vasquez, who led me to the pertinent cabinets and then left me to my own devices.

During this same swing along the East Coast, I visited the Niels Bohr Library of the American Institute of Physics, which has recently moved from New York City to College Park, Maryland. Extensive use has been made of its collection of interviews with major scientists of the twentieth century, as well as supporting documents, books, and periodicals. Director Spencer R. Weart, Joan N. Warnow-Blewett, and Bridget Sisk helped make my stay a productive one, and I am further indebted to the institute for a grant-in-aid which helped offset travel expenses.

My hat is also off to the staff of Cunningham Memorial Library at Indiana State University, especially the Interlibrary Loan Department, which provided access to many otherwise difficult-to-obtain articles. The university also granted me a leave, during which much of the research and writing was done, buffering me from the usual obligations and distractions of academic life. I am also grateful to the Faculty Research Committee for granting me financial support and to the students in my courses in the history of science and biography for their intellectual stimulation.

A National Endowment for the Humanities Summer Stipend made the costs of traveling far afield more manageable, and I wish to thank the NEH for its confidence in this project.

Others to whom I am indebted are my agent Michael Congdon, my editor John Glusman, editorial assistant Ethan Nosowsky, my indispensable copyeditor Jack Lynch and my designer Fritz Metsch.

Finally, I must express my unqualified gratitude to the John Simon Guggenheim Memorial Foundation for granting me a fellowship with which to complete the writing itself. Never before have I had such a block of unencumbered time with which to take up the pen. To have it provided by this august institution and to be able to join the ranks of the many distinguished Guggenheim Fellows is a heartfelt privilege.

As always, it is the members of one's own household who see one through. Thanks be to Rhonda and the Packydoodles—Mambo Manny, Snoote Rockne, and the ever elegant Count Blueski.

INDEX